OXFORD BIOGEOGRAPHY SERIES
Editors: A. Hallam, B. R. Rosen, and T. C. Whitmore

OXFORD BIOGEOGRAPHY SERIES

Editors
A. Hallam, Department of Geological Sciences, University of Birmingham.
B. R. Rosen, Department of Palaeontology, Natural History Museum, London.
T. C. Whitmore, Department of Geography, University of Cambridge.

The aim of the series is to publish a range of titles demonstrating the breadth of biogeography, from the biological to the geological ends of the spectrum. The subject is being revolutionized by plate tectonics, molecular phylogeny, and population models, vicariance, cladistics, and spatial classification analyses. For both specialist and non-specialist, the Oxford Biogeography Series will provide dynamic syntheses of new developments.

1. T. C. Whitmore (ed.): *Wallace's line and plate tectonics*
2. Christopher J. Humphries and Lynne R. Parenti: *Cladistic biogeography*
3. T. C. Whitmore and G. T. Prance (ed.): *Biogeography and Quaternary history in tropical America*
4. T. C. Whitmore (ed.): *Biogeographical evolution of the Malay Archipelago*
5. S. Robert Aiken and Colin H. Leigh: *Vanishing rain forests: the ecological transition in Malaysia*
6. Paul Adam: *Australian rainforests*
7. Wilma George and René Lavocat (ed.): *The Africa–South America connection*
8. Yin Hongfu: *The palaeobiogeography of China*
9. Alan J. Kohn and Frank E. Perron: *Life history and biogeography: patterns in* Conus
10. Anthony Hallam: *An outline of Phanerozoic biogeography*

Australian Rainforests

PAUL ADAM

*School of Biological Science,
University of New South Wales, Australia*

OXFORD   NEW YORK   TOKYO   MELBOURNE
OXFORD UNIVERSITY PRESS
1994

Oxford University Press, Walton Street, Oxford OX2 6DP
Oxford New York Toronto
Delhi Bombay Calcutta Madras Karachi
Kuala Lumpur Singapore Hong Kong Tokyo
Nairobi Dar es Salaam Cape Town
Melbourne Auckland Madrid
and associated companies in
Berlin Ibadan

Oxford is a trade mark of Oxford University Press

Published in the United States
by Oxford University Press Inc., New York

© Paul Adam 1992
First published 1992
First published in paperback 1994

All rights reserved. No part of this publication may be reproduced, stored in a retrieval system, or transmitted, in any form or by any means, without the prior permission in writing of Oxford University Press. Within the UK, exceptions are allowed in respect of any fair dealing for the purpose of research or private study, or criticism or review, as permitted under the Copyright, Designs and Patents Act, 1988, or in the case of reprographic reproduction in accordance with the terms of licences issued by the Copyright Licensing Agency. Enquiries concerning reproduction outside those terms and in other countries should be sent to the Rights Department, Oxford University Press, at the address above.

This book is sold subject to the condition that it shall not, by way of trade or otherwise, be lent, re-sold, hired out, or otherwise circulated without the publisher's prior consent in any form of binding or cover other than that in which it is published and without a similar condition including this condition being imposed on the subsequent purchaser.

A catalogue record for this book is available from the British Library

Library of Congress Cataloging in Publication Data
Adam, Paul, Ph.D.
Australian rainforests/Paul Adam.
p.cm.—(Oxford biogeography series; 6)
Includes bibliographical references.
1. Rain forests—Australia. 2. Rain forest conservation—
Australia. I. Title. II. Series: Oxford biogeography series; no. 6.
QH197.A3 1992   574.5' 2642' 0994—dc20   91-43261
ISBN 0 19 854872 9

Printed in Hong Kong

PREFACE

> I love a sunburnt country,
>   A land of sweeping plains,
> Of ragged mountain ranges,
>   Of droughts and flooding rains;
> I love her far horizons,
>   I love her jewel sea,
> Her beauty and her terror—
>   The wide, brown land for me!

This famous stanza by Dorothea Mackellar encapsulates many peoples' impressions of Australia—a wide, brown land, a continent dominated by the outback. Until recently, few would have associated rainforest with Australia, the international image of the country being an amalgam of desert and gum trees. Even within Australia, the presence of rainforest had barely registered on the public consciousness, as Australia has one of the most urban (or rather, suburban) populations of any country in the world; the majority of the population lives in one or other of the urban sprawls centred on State capitals.

It was thus a great surprise when, in the late 1970s and early 1980s, rainforest appeared on the front pages of major newspapers and the nightly television news showed confrontations between protesters and loggers. The conservation of rainforest rapidly became an issue of consequence in the political arena; in New South Wales it may have been a significant factor in determining the outcome of a State election. At the Federal level, the environmental credentials of political parties were assessed in the light of rainforest conservation policies in north Queensland, thus setting the stage for a major confrontation between the Federal Government and the Queensland State Government. The 1980s closed with rainforest logging virtually ended, with much of the rainforest included within National Parks, and large areas, in Tasmania, New South Wales, and Queensland, inscribed on the World Heritage List.

The widely felt public concern for rainforest conservation can be seen as the Australian expression of increased environmental awareness throughout the Western world and, in particular, of the much wider movement (focused on Amazonia and South-east Asia) opposed to rainforest destruction. However, the realization that Australia's rainforests are unique in many respects also served to stimulate public interest.

For many years, scientific and general public interest in natural history concentrated on what was regarded as the unique, quintessentially Australian elements in the biota. Rainforest was seen—by the relatively few who knew of it—as somehow 'un-Australian' and was interpreted as being an invasive element. The evidence for this was not strong but it was an interpretation which accorded with the prevalent 'cultural cringe' which held that little that was good originated in Australia. (A view which, perhaps, also finds expression in the vernacular use of 'brush' and 'scrubs' to refer to rainforest stands.) In the 1970s there was a major paradigm shift, with the recognition that rainforest had once been the dominant vegetation type across the continent, and that the biota of the present rainforests represented the survivors from these earlier times. Coming at a time of a search for an Australian national identity, this new evidence that the rainforests were Australian struck a chord with the general public, who overwhelmingly supported the protection of what remained. The awareness of rainforest also led to a rediscovery of earlier colonial history and an appreciation of the rainforest's role in the development of the eastern states.

PREFACE

My aim in this book is not to provide an introduction to rainforest ecology (for which the reader is referred to Whitmore 1990), but to give an outline of Australian rainforests: what they are, where they are, and how their present distribution evolved. I have also attempted to give a historical account of human interaction with the rainforest. From so close a perspective it is difficult to give a full picture of the turbulent 1980s: in the decades to come, it will be interesting to see what historians will make of the role of the environmental movement in general, and concern for rainforests in particular, in the political and social context of the late twentieth century. Some might suggest that I have devoted too much attention to recent history, but the interpretation and use of scientific evidence cannot be confined to ivory towers. What happened in Australia may offer important lessons to other countries. Public opinion may force politicians to make decisions in areas previously regarded as the domain of the professionals. As scientists, we have obligations to ensure that the public is as well informed as possible, but it cannot be denied that there is still much to learn: dogmatic assertions will not resolve controversies.

The change in the public image of rainforest was, in part, a response to new scientific findings, but has in itself been a major stimulus to new research. Australia is uniquely placed among the developed nations to carry out basic research on rainforest ecology and to provide guidance to developing nations on management of rainforest resources. Nevertheless, the steady stream of new publications on various aspects of Australian rainforests presents difficulties for anyone attempting to prepare a synthesis. What I hoped to do in this account was to provide a broad framework which will help readers place new results in context, while recognizing that some (probably much) of the detail will be refined and amended in future years. I have touched lightly, if at all, on some important topics, which, while essential to any understanding of rainforest functional ecology, do not appear to differ fundamentally in Australia from elsewhere; future research may change this perception (topics in this category include productivity and mineral cycling, and plant ecophysiology). My account concentrates on flora; this is not to deny that the fauna is an integral and essential part of the ecosystem, or that the Australian rainforest fauna is as uniquely Australian as the flora. Despite some excellent studies, much of the fauna (particularly the invertebrates) is still poorly documented, and it is easier to present a broad brush overview from a botanical perspective.

It would have been impossible to present any overview without a factual basis from which a synthesis could be drawn. Tribute is due to those few individuals who, during the period when rainforest was an unfashionable topic for research, nevertheless continued to collect the information on the distribution and diversity of rainforest which was to underpin the changed perspectives of the 1980s. Foremost amongst these are J. G. Tracey and L. J. Webb in Queensland and G. N. Baur and A. G. Floyd in New South Wales.

Some comments on the orthography of rainforest are called for. In Australia, the convention has been to use a single word, a practice advocated by Baur (1968)—'The spelling of "rainforest" as a single word, rather than the commonly used two words ("rain forest") is preferred as indicating the community's status as a fully independent plant formation and to avoid undue emphasis on rain as the sole determining environment factor'.

The concept of rainforest is not of great linguistic antiquity. The vegetation types to which it applies were foreign to northern hemisphere eyes and existing terminology was clearly inadequate to convey an impression of the structure. The term 'Regenwald' was introduced into the scientific literature by Schimper (1903) and in translation this became 'rain forest'.

In this account I have followed Australian convention, partly because it is the usage in much of the relevant literature, but also to acknowledge the strength of Baur's argument. This orthographic practice is at variance with current international usage—but the diversity of stands which are called 'rainforest' in Australia reinforces the importance of emphasizing the concept of 'rainforest' as a *vegetation type* rather than unduly stressing the significance of rain.

Nomenclature of vascular plants follows, as far as possible, that of Hnatiuk (1990). A taxonomic index is given separately. Many readers will be unfamiliar with Australia; many place names appearing in the text are shown on the map on page xiv and on the maps in Chapter 3; an index giving latitudes and longitudes is also included. Values in $ are in Australian dollars (worth somewhat less than the US $).

I am grateful to many people for their help and assistance, although I take responsibility for all errors and omissions remaining in the text. T. C. Whitmore made the initial invitation to write this book and provided the encouragement to ensure its completion. I would like to acknowledge the assistance of M. Archer, A. E. Ashford, P. Eby, M. D. Fox, H. Godthelp, S. Hand, M. M. Hindmarsh, D. Lunney, H. A. Martin, H. P. Ramsay, B. S. Wannan, J. F. Whitehouse, B. M. Wiecek, and N. C. Wilson in providing information and discussion. Joan Ratcliffe and Sally Durham deciphered my handwriting to produce the typescript, Sally Durham and Alaric Fisher produced many of the figures, and Sue Bullock, Ross Arnett, and Robyn Murphy provided photographic assistance.

For permission to reproduce figures, tables, and photographs acknowledgements are due to:

Jeff Allen ('Eric the Echidna' cartoon Chapter 7)
Prof. M. Archer (7.1.7.2)
Dr J. Ash (4.3)
Dr A. E. Ashford (3.31)
Dr D. Ashton (5.7)
Mr R. Atherton (3.40, Table 2.3)
Mr G. N. Baur (3.16, 5.3, Table 3.5)
Dr F. Bell (3.12, 3.40, Tables 2.3, 3.5)
Dr J. M. Bowler (6.4)
Dr M. Brown (Table 3.3)
Ms P. Eby (9.14)
Dr B. J. J. Embleton (6.1)
Dr R. G. Florence (4.9)
Dr H. A. Ford (Table 7.2)
Dr M. Fox (4.11, 9.15)
Dr R. S. Hill (6.7, Table 6.1)
Dr M. S. Hopkins (1.13, 5.1, 5.2, 5.6, Table 2.2)
Dr D. Jeans, editor, *Australia—A geography* (1.1, 1.5, Table 1.1)
Prof. A. P. Kershaw (6.14, Tables 6.1, 6.3)
Prof. J. Kikkawa (3.11)
Dr M. Lowman (5.9)
Dr H. A. Martin (6.5)
Matthew Martin, *Sydney Morning Herald* (cartoon, Chapter 9)
Mr W. McDonald (3.14)
Prof. H. Nix (1.4, 1.11, 6.15, 6.18)
Dr L. Pahl (3.12, 3.40, Table 2.3)
Dr J. Pickard (3.35)
Dr V. Semeniuk (8.6)
Prof. R. L. Specht (3.10, 4.7, Table 2.1)
Prof. G. C. Stocker (5.5)

Dr E. M. Truswell (6.6)
Dr J. Turner (4.10, Table 4.1)
Dr G. L. Unwin (4.4, 4.5)
Dr J. J. Veevers (6.1)
Prof. D. Walker (6.18)
Dr B. S. Wannan (3.22, 3.6, 9.1)
Dr L. J. Webb (2.3, 3.36–3.39, 6.16, key to rainforest types in Chapter 2)
Dr G. Werren (3.8)
Dr T. C. Whitmore (6.3, 7.3)
Mr J. B. Williams (2.1, 3.15)
Prof. M. A. J. Williams (6.11. 6.12, 6.13)
Mr N. Wilson (3.18, 3.27, 3.28)
Dr J. Winter (3.40)
Academic Press (London) (6.7)
Australian Geological Society (6.2)
Australian Institute of Marine Science (8.6)
Australian Journal of Botany (4.4, 4.5)
Australian Plants (Society for Growing Australian Plants) (3.38, key to rainforest plant communities in Chapter 2)
Biotropica (5.5)
Blackwell Scientific Publications (4.3)
Blackwell Scientific Publications (Aust) (1.13)
Colonial Living, Drummoyne (9.10, 9.11)
Commonweath of Australia (3.4, 3.9, 3.40, 6.10, Table 3.3)
CSIRO Publications (1.13, 2.2, 3.13, 4.4, 4.5, 5.1, 5.2, 5.6, 5.7, Tables 3.1 and 3.2)
Curtis Brown (Aust.) Pty Ltd on behalf of the estate of the late Dorothea Mackellar (lines from 'My Country' in preface)
Director, Royal Botanic Gardens, Sydney (1.10, 3.35)
Drafting Services Branch, Victorian Department of Conservation and Environment (3.26)
Ecological Society of America (3.37, 3.38, 3.39)
Elsevier Science Publishers (6.5, Table 4.1)
Forestry Commission of NSW (3.16, 5.3, 9.3)
Forestry Commission of Tasmania (Table 3.3)
Inkata Press (Table 2.2)
Kluwer Academic Publishers (1.4, 1.11, 3.36, 4.7, 6.16)
Landscope, Department of Conservation and Land Management, W.A. (3.1, 3.2)
Light Fantastic Photography, Adrian Pitman (3.32)
Linnean Society of New South Wales (4.9)
Maxwell MacMillan Publishing Australian Pty Ltd (5.1, 5.2, Table 2.1)
National Museum of Natural Sciences, Ottawa (3.11)
NSW National Parks and Wildlife Service (2.3, 3.8, 3.20, 4.10, 4.11, 6.4, 6.15, 7.4, 9.7, 9.8)
Peacock Press (6.4, 6.15)
Royal Society of Queensland (3.12)
Quaid Real Estate Pty Ltd (9.16)
Queensland Department of Primary Industries (3.14)
Surrey Beatty & Sons Pty Ltd (5.9, 6.14, Tables 7.1, 7.2)

Wet Tropics Management Authority (data on which 9.9 is based)
Working Group for Forest Conservation, Tasmania (3.29, Table 9.2)

Since completion of the manuscript a considerable number of new papers on Australian rainforests has appeared. In addition I have become aware of a number of older but important works. This new body of literature adds substantially to the available data on Australian rainforests and is listed below:

Archer, M., Hand, S. J., and Godthelp, H. (1991). *Riversleigh. The story of animals in ancient rainforests in inland Australia*. Reed, Sydney.
Barker, M. (1990). Effects of fire on the floristic composition, structure, and flammability of rainforest vegetation. *Tasforests* **2**, 117–20.
Baur, G. N. (1987). Rainforests of New South Wales. In *Forest management in Australia*, pp. 313–26. Institute of Foresters of Australia Conference Proceedings 1987, Perth.
Burrett, C., Duhig, N., Berry, R., and Varne, R. (1991). Asian and south-western Pacific continental terranes derived from Gondwana, and their biogeographic significance. *Australian Systematic Botany* **4**, 13–24.
Dale, J. A. and Johnson, G. T. (1987). Hoop pine forest management in Queensland. In *Forest management in Australia*, pp. 283–98. Institute of Foresters of Australia Conference Proceedings 1987, Perth.
Driscoll, P. V. and Kikkawa, J. (1989). Bird species diversity of lowland tropical rainforests of New Guinea and northern Australia. In *Ecological Studies Volume 69, Vertebrates in complex tropical ecosystems* (ed. M. L. Harmelin-Vivien and F. Bouliere), pp. 123–52. Springer, New York.
Dodson, J. R. (1988). The perspective of pollen records to study response, competition and resilience in vegetation on Barrington Tops, Australia. *Progress in physical geography* **12**, 183–208.
Duff, G. A. and Stocker, G. C. (1989). The effects of frosts on rainforest/open forest ecotones in the highlands of north Queensland. *Proceedings of the Royal Society of Queensland* **100**, 49–54.
Haynes, C. D., Ridpath, M. G., and Williams, M. A. J. (ed.) (1991). *Monsoonal Australia, Landscape, ecology and man in northern lowlands*. Balkema, Amsterdam.
Hickey, J. (1990). Change in rainforest vegetation in Tasmania. *Tasforests*, **2**, 143–50.
Hickey, J. E. and Felton, K. C. (1987). Management of Tasmanian cool temperate rainforest. In *Forest management in Australia*, pp. 327–42. Institute of Foresters of Australia Conference Proceedings 1987, Perth.
Hickey, J. E., Brown, M. J., Rounsevell, D. E., and Jarman, S. J. (1990). *Tasmanian rainforest research*. Proceedings of a Seminar on Rainforest Research, Tasmanian National Rainforest Conservation Program Report No. 1, Forestry Commission, Hobart.
Hill, R. S. (1990). Sixty million years of change in Tasmania's climate and vegetation. *Tasforests* **2**, 89–98.
Hill, R. S. and Read, J. (1991). A revised infrageneric classification of *Nothofagus* (Fagaceae). *Botanical Journal of Linnean Society* **105**, 37–72.
Jarman, S. J., Kantvilas, G., and Brown, M. J. (1986). The ecology of pteridophytes in Tasmanian cool temperate rainforest. *Fern Gazette* **13**, 77–86.
Jarman, S. J., Kantvilas, G., and Brown, M. J. (1991). *Floristic and ecological studies in Tasmanian rainforest*. Tasmanian NRCP Technical Report No. 3. Forestry Commission, Tasmania and Department of the Arts, Sport, the Environment, Tourism and Territories, Canberra.
Jones, R. L. (1990). Late Holocene vegetational changes on the Illawarra Coastal Plain, New South Wales, Australia. *Review of Palaeobotany and Palynology* **65**, 37–46.

Just, T. E. (1987). Management of tropical rain forests in north Queensland. In *Forest management in Australia*, pp. 299–322. Institute of Foresters of Australia Conference Proceedings 1986, Perth.

Kantvilas, G. (1990). Succession in rainforest lichens. *Tasforests* 2, 167–72.

Kodela, P. G. (1990). Modern pollen rain from forest communities on the Robertson Plateau, New South Wales. *Australian Journal of Botany* 38, 1–24.

Kodela, P. G. (1990). Pollen-tree relationships within forests of the Robertson–Moss Vale region, New South Wales, Australia, *Review of Palaeobotany and Palynology* 64, 273–9.

Laurance, W. F. (1991). Ecological correlates of extinction proneness in Australian tropical rainforest mammals. *Conservation Biology* 5, 79–89.

Martin, P. G. and Dowd, J. M. (1991). Application of evidence from molecular biology to the biogeography of angiosperms. *Australian Systematic Botany* 4, 111–16.

McKenzie, N. L., Johnston, R. B., and Kendrick, P. G. (ed.) (1991). *Kimberley rainforests*. Surrey Beatty, Sydney.

Melick, D. R. (1990). Ecology of rainforest and sclerophyllous communities in the Mitchell River National Park, Gippsland, Victoria. *Proceedings of the Royal Society of Victoria* 102, 71–87.

Melick, D. R. and Ashton, D. H. (1991). The effects of natural disturbance on warm temperate rainforests in south-eastern Australia. *Australian Journal of Botany* 39, 1–30.

Michaux, B. (1991). Distributional patterns and tectonic development in Indonesia: Wallace reinterpreted. *Australian Systematic Botany* 4, 25–36.

Morgan, G. J. (1988). Freshwater crayfish of the genus *Euastacus* Clark (Decapoda: Parastacidae) from Queensland. *Memoirs of the Museum of Victoria* 49, 1–49.

Nix, H. A. and Switzer, M. A. (1991). *Rainforest animals. Atlas of vertebrates endemic to Australia's wet tropics*. Australian National Parks and Wildlife Service, Canberra.

Podger, F. (1990). *Phytophthora*, fire and change in vegetation. *Tasforests* 2, 125–8.

Poore, D. (ed.) (1989). *No timber without trees. Sustainability in the tropical forest*. Earthscan Publications, London.

Russell-Smith, J. (1991). Classification, species richness, and environmental relations of monsoon rain forest in northern Australia. *Journal of Vegetation Science* 2, 259–78.

Summerbell, G. (1991). Regeneration of complex notophyll vine forest (humid subtropical rainforest) in eastern Australia—a review. *Cunninghamia* 2, 391–410.

Vanclay, J. K. (1990). Effects of selection logging on rainforest productivity. *Australian Forestry*, 53, 200–14.

Veevers, J. J. (1991). Phanerozoic Australia in the changing configuration of Proto-Pangea through Gondwanaland and Pangea to the present dispersed continents. *Australian Systematic Botany* 4, 1–12.

Whitehouse, J. F. (1991). East Australian rain-forests: a case-study in resource harvesting and conservation. *Environmental Conservation* 18, 33–43.

New South Wales
June 1991

P.A.

CONTENTS

1 Introduction: The Australian environment — 1
   1.1 Landforms — 1
   1.2 Climate — 7

2 Rainforest—Definition and classification — 13
   2.1 Defining rainforest — 13
   2.2 Australian vegetation classification — 13
      2.2.1 Incorporating dynamics into the definition of rainforest — 17
   2.3 Sclerophylly — 18
   2.4 Types of rainforest — 19
      2.4.1 New South Wales forest classification — 19
      2.4.2 Floristic classifications — 22
      2.4.3 Physiognomic classifications — 24
   2.5 Phytogeographical analysis of rainforest floras — 30
   2.6 Classification framework used in this book — 32

3 An overview of Australian rainforests — 33
   3.1 North-west Australia — 33
   3.2 The Northern Territory — 37
   3.3 Northern Cape York Peninsula — 43
   3.4 The wet tropics of north-east Queensland — 46
   3.5 Rainforests of provinces A_1 and A_2—southern Queensland and New South Wales — 51
   3.6 Rainforests of Victoria — 63
   3.7 Tasmanian rainforests — 66
   3.8 The rainforests of provinces C_1 and C_2 — 66
   3.9 Rainforests of the Great Barrier Reef Islands — 70
   3.10 Norfolk Island — 71
   3.11 Lord Howe Island — 72
   3.12 What is rainforest—the problem revisited — 73
      3.12.1 What is warm temperate rainforest? — 75
   3.13 Trends of variation in Australian rainforest — 79
   3.14 How much rainforest? — 83
   3.15 Australian rainforest in a world context — 84

4 Rainforest boundaries and the problem of mixed forests — 86
   4.1 North-east Queensland — 86
      4.1.1 Stability of boundaries — 87
   4.2 South-east Australia — 91
   4.3 Can mixed forest be stable? — 95
      4.3.1 Conclusions — 99
   4.4 Evolution of tall open forest — 99
   4.5 The definition of rainforest revisited — 100

5 Regeneration and responses to disturbance — 103
- 5.1 The gap-filling model of forest regeneration — 103
  - 5.1.1 Regeneration in rainforests with species-poor canopies — 110
  - 5.1.2 Nutrient distribution following disturbance — 114
- 5.2 Aspects of the life cycle of rainforest plants — 114
  - 5.2.1 Flowering, pollination, and fruit-set — 115
  - 5.2.2 Seed dispersal — 117
  - 5.2.3 The seed bank — 119
  - 5.2.4 Germination — 122
  - 5.2.5 Seedling establishment and mortality — 122
  - 5.2.6 Epiphytic tree establishment — 123
  - 5.2.7 Multistemmed trees in rainforest — 124
  - 5.2.8 The longevity of rainforest trees — 124
- 5.3 Disturbances to rainforest — 125
  - 5.3.1 Fire — 125
  - 5.3.2 Volcanoes — 128
  - 5.3.3 Cyclones and storms — 128
  - 5.3.4 Frosts and snow — 129
  - 5.3.5 Flood — 130
  - 5.3.6 Landslips — 131
  - 5.3.7 Pathogens — 131
  - 5.3.8 The role of insects — 133
- 5.4 The promotion of rainforest taxa in sclerophyll communities — 135

6 Origins and history of Australian rainforest — 137
- 6.1 Plate tectonic history of Australia — 137
  - 6.1.1 Topographic change — 139
- 6.2 Climatic history — 142
- 6.3 Vegetation history — 143
- 6.4 Interpretation — 148
- 6.5 Regional variation in rainforest vegetation — 150
  - 6.5.1 Temperate rainforest — 150
  - 6.5.2 Monsoonal forests — 155
- 6.6 Primitive angiosperms in Australian rainforests — 157
- 6.7 The Quaternary — 160
- 6.8 The evidence from present distribution — 166
- 6.9 A synthesis — 167
- 6.10 Rainforest refugia — 169
- 6.11 The vagility of rainforest taxa — 169
  - 6.11.1 New Zealand and New Guinea — 171
- 6.12 The sclerophyll flora — 176
- 6.13 The non-vascular flora — 178

7 The rainforest fauna — 180
- 7.1 Mammals — 180
  - 7.1.1 History of rainforest mammals — 184

	7.2	Avifauna	191
		7.2.1 Origins of the avifanua	193
	7.3	Herpetofauna	195
	7.4	Invertebrates	196
	7.5	Aquatic fauna	197
	7.6	Community organization	198
8	Mangroves		200
	8.1	Australian mangroves in a world context	201
	8.2	General features of mangroves	201
	8.3	Distribution of mangroves in Australia	202
		8.3.1 Distribution within estuaries	204
		8.3.2 Dynamics of mangroves	204
	8.4	Coping with the environment	204
	8.5	Natural disturbance in mangroves	206
	8.6	Human impacts	208
	8.7	Fauna of mangroves	209
	8.8	The ecological role of mangroves	211
	8.9	History	212
9	The human influence		214
	9.1	Aboriginal history	214
		9.1.1 Aborigines and rainforest	215
		9.1.2 The impact of Aborigines on rainforest	217
	9.2	European history	219
		9.2.1 Agriculture	222
		9.2.2 Forestry	225
		9.2.3 Other forest products	231
		9.2.4 Mining industry	233
	9.3	Extinctions and introductions	234
	9.4	Conservation	236
		9.4.1 The role of the federal government in conservation and land-use decisions	239
		9.4.2 The New South Wales rainforest nomination	244
		9.4.3 The north Queensland rainforests	246
	9.5	The forestry versus conservation debate	248
		9.5.1 The importance of the World Heritage List	257
	9.6	Australian rainforests in a world context	258
	9.7	The future of Australia's rainforests in a 'greenhouse' world	259
	9.8	Management issues	260
		9.8.1 Threats from development	264
	9.9	Australian rainforest—a paradigm for world rainforest	265

References 268

Index 295

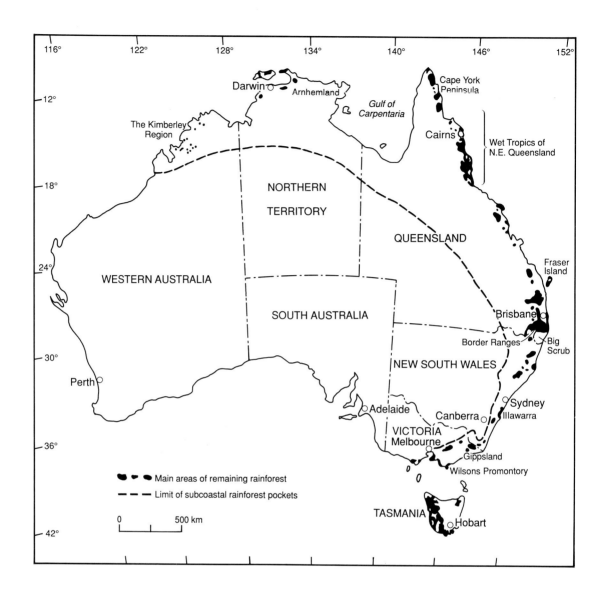

1 INTRODUCTION: THE AUSTRALIAN ENVIRONMENT

At the present day, rainforest in Australia is very limited in its distribution (opposite). It was once much more widespread, and the contraction to its present limits reflects environmental changes over many millions of years (Chapter 6). In order to provide the background for understanding why rainforest survives where it does, and is absent elsewhere, this chapter presents an overview of the environment of the Australian continent. More detailed recent reviews have been provided by Nix (1981), Bridgewater (1987), and in Jeans (1986).

1.1 LANDFORMS

The most outstanding feature of the Australian continent is its flatness. The mean altitude of Australia, although low, is marginally higher than that of Europe (Table 1.1): most of the continent consists of plateaux (Fig. 1.1) and coastal lowlands are limited in extent (Jennings and Mabbutt 1986).

The most extensive upland areas extend some 2500 km along the eastern margin of the continent and are generally referred to as the Eastern

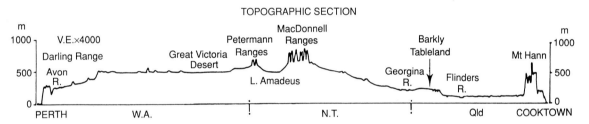

Fig. 1.1 Diagrammatic cross-section of Australia, from Perth in the south-west to Cairns in the north-east (from Jennings 1986).

Table 1.1 Physiographic comparison of the continents (from Jennings and Mabbutt 1986)

	Australia	Europe	Asia (including Europe)	Africa	North America	South America	Antarctica
Area (km² × 10⁶)	7.7	10	54.2	29.8	24.2	18	13.1
Area including continental shelf	11	10.8	65	31	31	20	15
Mean altitude (m)	330	290	860	660	780	650	2200
Highest point (m)	2228	5633	8848	5895	6133	6959	5140
Lowlands below 200 m (%)	39	57	25	10	30	38	6
Plateaux etc. above 1000 m (%)	2	7	30	23	27	13	86

Fig. 1.2 The dissected nature of the Blue Mountains, west of Sydney

Highlands or Great Dividing Range. The highest parts of the Great Dividing Range are in southeastern Australia and are often referred to as the Australian Alps (and include the highest point on the continent, Mt. Kosciusko—2228 m) but only limited segments reach 1500 m and in the north, in Cape York, the Eastern Highlands peter out into low ranges of only a few hundred metres elevation. Although the Eastern Highlands are rugged, the most impressive topographic features are not peaks but deeply incised valleys and gorges (Fig. 1.2).

Although the Eastern Highlands separate the short eastward-flowing rivers from the Murray–Darling Basin and drainage to the Gulf of Carpentaria (hence the Great Dividing Range), the actual watershed is rarely an impressive topographical feature. Although it can be delineated on maps, on the ground the actual divide is often barely perceptible on broad plateaux (Ollier 1986). As Jennings and Mabbutt (1986) observed, 'An aircraft leaving from Cooma airport starts at one side of the Great Dividing Range and takes off on the other. Whatever the official decrees, no self-respecting physical geographer can employ such a misnomer'. The eastern margin of the Eastern Highlands is, however, marked by a striking landscape feature—the Great Escarpment (Ollier 1986—see Fig. 1.3 and also Chapter 6).

Australia also has very old landforms. In much of northern Europe and North America the landscape was subject (mostly on more than one

Fig. 1.3 The Great Escarpment from Point Lookout, New England National Park, New South Wales

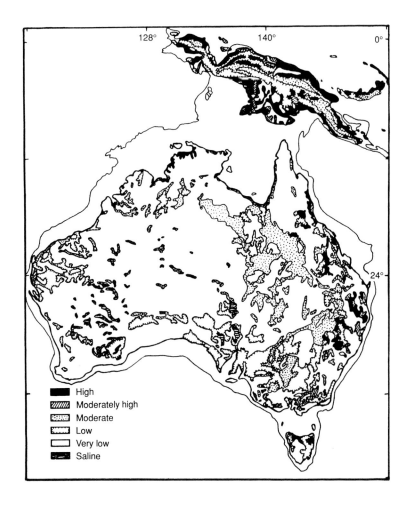

Fig. 1.4 Soil fertility in Australia (from Nix 1981).

occasion) to Quaternary glaciation. This means that over much of these areas soils are less than 20 000 years old and landforms were moulded very largely by glacial or periglacial processes. In Australia the last major glaciation was in the Permian, 250 Ma, although, during the Quaternary, limited areas of the highest parts of the Eastern Highlands and in Tasmania were subject to glacial or periglacial activity. Over much of the continent, landforms have evolved subaerially since the Permian, and, in limited areas which escaped the Permian glaciation, geomorphology may reflect even older times (Ollier 1986). Since the Permian, parts of the continent have been

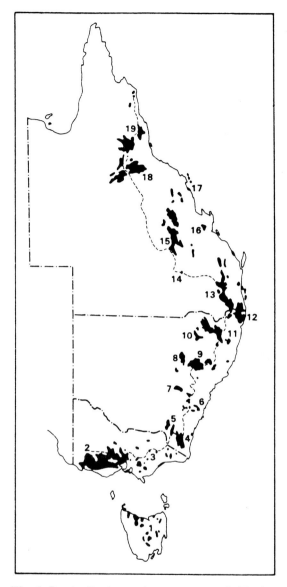

Fig. 1.5 Distribution of exposures of Tertiary volcanic rocks in eastern Australia in relation to the Great Dividing Range (----). 1 Tasmania; 2 newer volcanics of Victoria; 3 older volcanics of Victoria; 4 Monaro; 5 Snowy; 6 Abercrombie; 7 Orange; 8 Warrumbungles; 9 Liverpool; 10 Nandewar; 11 Inverell; 12 Tweed; 13 Toowoomba; 14 Mitchell; 15 Springsure; 16 Rockhampton; 17 Cape Hillsborough; 18 Nulla; 19 Mount Surprise (from Ollier 1986).

Fig. 1.6 Mount Warning in the caldera of the Tweed Volcano, north-east New South Wales. The rainforests on Mount Warning are included on the World Heritage List (see Chapter 9).

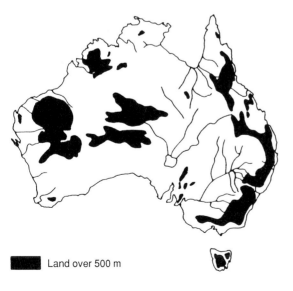

Land over 500 m

Fig. 1.7 Areas of Australia above 500 m elevation

periodically submerged under shallow seas (Chapter 6), and this has not only permitted deposition of marine sediments but subsequently gave a new start to the development of landforms.

Much of Australia is mantled by soils which are, by world standards, very old and which have not been rejuvenated by recent glaciation. In consequence, many of these soils are extremely infertile (Fig. 1.4) with, by world standards, exceptionally low phosphorus concentrations (Adam *et al.* 1989).

There are, however, extensive areas, concentrated along the Eastern Highlands, of more fertile soils (Fig. 1.4) and these correspond with areas of volcanic rocks (Fig. 1.5, see also Chapter 6). These volcanic areas also exhibit some spectacular erosional landforms (Fig. 1.6). The most recent volcanic activity, in western Victoria, was within the last few thousand years.

1.2 CLIMATE

The climate of Australia reflects both the position of the continent astride the Tropic of Capricorn and its topography.

The isohyets for mean annual precipitation are a close reflection of topography (Figs. 1.7, 1.8), particularly in the east. Much of the continent is arid or semi-arid (annual rainfall <500 mm). High annual rainfall (>1000 mm) occurs in northern Australia, along the eastern margin of the Eastern Highlands, in Tasmania, and in the south-west. Annual rainfall figures do not give any indication of seasonality (Fig. 1.9); rain in northern Australia and in the south-west is strongly seasonal in occurrence. Along much of the east coast, rain occurs throughout the year but with seasonal maxima in the summer (in the north, Fig. 1.9) or the winter (in the south). The pattern

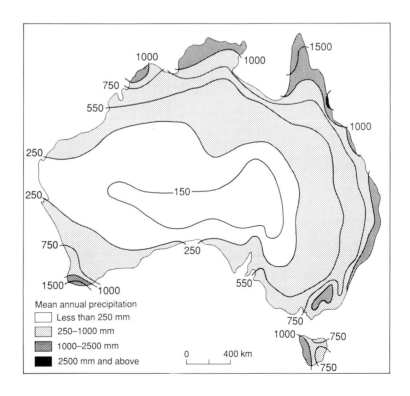

Fig. 1.8 Mean annual precipitation in Australia (Perrens 1982, from data from the Commonwealth Bureau of Meterology).

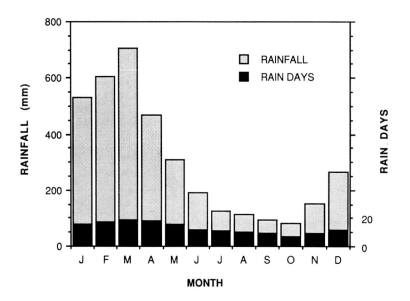

Fig. 1.9 Mean monthly rainfall and rain days at Innisfail, northeast Queensland (drawn from data in Tracey 1982), showing pronounced summer maximum in rainfall but less variation in the number of rain days per month

of rainfall distribution may, however, be a poor measure of availability of water for plant growth. Hopkins and Graham (1989) applied a water balance model to a lowland rainforest in north-east Queensland. Although the wettest season of the year is between November and April, the model predicted several short periods of shortage of available water. Conversely the model did not predict shortages during the 'dry' August to October period. Although the summer is the wet season, rain is not a daily occurrence; as the summer is the time of highest temperatures and radiation inputs, evaporation rates are high, leading to intermittent short periods of water shortage. Although the community peak litter-fall was in winter (patterns for individual species are not all synchronized) the availability of water for plant growth throughout the winter suggests that this leaf-fall peak was not induced by drought (Hopkins and Graham 1989).

As well as seasonality in water availability, there is also considerable variation in rainfall between years. Australia is notorious for extensive and prolonged drought, and very few parts of the continent have escaped intermittent drought during the period of recorded climatic data.

The variability in rainfall, even at sites supporting rainforest, is well illustrated in the record for Mt. Tomah between 1969 and 1985, shown in Fig. 1.10. Mt. Tomah, in the Blue Mountains west of Sydney, would be regarded by the general public as having a reliable and high annual rainfall.

Most precipitation is rainfall. On the higher peaks in the Eastern Highlands and in Tasmania, effective precipitation may be considerably augmented by cloud stripping (Ellis 1971). Regular winter snowfall is restricted to higher peaks in Tasmania and the Victorian and New South Wales Alps. Infrequent snowfall has been recorded on the Eastern Highlands as far north as Central Queensland.

Hailstorms are individually localized but are widespread in occurrence. Although making an insignificant contribution to precipitation, hail may cause considerable, although temporary, damage to vegetation canopies.

The patterns of temperature distribution

Fig. 1.10 Monthly rainfall 1969–85 at Mount Tomah, in the Blue Mountains west of Sydney (redrawn from Rodd 1987).

INTRODUCTION: THE AUSTRALIAN ENVIRONMENT

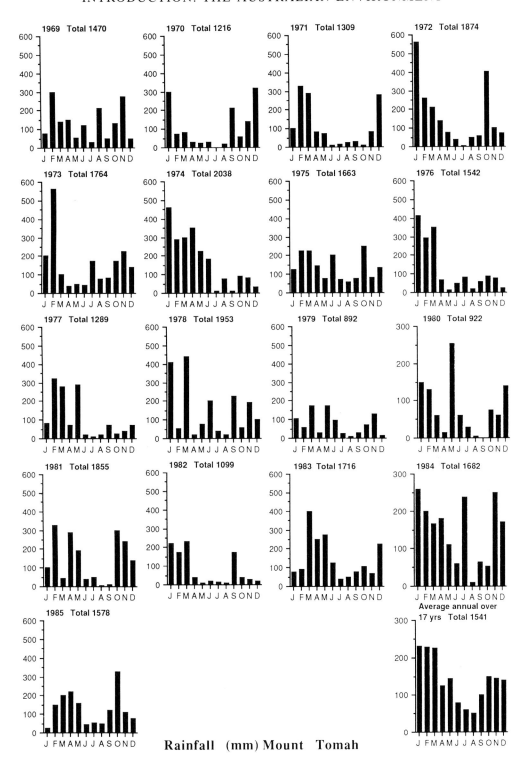

Rainfall (mm) Mount Tomah

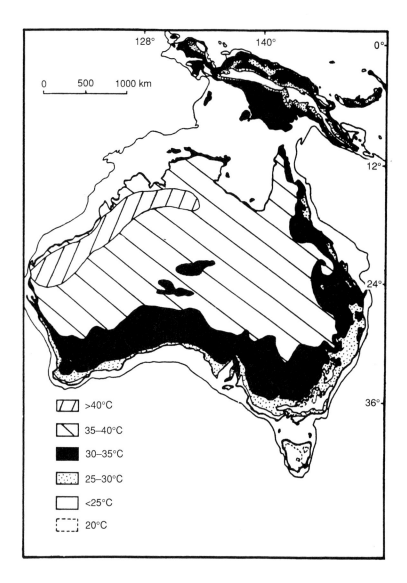

Fig. 1.11 Mean daily maximum temperature of the hottest week (from Nix 1981).

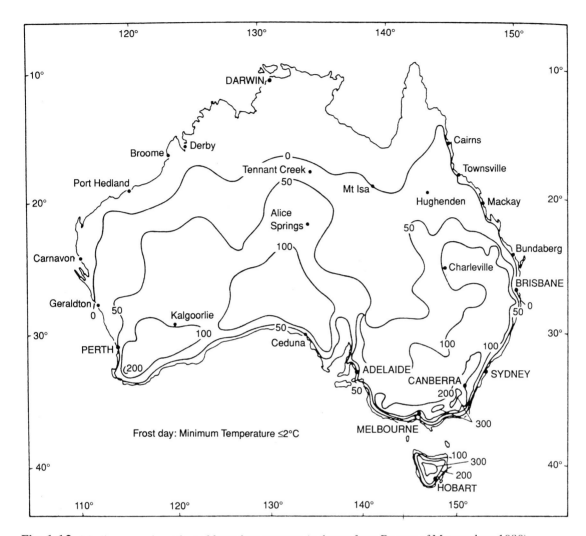

Fig. 1.12 Median annual number of frost days per year (redrawn from Bureau of Meteorology 1989).

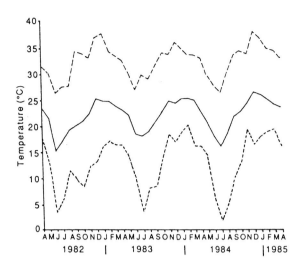

Fig. 1.13 Mean monthly temperatures (solid line) and monthly maximum (upper dashed line) and minimum (lower dashed line) temperatures recorded in a clearing within lowland complex mesophyll vine forest at Liverpool Creek in north-east Queensland (from Hopkins and Graham 1989).

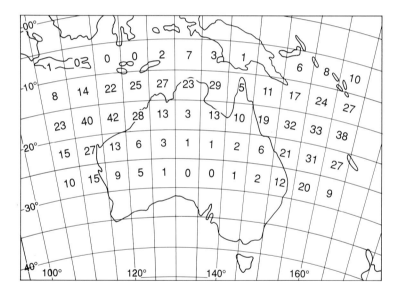

Fig. 1.14 Average decadal incidence of cyclones occurring in 5° latitude/longitude squares July 1959–June 1975 (redrawn from Lourensz 1977).

largely reflect latitude, but topography and cloud cover cause significant lowering of temperatures towards the east coast (Fig. 1.11). Frost is a predictable occurrence well north of the Tropic of Capricorn (Fig. 1.12). Even in the coastal lowlands of north-east Queensland, winter night temperatures, although not resulting in frost, may drop below 5° (Fig. 1.13).

Australia lies across the cyclone belt. The highest incidence of cyclones is off the north-western coast but cyclones are also frequent on the east coast (Fig. 1.14). As cyclones move across the coast and decline, they may become intense rain-bearing depressions responsible for high rainfall intensity events (for example Cyclone Peter dumped 1947 mm on Mt. Bellenden Ker, north-east Queensland, in 48 hours on 4–5 January 1979).

2 RAINFOREST—DEFINITION AND CLASSIFICATION

... in the coastal belt of Queensland, two contrasting types of vegetation are found side by side: the indigenous Australian 'open' forest and the rain-forest or 'jungle'. The distinction between them is sharp and definite, for they rarely intergrade. Seen from some high vantage-point the two formations stand out in the landscape as though they had been painted in different colours—the one sparse and pale, the other dark and close-grained ...

... Not only is the jungle utterly unlike the open forest in appearance; botanically it is both quite different and much more complex. In a square mile of open forest, it will often be impossible to find as many as half-a-dozen different species of trees. In a similar area of jungle there will probably be sixty. Moreover, quite a large proportion of this great variety of rain-forest species will bear fruit with some pretence of pulp and succulence ...

... Half the secret of its first appeal, I think, lies in the fact that the jungle at its best completely fulfils the city-bred westerner's idea of what tropical vegetation should be like. There is a luxuriance and gloom, broken here and there by shafts of light which filter through the dense canopy overhead and throw into silhouette the knotted vine-rope and the queerly sculptured trunks. The trees support a veritable garden of aerial ferns and orchids, which stick like brackets to the branches. In the clearings grow plants with large dark shining leaves, and palms, and treeferns. The silence is broken by the strange voices of invisible birds. (Ratcliffe 1938)

These quotations stress the sharp disjunction, structural and floristic, between rainforest and non-rainforest in Australia and illustrate the distinctive features of at least some types of rainforest. If rainforest is indeed so distinct then its definition should pose no problems. Nevertheless, in recent years the definition of rainforest has been much disputed and has even been the subject of formal inquiries and legal action (see Chapter 9). Some of the vegetation which Australians would agree is rainforest would scarcely seem so to a visitor from overseas. Ratcliffe (1938) highlighted a real phenomenon, but it is one that has proved surprisingly difficult to encapsulate within simple definitions.

In this chapter, two issues are addressed: distinguishing rainforest from non-rainforest and the recognition of different types of rainforest.

2.1 DEFINING RAINFOREST

In general discussion, the term rainforest normally refers to the evergreen wet forest of the lowland tropics—the Tropical Rainforest. The term was coined by the German phytogeographer, Schimper (1903), who employed a broad definition whereby the essential features of rainforest are that it is evergreen and hygrophilous. Tropical rainforest ('tropische Regenwald') was one category of rainforest, which fitted the diagnosis 'Evergreen, hygrophilous in character, at least 30 m high, but usually much taller, rich in thick-stemmed lianes and in woody as well as herbaceous epiphytes'. Outside the tropics, Schimper (1903) suggested that rainforest could occur in any temperate region with a mild winter and abundant rain (at least 1200 mm per annum), with no dry season and decribed as rainforest a number of vegetation types from various parts of the world.

2.2 AUSTRALIAN VEGETATION CLASSIFICATION

How is rainforest defined in Australia, and does the definition differ markedly from that originally proposed by Schimper? In order to answer these questions it is appropriate to review vegetation classification in Australia to see how the concept of rainforest has changed over the years.

When Australia was colonized by northern

Europeans, the vegetation was clearly markedly different from any previously experienced. Although a number of vegetation types could be distinguished, they differed in species composition, structure, and physiognomy from any North European vegetation types. Giving names to these types was clearly important, not as an adjunct to some scientific classification, but for everyday converse. Very few terms were taken up from Aboriginal languages; for the most part, words already existing in the English language were pressed into use. Even at this early stage of European contact, the vegetation which is now referred to as rainforest was recognized as distinct from the predominant eucalypt forest and woodland (see Chapter 6), although formal diagnosis of vegetation types was not attempted.

In the case of rainforest, the most frequently used terms were scrub and brush, and many rainforest stands are still known by these names. Both terms were in use in the early years of European settlement and were not exclusive to rainforest; rather they were applied to any vegetation type which was difficult to penetrate.

Why scrub and brush were applied to such complex vegetation as rainforest is unclear; Vader (1987) suggests that they derive from American usage. Whatever their origin, they fail to convey the nature of the forests. Hodgkinson (1845) was moved to write:

I must here make a digression to attempt to convey to the English reader some idea of the very peculiar appearance of that kind of vegetation to which the colonists have assigned the unmeaning name of brush. It grows on the richest alluvial land, and consists of trees of almost endless variety, and very large dimensions, totally differing in appearance from the ordinary Eucalypti and Casuarinae, which grow in the common open forests of Australia, for the brush trees in general possess a rich umbrageous foliage of bright shining green ... But the peculiar appearance of the brush is principally caused by the countless species of creepers, wild vines, and parasitical plants of singular conformation, which interlaced and entwined in inextricable confusion, bind and weave together the trees almost to their summits, and hang in rich and elegant flowering festoons from the highest branches.

When more scientific studies of vegetation were developed, their nomenclature built upon this vernacular tradition, rather than starting again from scratch, despite the lack of antiquity of the tradition. More recent classifications, which have proposed new terminologies, have achieved limited success—such is the attachment to earlier nomenclature. The result of attempts to produce more and more colours from a very limited palette are terminologies which, by permitting oxymorons, such as 'dry rainforest', are on first encounter confusing to any non-Australian. It is perhaps best to regard the names of plant communities as simply labels, the concepts to which they apply being explained by a formal diagnosis, rather than attempting to derive a definition from apparent etymology. Unfortunately, the tendency to read more into the connotations of names than was intended originally is almost irresistible.

The formal classification of Australian vegetation developed slowly, from a variety of local studies. A history of descriptive and classificatory studies is provided by Johnston and Lacey (1984). The majority of these studies could be regarded as intuitive, in that they recognized units which the authors considered to be visually distinct and then provided diagnoses of these units using a variety of attributes; habitat, structure, physiognomy, and floristics. Beadle and Costin (1952) proposed a classification system based on floristics and structure, which was intended to provide a uniform national framework for ecological studies. Structure was used to delimit major formations, which were essentially units which had long been recognized, such as sclerophyll forest, woodland and rainforest. Within these formations, subdivision was on the basis of floristics of the dominant stratum, and not total stand composition. The basic unit was the association which was defined as a 'climax community of which the dominant stratum has a qualitatively uniform floristic composition and which exhibits a uniform structure'. Associations were identified 'by eye' rather than on the basis of analysis of quantitative data. In survey work, the recognition of particular stands as being in a climax state must involve a degree of faith. Associations were amalgamated into alliances, defined as groups of 'floristically related associations of similar structure'.

The formations proposed by Beadle and Costin (1952) formed the units for several subsequent vegetation maps and the framework for numerous ecological studies. Rainforest was recognized as a single formation with four subformations; temperate rainforest, subtropical rainforest, tropical rainforest, and monsoon rainforest. Use of floristic criteria to recognize associations and alliances has occurred, but the diagnoses are difficult to apply. Some units, particularly relatively species-poor ones, are easily characterized and recognized; subdivisions within more species-rich stands are less readily defined and although given names, the lack of rigorous quantitative criteria for definition may make it difficult for other workers to apply such classifications with any degree of confidence.

These various 'intuitive' classifications accepted that there is an entity in Australia which can be called rainforest and provided, in some cases, means of subdividing rainforests. Other than by learning by experience, how was rainforest to be defined? Rainforest definition has proved elusive (Baur 1957; Werren and Allworth 1982), and has often been avoided. Even the broadest of definitions have not been immune from criticism (see M. Dale *et al.* 1980; M. Dale 1981; Felton 1981).

Such definitions as have been advanced clearly reflect an affinity to Schimper's definition of tropical rainforest. Beadle and Costin (1952) suggested that rainforest is a 'closed community dominated by usually mesomorphic meso- or mega-phanerophytes forming a deep, densely interlacing canopy in which lianes and epiphytes are invariably present, with mesomorphic subordinate strata of smaller trees, shrubs and ferns and herbs'; the inclusion of 'usually' permitting an unspecified degree of flexibility in interpretation, although the claim that lianes and epiphytes are invariably present would exclude some vegetation types which by 1952 were, by convention, regarded as 'rainforest'. Webb (1959) thought it easier to define rainforest negatively as being different from other communities, but in a positive diagnosis suggested that 'rain forest, excluding its transitions, is essentially a closed forest, with closed spaced trees generally arranged in several more or less continuous storeys, the uppermost of which (the canopy level) may be even or uneven. Rain forest is distinguished from other closed canopy forests by the prominence of life-forms, such as epiphytes and lianes, by the absence of annual herbs on the forest floor, and by its floristic complexity'. Despite this, Webb (1959) included within his classification some communities with low floristic diversity and an absence of lianes.

Baur (1965) offered a similar definition: '*Rainforest*. A closed moisture-loving community of trees, usually containing one or more subordinate storeys of trees and shrubs; frequently mixed in composition; the species typically, but not invariably, broadleaved and evergreen; heavy vines (lianes), vascular and non-vascular epiphytes, stranglers and buttressing often present and sometimes abundant; floristic affinities mainly with the Antarctic or Indo-Malaysian floras; eucalypts typically absent except as relics of an earlier community'. This diagnosis has sufficient flexibility ('usually', 'typically, but not invariably') to encompass virtually all communities conventionally regarded as rainforest in New South Wales, but this very flexibility is potentially confusing for the novitiate.

These definitions indicate that the diversity of life-forms and structural complexity of the stand were important features distinguishing rainforest from other vegetation types.

Although classifications derived from the Beadle and Costin scheme are still widely used, there has been a more recent tendency to favour the classification developed by Specht (1970, 1981*a*). This scheme partitions structure (the spatial distribution of biomass) into categories using two attributes, the projective foliage cover of the tallest stratum and the life-form and height of the tallest stratum. The 1981 version of the

Table 2.1 The structural classification of Australian vegetation proposed by Specht (1981a)

Life-form of tallest stratum		Foliage projective cover of tallest stratum				
		100–70% (4)†	70–50% (3+)	50–30% (3−)	30–10% (2)	<10% (1)
Trees*>30 m	(T)†	Tall closed forest	Tall forest	(Tall open forest)‡	(Tall woodland)‡	—
Trees 10–30 m	(M)	Closed forest	Forest	Open forest	Woodland	Open woodland
Trees <10 m	(L)	Low closed forest	Low forest	Low open forest	Low woodland	Low open woodland
Shrubs*>2 m	(S)	Closed scrub	Scrub	Open scrub	Tall shrubland	Tall open shrubland
Shrubs 0.25–2 m						
Sclerophyllous	(Z)	Closed heathland	Heathland	Open heathland	Shrubland	Open shrubland
Non-sclerophyllous	(C)	—	—	Open shrubland	Low shrubland	Low open shrubland
Shrubs <0.25 m						
Sclerophyllous	(D)	—	—	— (fell-field)	Dwarf open heathland (fell-field)	Dwarf open heathland
Non-sclerophyllous	(W)	—	—	—	Dwarf shrubland	Dwarf open shrubland
Hummock grasses	(H)	—	—	—	Hummock grassland	Open hummock grassland
Herbaceous layer						
Graminoids	(G)	Closed (tussock) grassland	(Tussock) grassland	(Tussock) grassland	Open (tussock) grassland	Very open (tussock) grassland
Sedges	(Y)	Closed sedgeland	Sedgeland	Sedgeland	Open sedgeland	Very open sedgeland
Herbs	(X)	Closed herbland	Herbland	Herbland	Open herbland	Very open sedgeland
Ferns	(F)	Closed fernland	Fernland	Fernland	—	—

* A tree is defined as a woody plant usually with a single stem; a shrub is a woody plant usually with many stems arising at or near the base.
† Symbols and numbers in parentheses may be used to describe the formation, e.g. a tall closed forest—T4, hummock grassland—H2.
‡ Senescent phases of Tall forest.

classification is presented in Table 2.1, it differs from the earlier (1970) model in having five, rather than four, categories of cover class.

There are certain practical difficulties in applying this classification exactly as prescribed by Specht (1970, 1981a). The cover parameter is projective foliage cover, the percentage of the sample site occupied by the vertical projection of foliage. In forest vegetation this can be estimated using an optical point quadrat, but this can be a tedious, time-consuming operation. Frequently the line of sight to the upper stratum is obscured by lower foliage and movement of leaves in the breeze may restrict measurement to the early morning or evening. If the upper stratum contains deciduous species, the method is inappropriate for part of the year. Specht (1970) indicated that communities with deciduous canopy species are to be classified on the basis of foliage projective cover at the time of maximum leaf development but although he stressed the relationship between foliage projective cover and communities' use of water and energy (relationships questioned by Johnston and Lacey 1984), most users of the classification are not, in the first instance, interested in prediction of ecosystem functions, but rather require simple pigeon-holes for communication and purposes such as vegetation mapping.

Many of the applications of the scheme have not involved precise measurement of projective foliage cover but a quick visual estimate which probably approximates to crown cover (in which tree canopies are regarded as more or less opaque). The structural classification developed by Walker and Hopkins (1984) has similarities with that of Specht, but formally adopts crown cover, rather than projective foliage cover, as the measure of cover, and treats rainforest separately from other vegetation types.

Difficulties can also arise with the definition of the upper stratum. Do a few emergents (say 2–3 per hectare) constitute a very open upper stratum? This question is of some significance in defining the limits of rainforest when eucalypts are emergents over a closed lower canopy of rainforest species.

Specht (1970) in both text and the accompanying vegetation map synonymized closed forest with rainforest, thus apparently solving the problem of rainforest definition. However, there are closed canopy communities (falling into Specht's low closed-forest category) which on other criteria would not be regarded as rainforest. Seddon (1984) observed that 'The tendency in Australia for botanists in their weaker moments to used "closed-forest" as if it were synonymous with "rainforest" really indicates rather clearly that this is the only forest form that is at all common in Australia that *does* cast a dense shade. It is so different from the familiar, light-drenched open-forest dominated by eucalypts— what instinctively springs to the mind's eye when we think of "the bush". The closed canopy is, moreover, the key feature of rainforest in that it then makes possible all the special life-forms we have discussed, and these are the unique features of rainforest.'

We can, therefore, terminate the debate as to what are rainforests by accepting that, in the Australian context, they fall into the closed canopy category defined by Specht (1981a), with the proviso that some species are deciduous and so do not always have a closed canopy, and with the further qualification that there are some closed canopy forests dominated by species which by convention are not rainforest species.

The definition appears, at least potentially, to differ from that of Schimper (1903). The definitions both refer to closed woody vegetation, but do the Australian communities meet Schimper's other criteria of being evergreen, hygrophilous, and occurring at sites with at least 1200 mm annual rainfall? As will become clear in Chapter 3, most, but not all, of those communities which are rainforests in the Specht scheme also meet Schimper's criteria.

2.2.1 Incorporating dynamics into the definition of rainforest

Most definitions of plant communities are static, that is, they apply to descriptions of stands at particular moments in time. However, it is well known that different aspects of any area of vegetation are changing, at various rates, over time. One of the main drawbacks of most descriptive ecology is the failure to address the dynamic nature of vegetation.

In Australia, there are major differences between rainforest and sclerophyll forests in their response to various disturbances and in the role of disturbance in the normal processes of stand regeneration (see Chapters 4 and 5).

The recently developed classification of Tasmanian rainforest (Jarman and Brown 1982, 1983; Jarman *et al.* 1984, 1987) adopts a formal definition of rainforest which relies on the reproductive characteristics of the key species. Jarman and Brown considered the problem of defining rainforest in the Tasmanian context at some length, and concluded that a floristic approach was most appropriate. Rainforest was considered to be forest vegetation (with trees greater than 8 m high) dominated by species of *Nothofagus*, *Eucryphia*, *Atherosperma*, *Athrotaxis*, *Lagarostrobos*, *Phyllocladus*, or *Diselma*. These are able to regenerate below undisturbed canopies, or by exploitation of local recurring disturbance, such as that produced by windthrow or flooding; species depending on fire for regeneration were excluded from the rainforest species group.

The definition excludes lowland mixed forest where eucalypts occur over a rainforest understorey. However, whether upland mixed forest should be regarded as rainforest is unresolved, as there is some evidence that the eucalypts in that situation may regenerate successfully in the absence of fire (Jarman *et al.* 1987).

The definition could not be applied outside Tasmania, as several of the defining species are, at the present day, absent from the mainland. It would be possible to define mainland rainforest in terms of rainforest species (which themselves could be defined by regenerative characteristics), but the list would be so extensive as to render the definition impractical except in cool temperate rainforest (microphyll moss and fern forest).

A definition essentially similar to that developed in Tasmania has been adopted in Victoria: 'Rainforest is defined ecologically as closed broadleaved forest vegetation with a more or less continuous rainforest tree canopy of variable height, and with a characteristic composition of species and life forms.

Rainforest canopy species are defined as shade tolerant tree species which are able to regenerate below an undisturbed canopy, or in small canopy gaps resulting from locally recurring minor disturbances, such as isolated windthrow or lightning strike, which are part of the rainforest ecosystem. Such species are not dependent on fire for their regeneration.' (Department of Conservation, Forests and Lands 1987.)

The issue of whether communities which may be seral to rainforest should, in themselves, be classified as rainforest has aroused considerable controversy. Until recently, such vegetation was, by convention, excluded from the definition of rainforest and either recognized as a distinctive vegetation type or treated as a variant of sclerophyll forest. Dale *et al.* (1980) argued that the definition of rainforest should be extended to include 'transitional and seral communities with sclerophyll emergents that are of similar botanical composition to mature rainforests in which sclerophylls are absent' (see Chapter 4). This broader definition has been adopted by many conservationists but is disputed by some foresters.

2.3 SCLEROPHYLLY

There is a striking contrast between the closed-canopied rainforest and the open eucalypt-dominated canopies of much of the remaining forest cover of Australia. The two forest types differ in structure, range of life-forms, floristic composition, and response to disturbance (particularly fire).

Those forests which are not regarded as rainforest are frequently categorized as sclerophyll forest. Like rainforest, the concept of sclerophylly was introduced into phytogeographical literature by Schimper (1903): 'The mild temperate districts with winter-rain and prolonged summer-drought are the home of evergreen xerophilous woody plants, which, owing to the stiffness of their thick, leathery leaves may be termed sclerophyllous woody plants'. In Australia, the term was first used by Diels (1906) in a description of vegetation in south-west Australia, a region which climatically fits Schimper's diagnosis. Subsequent to Diels, sclerophyll has been taken up as a term to apply to almost all vegetation types in Australia which display even slight xeromorphic characters. Seddon (1974) has provided a detailed analysis of the changing usage and definitions of the term sclerophyll. It is difficult to establish criteria by which sclerophylly may be recognized, although Gillison (1981) suggested that a sclerophyllous leaf is one which cracks when wrapped around a pencil. Among vegetation which would be generally agreed to be rainforest, there are many species with leaves which possess some xeromorphic characteristics. Sometimes these can be categorized as 'coriaceous', a term as difficult to define as sclerophyllous, but, by example at least, one can usually recognize a coriaceous leaf *Gestalt* which is different from that of sclerophylls. Even this distinction can fail, and there is then a tendency to resort to the circularity illustrated by Walker and Hopkins (1984): 'Hard-leaved plants can occur in rainforest, and commonly occurring genera are *Agathis*, *Podocarpus* and *Araucaria*; by definition these are non-sclerophylls'. The sclerophyllous nature of some rainforest leaves was discussed by Webb (1959, 1968, 1978a); it is of interest that Richards (1952) regarded sclerophyllous leaves as characteristic of tropical rainforest (see P. W. Richards, p. 80).

2.4 TYPES OF RAINFOREST

For most purposes recognition of a single unit—rainforest—is of limited value. This vegetation type is too widely distributed and too floristically variable. We need to be able to recognize units about which certain (if only limited) generalizations and predictions can be made. This requires us to subdivide rainforest into more homogeneous categories of vegetation.

The description and classification of vegetation are topics which tend to arouse great controversy. Many classifications are possible, and to ask, in an absolute sense, whether one classification is better than another is inappropriate. The usefulness of a classification must be judged in the context for which it was designed. Classifications developed for a specialist purpose may be applied in other circumstances, but various limitations may become obvious (a forest typing developed in the context of timber harvesting may be used as a framework for ecological discussion but, if other attributes had been used, a different classification might have emerged).

The account below is not a comprehensive discussion of all classifications of Australian rainforests which have been proposed, but it does introduce examples of the major approaches.

2.4.1 New South Wales forest classification

The classification of the New South Wales forests (Baur 1965; Forestry Commission of NSW 1989) is an example of a traditional intuitive structural/selective floristic approach. Rainforest is divided into four subforms, referred to as 'leagues' by Baur (1965), and characterized primarily by structural features and physiognomy. The features of the four leagues, subtropical, warm temperate, cool temperate, and dry are illustrated in Fig. 2.1. The nomenclature is unfortunate, subtropical, warm temperate, and cool temperate might imply a latitudinal sequence which is not the case while dry rainforest is a source of considerable confusion (the concept of warm temperate rainforest is examined further on p. 75). Notwithstanding these names, the vegetation types to which they refer are distinct in structure and, at least in part, floristics.

Within the leagues, subdivision is on the basis of floristics into types. Recognition of types is intuitive with particular attention being given to potentially commercial species, and some of the

SUBTROPICAL RAINFOREST

DRY RAINFOREST

Trees –2 or 3 strata of trees
  – diverse: 10–60 species in canopy
  – leaf size large: notophylls and mesophylls common
  – leaves often compound, and mostly with entire margins
  – stranglers (figs) often common
  – palms often common
  – plank-buttresses common
  – uneven, non-uniform canopy

Vines – large, thick-stemmed vines common and diverse
Large epiphytes – (orchids, ferns, aroids) common and diverse
Special features – large-leaved herbs and ground-ferns common

Habitat – high rainfall (> 1300 mm)
  – eutrophic parent rocks (basalt, rich shales, some granodiorites, etc.)
  – favoured by shelter e.g. in gullies
  – warmer areas: sea level to 900 m

Note – although the tree species are very diverse the commoner ones often include: *Argyrodendron* spp. (booyongs), *Sloanea woollsii* (yellow carabeen), *Dysoxylum fraseranum* (rosewood), *Ficus* spp. (figs), *Acmena*, *Syzygium* (lilly pillies).

Trees – mostly 2 strata with the *lower* stratum 6–18 m tall and continuous, the upper one mostly of scattered tall *Araucaria* or semideciduous spp.
  – diverse range of species : 10–30 species in lower layer
  – leaf size small, mainly microphyll
  – leaves often hard and with blunt tips
  – plank-buttressing rare
  – palms ± absent
  – stranglers rare or frequent

Vines – large vines very common and diverse
Large epiphytes – rare or sometimes common, but few species
Special features – mosses and ground ferns rare, no tree ferns, prickly shrubs common

Habitat – warm areas with fairly low rainfall (630–1100 mm). Eutrophic parent rocks (basalt etc.) and/or good shelter

Note – the common emergent trees include: *Araucaria cunninghamii*, (hoop pine), *Brachychiton discolor*, (lace-bark tree) *Flindersia australis* (teak).

  – the canopy layer is often dominated by trees of a few families: Sapindaceae, Euphorbiaceae, Celastraceae, Oleaceae, Anacardiaceae, Moraceae.

Fig. 2.1 Features of the four major forms (leagues—Baur 1965) of rainforest recognized in the intuitive structural/floristic classification frequently used in New South Wales and southern Queensland (from Williams *et al.* 1984).

WARM-TEMPERATE RAINFOREST

COOL-TEMPERATE RAINFOREST

Trees–2 strata of trees
 – less diverse than STRf: canopy layer of 3–15 species
 – leaf size medium: notophylls and microphylls common
 – simple leaves with toothed margins most common
 – stranglers and palms rare or absent
 – plank-buttresses rare or absent
 – fairly even canopy

Vines – large vines sparse to fairly common
Large epiphytes – sparse to fairly common, but few species
Special features – slender rather uniform tree trunks
 – many trunks with whitish covering of lichens
 – ground ferns frequent

Habitat – fairly high rainfall (over 1300 mm)
 – medium to high altitude (450–1200 m)
 – poorer soils on oligotrophic rocks e.g. rhyolite, trachyte, slates, but on eutrophic rocks (e.g. basalt) in southern localities

Note – the commonest trees are usually *Ceratopetalum apetalum*, (coachwood), sassafras, (*Doryphora sassafras*), and various Lauraceae

Trees – mostly 2 strata, sometimes 1
 – very low diversity: canopy even and uniform, with 1–3 species
 – leaf size small: microphylls and nanophylls most common
 – simple leaves with toothed margins most common
 – stranglers and palms absent
 – plank-buttresses absent but base of trunks sometimes massive

Vines – large vines rare or absent, but thin wiry vines may be common
Large epiphytes – rare or absent but a few small ferns and orchids frequent
Special features – mossy epiphytes and lichens very abundant
 – ground ferns and treeferns often very common

Habitat – very high rainfall (1750–3500 mm)
 – high altitude (900–1500 m)
 – frequent mists
 – varied soil parent materials, e.g. trachyte, basalt, slates, granodiorite, rhyolite

Note – the commonest and often the only dominant tree is *Nothofagus moorei* (southern beech)

types exhibit considerable variability both within and between sites. Details of the 24 rainforest types recognized are given in Forestry Commission of NSW (1989). (The earlier account (Baur 1965) described 19 types.) While the classification was developed as an adjunct to forest management it has proved useful as a general-purpose scheme.

2.4.2 Floristic classifications

The forest types in the New South Wales scheme are defined floristically, but only in terms of certain key species, particularly those of high commercial value. Attempts have also been made to produce floristic classifications for general ecological purposes.

In the Beadle and Costin (1952) tradition, the basic floristic unit is the association, with associations being grouped into alliances. In the case of rainforest, however, Beadle (1981) argued that the species diversity and variability within and between stands are such as to render impossible the recognition of associations. For rainforest, the appropriate unit is the alliance, although in some instances suballiances may be recognized. (Some of these suballiances, for example the *Ceratopetalum apetalum–Doryphora sassafras* suballiance (the direct equivalent of Forest Type 12 in the Forestry Commission of NSW 1989 scheme) have a relatively constant species composition in the canopy and encompass a range of floristic variation comparable with that within associations recognized in other major vegetation types.)

Floyd, who has carried out extensive floristic surveys of rainforest stands throughout New South Wales, has developed a detailed classification involving 14 alliances and 43 suballiances. A listing of this classification appears in Floyd (1987), and a full account in Floyd (1990). The alliances listed in Floyd (1987) are grouped into the major structural subforms adopted by Baur (1965). The correspondence between Floyd's scheme and the Forestry Commission of NSW (1989) classification is documented in Baur (1989).

One of the major problems with alliance classification is that their derivation is intuitive. The diagnoses do not clearly describe the range of variation within units and it can be difficult for inexperienced workers to assign particular stands to an alliance. Some units are sufficiently distinctive as to be readily recognizable, but others are less clear cut. This difficulty in recognition reflects not only the variation within rainforest but also the problem of identifying canopy trees and obtaining estimates of their relative abundance. This is not to argue that the units are not valid communities, rather it emphasizes the difficulty of encapsulating the essence of complex plant communities. As it stands, the best way of learning to apply an alliance classification is by example; visiting the 'type site' and then recognizing similar vegetation elsewhere.

Although the total floristic composition of alliances contributes to their overall *Gestalt*, normally only a few key canopy species are required for their diagnosis. There have been few attempts, anywhere in the world, to develop phytosociological classifications involving recording of total floristic composition of rainforest stands. Webb *et al.* (1967*a,b*) explored the possibility of such an approach.

The data set was from eighteen 0.1 ha sites in north-east Queensland and Cape York, defined by 818 species. For the analyses only qualitative (i.e. presence/absence) data were used. Association analysis of the total data set (Webb *et al.* 1967*a*) indicated that the major dichotomy was between sites with strongly seasonal (monsoonal) climates and those with significant rainfall throughout the year (although still with a late summer maximum). Subsequent divisions could be interpreted as reflecting altitude and edaphic factors. The classification recovered only imperfectly the physiognomic classification (see p. 24) which had been employed to guide choice of sampling sites.

Identification of all species in the sample plots was a time-consuming operation. The concordance between the classification based on the full data set and those produced from various subsets of the total species list was explored by Webb et al. (1987b). The full data set classification was exactly recovered from the subset of 269 large tree species. Many of these species were restricted to single sites, and, as such, contributed little to the classification. Reduction of the data to only 65 species still allowed recovery of the major divisions of the full data set classification. The liane subset also displayed most of the features of the original classification.

Webb et al. (1967b) suggested that these results indicated that the rainforests were comprised of two elements. The canopy element, composed of big trees, large vines, and possibly vascular epiphytes responds to the macroclimate and defines the floristic structure of the forest. The second element, all the plants in the understorey, is buffered from the macroclimate by the canopy and the distribution of understorey species reflects edaphic factors and the local disturbance history.

The large proportion of single-site species in the data set suggested to Webb et al. (1967b) that floristic classifications were unlikely to be of value in rainforest studies. If the number of sites increased, Webb et al. (1967b) predicted that the number of single-site species would also increase and that any classification would be influenced primarily by species richness rather than by joint occurrences of species. For the purposes of devising a regional classification, recording full floristic site inventories would involve collecting much unnecessary data; Webb et al. (1967b) suggested that a physiognomic–structural approach was likely to be more profitable.

The above study was conducted when numerical approaches to classification were in their infancy, and there have been considerable advances since 1967. The limited number of sites made it impossible to develop a practical regional floristic classification. The study did, however, demonstrate that numerical techniques could be used to reveal features of a complex data set that would not be apparent on simple visual inspection. It also raised a number of interesting questions. For example, is it generally the case that the canopy composition determines the floristic structure of the forest, to what extent do synusia behave independently? If, at the regional scale, the major 'message' resides in the canopy, then this would suggest that the traditional alliance level of classification would remain the most appropriate.

W. T. Williams et al. (1973) subsequently investigated the use of quantitative data and the effects of plot size in a study of forests on the Nightcap Range in northern New South Wales. There is a sharp distinction (floristically and structurally) between complex rainforests (subtropical rainforest *sensu* Baur 1965) on basalt and structurally and floristically simple rainforest on rhyolite (warm temperate rainforest *sensu* Baur 1965). The success of various sampling and analytical approaches was judged by their ability to recover this clear a priori pattern. The results suggested that the most appropriate sampling unit was a relatively large quadrat (0.1 ha) for which a complete inventory of trees was recorded. Subsequent classifications by a range of techniques differed little, and quantitative data did not add to the results of using presence/absence data. Williams et al. (1973) recognized that collecting data for large plots was an onerous task; if the sample unit was smaller, then they recommended plotless sampling, recording a group of twenty or twenty-five contiguous trees (such data being quantitative, recording not only species but also the number of individuals of each species).

For floristic methods to produce useful regional classifications, it is necessary that at least some of the groupings recognized are not restricted to single sites but are recurrent over the region. In the case of species-rich rainforest, there has been some debate as to whether species' distribution is such as to result in recurrent patterns. Dale and Webb (1975) described a method that would permit recognition of species/site groups (associations in a phytosociological sense), but the test data that they used, from three rainforest localities in north-east Queensland, were found to segregate on the basis of localities and the units distinguished were site specific.

W. T. Williams and Tracey (1984) carried out an analysis of a data set from 146 sites defined by the presence/absence of 740 tree species in north-east Queensland. The two-neighbour network analysis resulted in the recognition of nineteen groups, which were not only floristically distinct but made ecological sense and were strikingly similar to the groupings recognized intuitively by Tracey and Webb (1975) and Tracey (1982). This study is the most successful example to-date of the use of numerical methods in the classification of complex floristic sets. The extensive analysis reported by Webb *et al.* (1984) (see p. 30) illustrates the ability of modern numerical methods to accommodate very large data sets but was not intended to produce a community classification *per se*.

The most extensive use of a phytosociological approach (Zurich-Montpellier school) to Australian rainforest vegetation is Busby's (1984) classification of *Nothofagus cunninghamii* vegetation in Victoria and Tasmania (see also Busby and Bridgewater 1977). These cool temperate rainforests have relatively low species diversity of woody plants compared with other rainforest types and this may be one of the reasons why a phytosociological treatment was successful. There is a rich and diverse bryoflora associated with cool-temperate rainforest but this was not included in Busby's database; the extent to which bryophyte distribution can be related to that of communities defined by vascular species remains to be explored. The vegetation sampled by Busby was selected on the basis of the presence of *Nothofagus cunninghamii* rather than on structure, and included stands which in terms of the structural categories of Specht (1981*a*) are not forest. The concordance between boundaries of floristic units and change in structure was poor. Full floristic tables were used to define the floristic units. The classification suggests two major groupings (Busby 1984, 1987); one restricted to high-altitude sites in south western Tasmania, the *Tasmannia–Astelia* order (with characteristic species including *T. lanceolata* and *A. alpina*); the other, the *Atherosperma–Blechnum* order (named after *A. moschatum* and *B. wattsi*) occurring in Victoria and Tasmania. Within this latter order are two alliances (alliances here being based on total floristic composition and not biased towards canopy species as in the Australian definition), the *Dicksonia–Polystichum* alliance (after *D. antarctica* and *P. proliferum*) found at lower altitudes in Tasmania and towards high altitudes in Victoria, and the *Anodopetalum–Anopterus* alliance (*Anodopetalum biglandulosum*, *Anopterus glandulosus*) which is restricted to Tasmania. Within the *Dicksonia–Polystichum* alliance, the constituent associations are also found in both Victoria and Tasmania.

A similar phytosociological approach might be developed for some other relatively species-poor rainforests, but while in theory the approach could be applicable to all rainforests, the difficulties of recording total floristics and the complexity of the resulting data matrices, are likely to militate against its implementation. In more complex rainforests, the question of whether synusia should be treated independently would require consideration; in the *Nothofagus cunninghamii* forests under consideration, vines and vascular epiphytes were insignificant so that the question did not arise.

2.4.3 Physiognomic classification

The distribution patterns of species are unstable (in that they are subject to change, possibly with a lag phase, in response to environmental change), many individual species have extremely localized or disjunct distributions, so that a large portion of the flora in any stand may not contribute to recurrent codistribution patterns which define floristic communities (indeed, for the more species-rich subtropical and tropical rainforest, it is still unclear whether or not floristic units of more than local occurrence can be defined), and there are major problems with floristic sampling and identification. Webb (1959, 1968, 1978*a*; Webb *et al.* 1970; Webb *et al.* 1976; Webb and Tracey 1981*a,b*) has argued that a physiognomic/structural approach may be more appropriate than floristics for achieving a synoptic overview of variation in Australian rainforest. Physiognomic/structural characters may be applicable over wider geographical areas than floristic ones, and can be recorded rapidly by workers lacking taxonomic expertise and without the need for subsequent recourse to herbaria.

It had long been recognized that there were recurrent patterns of variation in physiognomic/structural features of rainforest vegetation that could be correlated with environmental conditions (see Schimper 1903; Warming 1909, and, for a review of more recent data from the Far East, Whitmore 1984). Structural features have long been used in Australia to define formations which are subsequently subdivided on the basis of partial floristics. The importance of structure and physiognomy for rainforest classification was emphasized by Richards *et al.* (1940) and Beard (1955). Webb's approach builds on these earlier studies but takes the use of physiognomy and structure much further.

The early evolution of the classification (Webb 1959, 1968) involved an intuitive, empirical approach to data analysis. Later (Webb *et al.* 1970, 1976), it was shown that the data could be subject to numerical analysis.

The best guides to the classification are provided in Webb (1978*a*) and Webb *et al.* (1976); this latter paper contains a *pro forma* for field use and a detailed explanation of the various characters to be recorded.

One of the first issues to be faced in developing a structural/physiognomic classification is the choice of attributes to be recorded. In a floristic classification an a priori decision is made to record all species or all species in some defined subset of the total flora (e.g. all vascular plant or all canopy species); once that decision is made, the quantity of data recorded is determined by the features of each site. In the case of physiognomic features, the potential number which could be recorded is very large, and the decision as to which to record is essentially arbitrary. The amount of data to be collected is largely a function of the number of attributes selected.

The features utilized by Webb fall into four categories:

1. height and depth of canopy closure (prominence of different tree storeys or layers, evenness of canopy surface, presence and type of emergents, regularity of distribution of trunks of upper canopy trees, crown shapes, vertical extension of crowns in relation to length of trunks, etc.);
2. leaves (size, shape, texture, periodicity of leaf-fall, etc);
3. trunks (texture and colour of bark, extent of covering by epiphytes, cauliflory, stilt roots, spreading surface roots, club-like base, etc.)
4. special life-forms or growth forms (types of palm-like, fern-like, pandan-like, banana-like, bamboo like, grass-like plants; stranglers; banyans; thorns, prickles and hooks; robust, slender and wiry vines; types of epiphytes and their vertical extension, etc.) (Webb 1978*a*).

These features are discussed at greater length in Webb *et al.* (1976).

Quantification is achieved by estimating the prominence of features on a four-point scale (0—not evident, 1—rare or inconspicuous, 2—occasional or rare but conspicuous, 3—common, either uniformly or in patches). Webb *et al.* (1970) mentioned that recorders new to rainforest, and not familiar with the total range of variation shown in the presence of some features, may be overly impressed by them. Features like plank buttresses or robust lianes are so striking that their occasional presence may elicit a score of 3 'whereas an ecologist familiar with the luxuriance of these life-forms in more favourable conditions would have been content with a score of 2 or even 1' (Webb *et al.* 1970).

Leaf size is an important attribute both in defining groups and in their subsequent recognition. The leaf size classes are based upon those established by Raunkiaer (1934, see Fig. 2.2 this volume). In 1959, Webb recognized the particular prominence in Australia of leaves at the lower end of the range of size in Raunkiaer's mesophyll

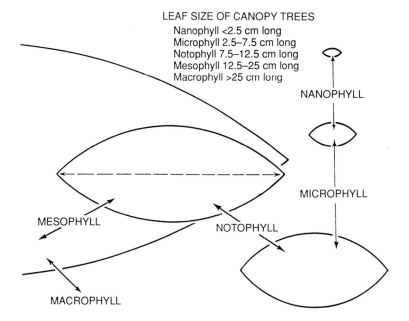

Fig. 2.2 Leaf-size categories utilized in the structural/physiognomic classification of Australian rainforests (from Tracey 1982).

category and formalized this distinction by proposing a new leaf size category notophyll (from the Greek 'notos'—southern).

In determining the leaf size characteristic of a stand only the trees are considered and, in the case of compound leaves, leaflets are considered as leaves. Within the crown of many species there is considerable variation in leaf size with leaves exposed to the sun being smaller than leaves in shade; only the sun leaves are considered. Where the average leaf size appears to be intermediate between two size classes, the larger class is nominated. Walker and Hopkins (1984), whose classification of rainforest has many features of Webb's scheme, suggested that estimates of leaf size should be standardized by sampling ten adjacent canopy trees and using a shot-gun or a catapult to collect sun leaves, stands are then categorized by leaf size according to the classes listed in Table 2.2. Stocker and Unwin (1989) have stressed that cursory assessment of dominant leaf size can be deceptive, but quantitative checks have rarely been made in most applications of the scheme.

The classification separates rainforests into three major categories (Webb 1968): vine forests characterized by robust woody vines (lianas), fern forests in which tree ferns and terrestrial ferns are conspicuous and vines are generally wiry and slender, and mossy forests characterized by abundance of epiphytic bryophytes. Webb (1978a) later developed a key to the forest types currently recognized and this is presented below.

1. Mesophylls and notophylls most common

    2. Robust lianes, vascular epiphytes, plant buttresses, macrophylls and compound mesophylls prominent; trunk spaces generally obscured by aroids and palms; stem diameter irregular, many av. 60–120 cm; canopy level av. 21–42 m

        3. Deciduous emergent and top canopy trees rare

            4. Palm trees not prominent in canopy ... *Complex mesophyll vine forest* (CMVF)

            4. Feather palm trees prominent in canopy ... *Mesophyll feather-palm vine forest* (MFPVF)

Table 2.2 Categorization of rainforest stands in terms of leaf size in the tallest stratum (from Walker and Hopkins 1984)

Term describing leaf size of forest stand	Number of individual trees (maximum 10) with specified leaf sizes	Percentage of individuals in tallest stratum with specified leaf sizes
1 Macrophyll	>5 macro	>50% macro
2 Macrophyll–mesophyll	3–5 macro and 1–4 meso	30–50% macro and 10–40% meso
3 Mesophyll	>5 meso	>50% meso
4 Mesophyll–notophyll	3–5 meso and 1–4 noto	30–50% meso and 10–40% noto
5 Notophyll	>5 noto	>50% noto
6 Notophyll–microphyll	3–5 noto and 1–4 micro	30–50% noto and 10–40% micro
7 Microphyll	>5 micro	>50% micro
8 Microphyll–nanophyll	3–5 micro and 1–4 nano	30–50% micro and 10–40% nano
9 Nanophyll	>5 nano	>50% nano

    3. Deciduous and semi-deciduous emergent and top canopy

        4. Mostly mesophylls ... *Semi-deciduous mesophyll vine forest* (SDMVF)

        4. Mostly notophylls ... *Semi-deciduous notophyll vine forest* (SDNVF)

  2. Robust lianes and vascular epiphytes not conspicuous in upper tree layers, which are simplified; spur rather than plank buttresses prominent; trunk spaces open, stem diameters (except for evergreen emergents) generally regular, av. 60 cm; canopy level av. 24–36 m. Simplification of structural features does not, however, approach that of simple notophyll evergreen trees. Sclerophylls (e.g. *Acacia*) may be scattered in canopy

    3. Deciduous emergents and top canopy trees rare or absent. Mostly mesophylls

        4. Palm trees not prominent in canopy ... *Mesophyll vine forest* (MVF)

        4. Fan palm trees prominent in canopy ... *Mesophyll fan-palm vine forest* (MFAPVF)

1. Notophylls and micropylls most common

  2. Robust and slender woody lianes, vascular epiphytes, plank buttresses, and compound entire leaves prominent; trunk spaces generally obscured by the aroid *Pothos*; stem diameters irregular, many av. 60–120 cm

    3. Canopy level uneven, av. 21–45 m emergents mostly evergreen and umbrageous ... *Complex notophyll vine forest* (CNVF)

    3. Canopy level uneven, av. 15–36 m, occasional deciduous species with common emergent *Araucaria* or *Agathis*, reaching av. 36–51 m ... *Araucarian notophyll vine forest* (ANVF)

  2. Robust lianes and vascular epiphytes inconspicuous in tree tops; slender woody and wiry lianes prominent in understorey; plank buttresses inconspicuous; simple toothed leaves prominent; trunk spaces open; stem diameters (except for emergents) generally regular, av. 60 cm; tree crowns evergreen and generally sparse and narrow; strong tendency to single species dominance (e.g. *Ceratopetalum*) in upper tree layers; canopy level even, av. 21–33 m, often with sclerophyllous emergents and codominant ... *Simple notophyll evergreen vine forest* (SNEVF)

2. Robust lianes, vascular epiphytes and plank buttresses present, but not so prominent as in complex types; tree crowns mostly evergreen, but with a few semi-evergreen or deciduous species; i.e. structural features are intermediate between simple and complex types ... *Notophyll vine forest* (NVF)

2. Robust and slender lianes generally present, wiry lianes (climbing ferns) generally conspicuous in understorey; vascular epiphytes and plank buttresses inconspicuous; feather palms generally conspicuous; tree crown evergreen; canopy level av. 20–25 m ... *Evergreen notophyll vine forest* (ENVF) ± *feather palms*

2. Robust, slender and wiry lianes generally inconspicuous; fleshy vascular epiphytes may be prominent on trunks; plank buttresses inconspicuous; simple entire leaves prominent; deciduous species generally absent but many tree crowns become sparse during the dry season, i.e. semi-evergreen; typically mixed with sclerophyllous emergents and codominants

   3. Canopy level av. 10–20 m *Simple semi-evergreen notophyll vine forest* (SSENVF)

   3. Canopy level av. 3–9 m, generally even, and canopy trees often branched low down (shrub-like) ... *Simple semi-evergreen notophyll vine thicket* (SSENVT)

1. Microphylls most common

   2. Mossy and vascular epiphytes inconspicuous in top tree layers; robust lianes generally prominent; plank buttresses absent; prickly and thorny species frequent in usually dense shrub understorey; ground layer sparse; compound leaves and entire leaf margins common.

      3. Canopy level uneven, av. 9–15 m with mixed evergreen and semi-evergreen emergent and upper tree layer species; araucarian and deciduous emergents rare or absent ... *Low microphyll vine forest* (LMVF)

      3. Canopy level uneven, av. 4–9 m, with mixed evergreen, semi-evergreen and deciduous emergents to av. 9–18 m, swollen stems ('bottle trees') ... *Semi-evergreen vine thicket* (SEVT)

   3. Canopy level uneven and discontinuous, av. 4–9 m; practically all emergents are deciduous, and many understorey species are deciduous or semi-evergreen swollen stems ('bottle trees' and other species) may be common ... *Semi-evegreen vine thicket* (SEVT)

   2. Mossy and vascular epiphytes usually present in top tree layers; robust lianes inconspicuous; slender and wiry lianes generally prominent; plank buttresses absent; prickly and thorny species absent; simple leaves with toothed margins common; strong tendency to single species dominance (*Nothofagus, Eucryphia*) in tree layer; tree ferns and ground ferns prominent; sclerophyll emergents generally present in marginal situations

      3. Canopy level tall, even except for sclerophylls, av. 20–45 m ... *Microphyll fern forest* (MFF)

      3. Canopy level stunted, generally even and mixed with sclerophylls, av. 6–9 m ... *Microphyll fern thicket* (MFT)

1. Nanophylls most common

   2. Mossy epiphytes conspicuous; robust lianes and true prickles and thorns absent or rare; plank buttresses absent; simple leaves with toothed margins common; strong tendency to single species dominance (*Nothofagus*) in tree layer; tree ferns and ground ferns prominent; floor often peaty and covered by mosses; sclerophyll emergents generally present

      3. Canopy level tall, except for sclerophylls, av. 18–40 m ... *Nanophyll fern forest* (NFF) *and mossy forest* (NMF)

      3. Canopy level stunted, uneven, often with sclerophylls, av. 6–9 m ... *Nanophyll fern thicket* (NFT) *and mossy thicket* (NMT)

Given the continuous variation within rainforests, attempts to place every single stand using the key are likely to be unsuccessful; some stands will fall between existing categories. Nevertheless, in most cases, matching with a previously recognized type can be achieved. Webb *et al.* (1970) noted that one of the difficulties with a physiognomic classification was comunication. The concepts and terminology were unfamiliar to those accustomed to floristic or simple structural classifications. This problem has, with the passage of time, lessened, and the units recognized in the key are increasingly used in general discussions of Australian rainforest, but have not caught on elsewhere. The majority of categories previously recognized at formation or subformation level can be related to structural/physiognomic units (Table 2.3) although some subformations (e.g. monsoon forest) may be represented by a number of structural/physiognomic units.

The classification in its present form provides an appropriate framework for discussion of rainforests at a national scale. It has been argued (for example by Helman 1987) that for detailed regional studies the units are too broad and need further qualification using other information, for example floristics. While recognizing the need for more narrowly defined units in local studies, it is

Table 2.3 Structural/physiognomic categories of rainforest and relationship to earlier named categories (modified from Winter *et al.* 1987*b*)

Structural type	Other names
Complex mesophyll vine forest (CMVF)	Tropical rainforest
Mesophyll vine forest (MVF)	Tropical rainforest
Semi-deciduous mesophyll vine forest (SDMVF)	Tall monsoon forest
Semi-deciduous notophyll vine forest (SDNVF)	Monsoon forest
Evergreen notophyll vine forest (ENVF)	Monsoon forest
Simple semi-evergreen notophyll vine forest (SSENVF)	Monsoon forest
Simple notophyll semi-evergreen vine thicket (SNSEVT)	Monsoon forest
Deciduous vine thicket (DVT)	Monsoon forest
Complex notophyll vine forest (CNVF)	Subtropical rainforest
Notophyll vine forest (NVF)	Subtropical rainforest
Simple notophyll evergreen vine forest (SNEVF)	Subtropical rainforest
	Warm temperate rainforest
	Coachwood forest
Araucarian notophyll vine forest (ANVF)	Hoop pine rainforest
Araucarian microphyll vine forest (AMVF)	Hoop pine rainforest
Low microphyll vine forest (LMVF)	Dry rainforest
Semi-evergreen vine thicket (SEVT)	Bottle tree scrub
Microphyll fern forest (MFF)	Warm temperate rainforest
Microphyll fern thicket (MFT)	Warm temperate rainforest
Nanophyll fern forest (NFF)	Cool temperate rainforest
Nanophyll mossy forest (NMF)	Cool temperate rainforest
	Beech forest
	Myrtle forest
Nanophyll fern/moss thicket (NFT/NMT)	Cool temperate rainforest
	Beech forest
	Myrtle forest

possible that this may be achieved through expanding the number of attributes used in physiognomic classification. This remains to be tested.

2.5 PHYTOGEOGRAPHICAL ANALYSIS OF RAINFOREST FLORAS

Webb *et al.* (1984) have carried out a numerical analysis of species distributions throughout the range of rainforest in Australia, which supersedes earlier accounts in Webb and Tracey 1981*a,b*. The data set consisted of presence/absence records of 1316 tree species (from 406 genera) from 561 rainforest stands, ranging from the Kimberley region to Tasmania. Only the upper levels of the agglomerative cluster analysis were presented in the paper. At the highest level in the dendogram, three groups were recognized as ecofloristic regions. At a lower level in the hierarchy, eight groups which are termed ecofloristic provinces, were delimited.

It is important to note that the units recognized by Webb *et al.* (1984) were described as ecofloristic rather than being phytogeographical units in any traditional sense. Thus, while there is a spatial (geographical) component to the segregation of units at any particular level in the hierarchy, there is also a strong ecological (climatic/edaphic) component. The site groupings segregated at the ecofloristic region level are, for the most part, geographically distinct and the broad distribution of the regions can be portrayed clearly on continental scale maps (Fig. 2.3). The ecofloristic provinces are not necessarily segregated one from another at a broad geographical scale; rather, sites assigned to different provinces may occur in close proximity separated along altitudinal or other gradients. Webb *et al.* (1984) considered the provinces to be equivalent to the class level of organization in the phytosociological (Zurich–Montpellier) hierarchy. However, the qualitative nature of the original data and variation in sampling strategies between sites, make it difficult to make formal phytosociological diagnoses. Within the provinces there is considerable heterogeneity in the rainforest vegetation, and several communities, as recognized by other authors, may occur. Nevertheless, the regions and provinces provide a valuable integrative framework for discussion of Australian rainforests, a framework which is explicable both in terms of present environmental factors and the palaeobotanical history of the continent (see Chapter 6).

The primary division in the hierarchy separates the northern forests (with megatherm climate) from the southern mesotherm/microtherm forests. The plant thermal response groups are those defined by Nix (1982):

megatherm dominant—mean annual air temperature $>24°C$

megatherm/mesotherm—mean annual air temperature $>20°C<24°C$

mesotherm dominant—mean annual air temperature $>14°C<24°C$

mesotherm/microtherm—mean annual air temperature $>12°C<14°C$

microtherm dominant—mean annual air temperature $<12°C$

The southern group of sites constitute region A, the northern group is split into two regions; B is entirely tropical and C is mainly subtropical but extends into the subhumid tropics (see Fig. 2.3). The sites in region A experience relatively high, evenly distributed annual rainfall, while rainfall in regions B and C varies from mildly to strongly seasonal. The differentiation between the regions is most strongly expressed at species level and is clear cut at higher taxonomic levels (Webb and Tracey 1981*b*). Within region A, three provinces are recognized; three provinces are delimited in region B and two in region C.

Province A_1 has its core area centred on subtropical coastal south Queensland and northern New South Wales, a region which experiences the optimal humid mesotherm climatic regime (Nix 1982). The floristic elements which define the A_1 province extend (in part) to elevations of $c.900$ m and also on to poorer soils, and southwards to the

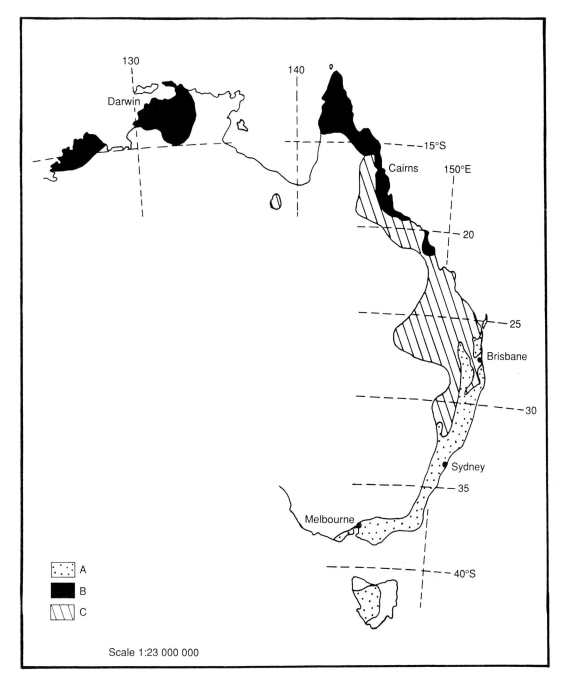

Fig. 2.3 Distribution of ecofloristic regions within Australia (redrawn from Adam 1987b, after Webb *et al.* 1984).

south coast of New South Wales (e.g. Mt. Dromedary). Province A_2 is subtropical—lower montane with a microtherm/mesotherm humid climate, and encompasses the *Nothofagus moorei*- and *Eucryphia moorei*-dominated microphyll fern forests and some *Ceratopetalum apetalum*-dominated stands. Province A_3 encompasses the rainforests of Tasmania and Victoria.

Provinces B_1 and B_2 both experience strongly seasonal rainfall with B_2 being drier than B_1 (B_1 and B_3 are closely related but the numbering of provinces in Webb *et al.* 1984 separated them in order to preserve conformity with Webb and Tracey 1981*a*). The B_2 province encompasses the large north-east Queensland block of rainforest between Ingham and Cooktown but with significant outliers northwards on the relatively humid tops of the McIlwraith Range and southwards on a number of mountain ranges as far as the Eungella Range inland from Mackay.

Province C_1 has as its core area the coastal lowland north of Brisbane, extending to scattered moist uplands, and experiences a subhumid–humid warm subtropical climate. The optimal structural type is araucarian notophyll vine forest, under low rainfall (c.800–900 mm per annum) it merges with province C_2, while under higher rainfall (c.1500 mm) it is replaced by A_1. Province C_2 experiences a subhumid warm subtropical (mesotherm/megatherm) climate. The rainforest stands are mostly scattered semi-evergreen vine thickets or, in north Queensland, deciduous vine thickets.

2.6 CLASSIFICATION FRAMEWORK USED IN THIS BOOK

As the basic descriptive framework, the structural/physiognomic classification of Webb as described in the key on p. 26 will be used. Where other local classifications have been developed, these will be referred to but the structural/physiognomic approach provides a uniform descriptive system applicable to all Australian rainforests. Where appropriate, reference will also be made to the ecofloristic framework proposed by Webb *et al.* (1984).

3 AN OVERVIEW OF AUSTRALIAN RAINFORESTS

The aim in this chapter is to present a broad-brush description of Australian rainforest, starting in the north-west and travelling clockwise around the continent. More comprehensive accounts are found in Beadle (1981), Figgis (1985), and Werren and Kershaw (1987).

This geographical approach is chosen as a matter of convenience, but other frameworks (for example structural/physiognomic) also would have been possible. Most of the structural–physiognomic forest types recognized by Webb are geographically widespread and variable in floristic composition. While relating structural–physiognomic variation to environmental gradients (p. 79) allows valuable insight into the factors controlling rainforest distribution, the internal floristic variability of the units would make an overview extremely complex. Although the boundaries of the geographical regions are in part arbitrary, the regions are distinct in terms of total floristics and the range of structural–physiognomic types represented in the rainforests.

One distinctive habitat-restricted rainforest is littoral rainforest occurring on sand dunes from north-west Australia to southern New South Wales. These stands have strong similarities in structure, and to a considerable extent in floristics, not only around Australia but also on tropical and subtropical shores elsewhere. Although littoral rainforests are discussed here on a regional basis, it would be equally appropriate to treat them separately. Littoral rainforest on sand dunes has similarities with rainforest on exposed headlands which is often also referred to as littoral rainforest. However, there appears to be greater regional variation in headland stands which indicates a stronger relationship to communities further inland.

3.1 NORTH-WEST AUSTRALIA

It is only recently that the occurrence of rainforest in north-west Australia has been widely recognized.

The first collections of specimens from the rainforests of the Kimberley region were made by Allan Cunningham, between 1819 and 1822 (McMinn 1970), but while the specimens and the unpublished journals clearly demonstrate the occurrence of rainforest (Kenneally and Beard 1987), the published accounts of the expeditions make no mention of rainforest. Gardner (1923) commented on the absence of rainforest in the Kimberley region, but later he did recognize a distinctive river plain forest which contained species 'which are more properly those of the rain forest' (Gardner 1944).

In the 1950s, a survey of Kimberley was carried out by CSIRO, although this was limited by the poor accessibility of the region. No rainforest stands were recorded and Speck (1960) specifically commented upon their absence.

The first formal recognition of the rainforests came, not from biologists, but from geologists who, in the mid-1960s, recorded the occurrence of vine thickets in the notes accompanying the geological map (Allen 1966). Following this, rainforests were recorded in a number of surveys (George 1978: Kenneally and Beard 1987). In 1987, as part of the Federal-State national rainforest conservation programme, a detailed survey on the region was carried out (McKenzie *et al*. 1987). Initially, LANDSAT TM (thematic mapper) satellite imagery was used and 400–500 stands were detected, mostly small but some nearly 100 ha in area. Subsequently, 83 of them were visited by helicopter (Fig. 3.1). Although the total area of rainforest is small, the habitat contributes considerably to the diversity of the

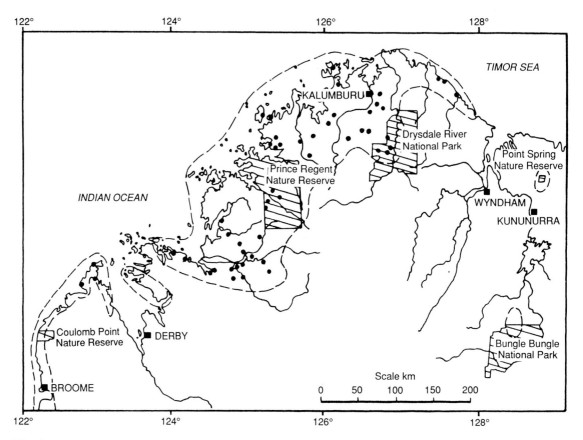

Fig. 3.1 Distribution of rainforest stands in the Kimberley region (from McKenzie *et al.* 1987). — — — limits to distribution of rainforest stands. ● sites surveyed by McKenzie *et al.* (1987).

Western Australian flora. To date, over 300 species have been recorded, eighty per cent being restricted to rainforest. The 1987 survey added several new genera and more new species to the Western Australian list (McKenzie *et al.* 1987); given the small number of sites studied compared with the total rainforest stands, it is likely that further recording will add yet more species.

The rainforest patches occur within 150 km of the coast and experience between 1500 and 600 mm rain per annum (Fig. 3.2). All the rain is received during the summer wet period, which for much of the Kimberley lasts for only about four months although on the Mitchell Plateau it extends for six months (Beard 1976).

Kenneally and Beard (1987) recognized four types of rainforest in north-west Australia on the basis of structure and physiognomy and provide species lists for the three types occurring in the Kimberley region.

The tallest and most luxuriant type, complex mesophyll vine forest, occurs mainly on creek and river banks extending out on to the floodplains. Trees are up to 20 m tall and form a dense canopy. The vine forest is normally separated from the river by a narrow fringe of *Melaleuca leucadendra* and *Pandanus aquaticus*.

Semi-evergreen mesophyll vine forest occurs principally in the higher rainfall areas. There is a discontinuous upper stratum, 12–15 m high, of

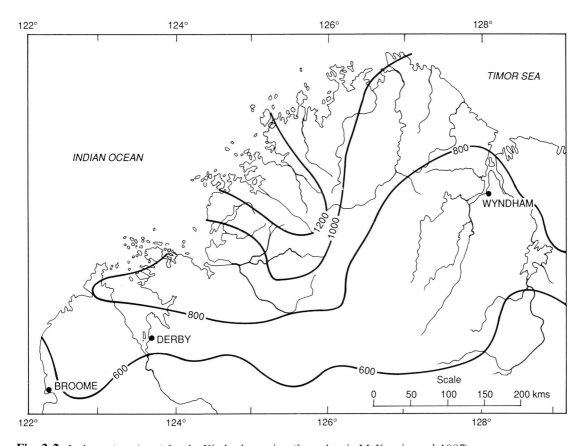

Fig. 3.2 Isohyets (mm/year) for the Kimberley region (from data in McKenzie *et al.* 1987).

mostly deciduous trees, over a dense understorey of shrubs and vines. The emergents tend to be clothed in vines (McKenzie *et al.* 1987) and vascular epiphytes are rare, except for occasional orchids. There is little or only sparse ground cover.

Deciduous vine thickets occur under lower rainfall conditions. Emergents are scattered, 6–9 m tall and virtually all deciduous, over a low shrub layer in which many species are deciduous or semi-evergreen. The majority of species are shared with the semi-evergreen vine forest. The most common tree is *Ziziphus quadrilocularis* (Rhamnaceae).

On the Dampier Peninsula, south of the north-west Kimberley, vine forests and thickets occur on deep coastal dune sands, a habitat which is very rare along the coasts to the north.

Most occurrences of both vine forests and thickets are associated with rocky habitats, on basalt, sandstone scree, or on rubble-strewn slopes below duricrusted plateau surfaces. (The duricrust on the Mitchell Plateau and Bougainville Peninsula is bauxite and exploration of this resource by mining companies helped to provide access for botanical studies.)

The factors controlling the distribution of rainforest in the Kimberley have been the subject of some discussion. Allen (1971) suggested that the occurrence of rainforest patches was correlated with the distribution of springs, but subsequent studies have not provided evidence of this. Beard (1976) thought that the occurrence of rainforest on the Mitchell Plateau was related to the high rainfall of the area and to runoff. However, the existence of numerous areas of vine thicket under low rainfall conditions on the Bougainville Peninsula (Beard *et al.* 1984) and on other areas indicates that the current distribution of rainforest is not a reflection of local high rainfall. Beard *et al.* (1984) and Clayton-Greene and Beard (1985) argued in favour of the hypothesis that fire is the major factor determining the distribution of rainforest. Fire scars on trunks show that fires regularly enter rainforest stands but normally do little damage. However, seedlings of rainforest species growing in open communities adjacent to rainforest are killed by fire. Clayton-Green and Beard (1985) suggested that fires slowly destroy rainforest stands, each fire killing the outer trees. On the Bougainville Peninsula and the Osborn Islands rainforest seems to be actively invading eucalypt woodland (Beard *et al.* 1984). These areas have a lower fire frequency than the Mitchell Plateau. The whole region was previously subject to systematic burning by Aborigines but this ceased in the 1940s, although there were intermittent visits by Aborigines until the 1960s. When mining exploration occurred in the late 1960s and 1970s, the Mitchell Plateau was burnt every year (Clayton-Greene and Beard 1985). Exploration of the Bougainville Plateau was less extensive and that area has been free of regular burning for several decades, permitting the expansion of vine thickets. The bouldery nature of most rainforest stands confers a degree of protection from fire.

In more recent years, the extension of the pastoral industry has resulted in damage to a number of stands by cattle. The damage to the vegetation has been associated with a loss of some bird species and land snails (McKenzie *et al.* 1987; Cribb 1987).

In Webb *et al.*'s (1984) analysis, the complex mesophyll vine forests are representative of the B_1 province. The semi-evergreen mesophyll vine forest and the deciduous vine thickets are included within the B_3 province.

Before leaving the north-west, it is appropriate to mention briefly the boab *Adansonia gregorii* which is one of the striking features of the region (Fig. 3.3). *Adansonia digitata* is a widespread species in sub-Saharan Africa and there are seven *Adansonia* species in Madagascar (Guillaumet 1984). *A. gregorii* appears to be more closely related to *A. digitata* than to the Madagascan species (Armstrong 1979). It is tempting to view the occurrence of *Adansonia* in Australia as a Gondwanan relict but in view of the absence to date of fossil records from Australia, the possi-

Fig. 3.3 *Adansonia gregorii*, near Wyndham, Kimberley Region, Western Australia.

bility of more recent long-distance dispersal cannot be ruled out. *A. gregorii* is well adapted to a seasonally dry climate because of the storage of large quantities of water in the swollen trunk. The bottle trees (*Brachychiton* spp.) display a similar growth form in the so-called softwood scrubs in Queensland.

3.2 THE NORTHERN TERRITORY

The rainforests of the Northern Territory have been subject to a number of studies; indeed Bowman (1988) suggested that they have received attention disproportionate to their area or species richness. A comprehensive review has been published by Russell-Smith and Dunlop (1987) and a more general account by A. Fox (1985). Details of the species composition of rainforest are provided by Taylor and Dunlop (1985), Brennan (1986), Russell-Smith and Dunlop (1987), and Brock (1988). Most of the information refers to western Arnhemland, much less is known about east Arnhemland.

The rainforest stands in the Northern Territory

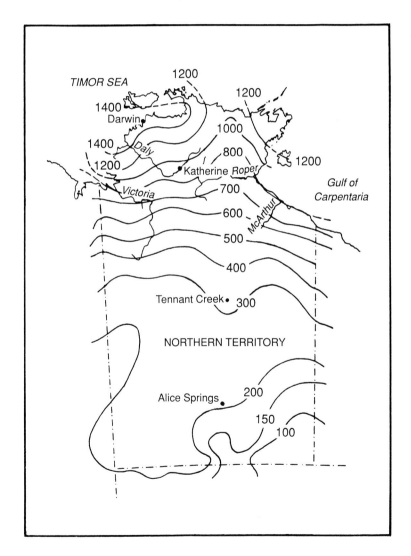

Fig. 3.4 Isohyets in the Northern Territory (from Bureau of Meteorology 1977).

are mostly small although there are some examples covering several hundred hectares. Rainforest is found from the offshore islands to as far inland as Tennant Creek, over a rainfall gradient from in excess of 2000 mm/yr. on Melville Island to less than 500 mm at Tennant Creek (Fig. 3.4). Parallel to this gradient is a decline in floristic and structural complexity. Rainfall is almost totally restricted to the summer wet season (Fig. 3.5). Temperatures are high throughout the year, with a mean annual temperature exceeding 24°C.

Russell-Smith and Dunlop (1987) and A. Fox (1985) recognized six types of rainforest, but Russell-Smith (1988) adopted a simpler classification of four types, which is outlined here. The types are categorized in terms of habitat, but also differ in terms of floristics and structure. The four types are: vine forests on sandstones, vine forests on rock outcrops (other than sandstone), vine forests of the lowland river plains, and coastal vine forests.

Sandstone outcrops are extensive across northern Australia, including the Arnhemland escarpment and plateau and offshore islands. Rainforests associated with sandstone regions may be divided into two subtypes—those occurring in sites with a constant availability of surface water and those occurring in seasonally dry sites.

Perennial moisture is found along streamlines in protected gorges, as seepage zones on steep slopes or the bases of cliffs, or at headwater springs. Vegetation associated with streamlines may form stands kilometres long and 100 m wide, while around springs stands may be less than 1 ha. The rainforests are evergreen and often around 30 m tall, with species such as *Calophyllum sil*, *Elaeocarpus angustifolius*, *Horsfieldia australiana*, *Ilex arnhemensis*, *Planchonella xerocarpa*, *Syzygium* spp., *Xanthostemon eucalyptoides*, and the Northern Territory endemic palm *Carpentaria acuminata*. Vines are not particularly prominent and epiphytes are uncommon. Both structurally and in terms of species composition, these stands have strong similarities to the complex mesophyll riverine forests of the Kimberley region.

Vine forests on seasonally dry sandstone sites are widespread but are particularly prominent in western Arnhemland, where stands up to hundreds of hectares in extent are dominated by a single evergreen species, *Allosyncarpia ternata*, which is endemic to the Arnhemland escarpment (Fig. 3.6) and in favourable situations may reach heights of 35 m or more. In more exposed sites or on shallow skeletal soils, *A. ternata* takes the form of a short-trunked tree (less than 10 m) with a wide spreading canopy. The majority of other tree

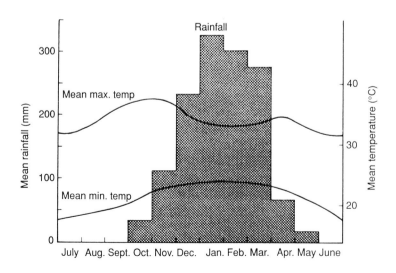

Fig. 3.5 Temperature and rainfall record for Oenpelli, the Alligator Rivers region (from Christian and Aldrick 1977).

Fig. 3.6 *Allosyncarpia ternata*. Twin Falls, Kakadu, Northern Territory. (Photo B. S. Wannan).

species associated with *A. ternata* are also evergreen. In the absence of *Allosyncarpia*, its role is often taken by species of *Xanthostemon*. Bowman *et al.* (1990) recognized *Allosyncarpia* forests as a distinctive vegetation type, restricted to siliceous substrates. Although over half the species they recorded from *A. ternata* stands were shared with communities regarded as rainforest, Bowman *et al.* (1990) did not consider *A. ternata* stands to be rainforest.

Vine forest occurs on a variety of rock types (including basalt, granite, metamorphic rocks, and limestone) away from the sandstone. At most of these sites, soils are poorly developed and the woody plants are rooted between rocks. Most stands are small, although there are some more extensive examples. The vegetation is very dense and normally low with abundant thorny vines and the majority of the canopy consists of deciduous trees (including *Allophyllus cobbe*, *Canarium australianum*, *Croton* spp., *Ficus* spp., *Sterculia quadrifida*, *Strychnos lucida*, and *Wrightia pubescens*), although a few evergreens (including *Celtis philippinensis*, *Diospyros humilis*, and *Drypetes lasiogyna*) also occur.

Rainforests are found in a number of situations in the lowlands, including both sites with perennial availability of water and seasonally dry habitats.

Evergreen rainforest is found around springs and seepage zones and on the banks of some permanent streams. In these sites, the soils are normally waterlogged. There are a number of examples in the Darwin region, known locally as jungles (for example Holmes Jungle, Black Jungle). Canopy heights are up to 25 m and are dominated by evergreen species (including *inter alia*, *Buchanania arborescens*, *Calophyllum soulattri*, *Ficus racemosa*, *Horsfieldia australiana*, *Syzygium nervosum*, and the palms *Carpentaria acuminata* and *Livistona benthamii*), although a small number of deciduous species may be present. These stands, both structurally and floristically, have similarities to the stands associated with perennial moisture on sandstones.

On seasonally dry habitats in the lowlands, rainforest stands have a much larger deciduous component (Fig. 3.7). Sites on which rainforest occur include the banks of seasonal streams, on the levees of some of the major rivers and on the landward fringes of the riverine floodplains. At many localities inundation by floodwaters occurs for varying periods of the wet season. Some of these stands are extensive, Russell-Smith (1988) refers to one on the floodplain of the Daly River over 10 km long and up to 1 km wide.

Where the water table remains within the rooting zone throughout the dry season, the canopy may be up to 20 m and dominated by evergreen species (many characteristic species are shared with the stands associated with springs but there are some floristic differences between the two habitats). With reduced moisture availability during the dry season, the canopy is lower and dominated by deciduous species (including *Bombax ceiba*, *Ficus virens*, *Sterculia quadrifida*, and *Terminalia seriocarpa*). The understorey is often open but may have dense clumps of *Bambusa arnhemica*.

Further inland, south of Katherine, closed scrubs dominated by *Acacia shirleyi*, *Lysiphyllum cunninghamii*, and *Macropteranthes keckwickii* occur in a variety of situations. Although it is perhaps stretching the definition to regard these stands as true rainforest they do have a distinct rainforest element in their flora and provide inland localities for species frequent in coastal rainforests, such as *Abrus precatorius*, *Alphitonia excelsa*, *Grewia breviflora*, *Gyrocarpus americanus*, and *Strychnos lucida*.

Coastal vine forests are found on a variety of substrates on the coast and experience seasonally dry conditions. The canopy is generally low although deciduous emergents may extend considerably above the general canopy level. On exposed sites, which experience strong salt-laden winds during the dry season, the canopy may be only a few metres high and be laced together by a dense tangle of vines. Although many of the canopy species in these coastal sands are deciduous, there is also an evergreen component, including the tamarind tree *Tamarindus indica* which might be one of the first introductions in the Australian flora. It is suspected that this species was brought to Australia from Indonesia by Macassan fishermen and merchants who annually visited northern Australia to harvest trepang for about 400 years prior to European settlement.

The deciduous and most semi-deciduous rainforests are representative of Webb *et al*.'s (1984) B_3 province, but the taller evergreen stands form part of the B_1 province.

Russell-Smith and Dunlop (1987) argued that the flora of the Northern Territory rainforest patches consists of two distinct elements, one of widely distributed actively colonizing species and the other consisting of species restricted to refugia and showing disjunct distributions. As evidence of the colonizing ability of rainforest, they pointed to the early European settlement of Victoria at Port Essington on the Cobourg Peninsula, which was occupied between 1839 and 1849, and whose ruins are now within a species-rich stand of semi-deciduous vine forest. In addition, many stands are on geomorphologically young sites, such as beach ridges and river levees. Some 70 per cent of the obligate rainforest flora can be found in sites which must be of relatively recent origin.

While there is clear evidence for recent colonization by rainforest, there is also evidence which can be interpreted as showing that rainforest has been lost from other sites. The jungle fowl (*Megapodes freycinet*) constructs large nest mounds in which eggs are incubated and old mounds are persistent. At the present time, jungle fowl are largely restricted to rainforest stands. Stocker (1971), working on Melville Island, and Russell-Smith (1985), working in the Alligator Rivers region, both interpreted the presence of old nest mounds in eucalypt woodland as an indication of former rainforest, although Bowman and Dunlop (1986) suggested that not enough is known about jungle fowl ecology to permit rejection of other hypotheses. Organic matter from the mounds can be radio carbon dated. On Melville Island, the dates from five abandoned mounds ranged from 8200 BP to 1600 BP (Stocker 1971). If the assumption that the mounds were originally established in rainforest is correct, then what might account for the loss of rainforest? A major factor is likely to be fire. Russell-Smith and Dunlop (1987) pointed out that the vine forest flora contains many species which regenerate vegetatively after fire. A single fire is unlikely to have a catastrophic effect except on stands growing in sheltered permanently moist sites, which are, in any case, unlikely to burn. However, if fires are frequent, the ability of species to respond may be much reduced and the rainforest may be invaded by other species. Sites on rock outcrops may be particularly vulnerable to changes in fire frequency, even though the habitat itself confers considerable protection from the spread of fire. Seedling establishment on these rocky sites is poor in most years, so that canopy loss from frequent fire may not be replaced by recruitment.

Considerable damage has been done to some sites, particularly those with permanently available water, by feral pigs and buffalo, which impair regeneration and spread exotic weeds. Through

Fig. 3.7 Rainforest stand at the foot of Obiri Rock, a sandstone outlier on the edge of the Alligator Rivers floodplain, at the start of the dry season. *Pandanus* sp. in foreground

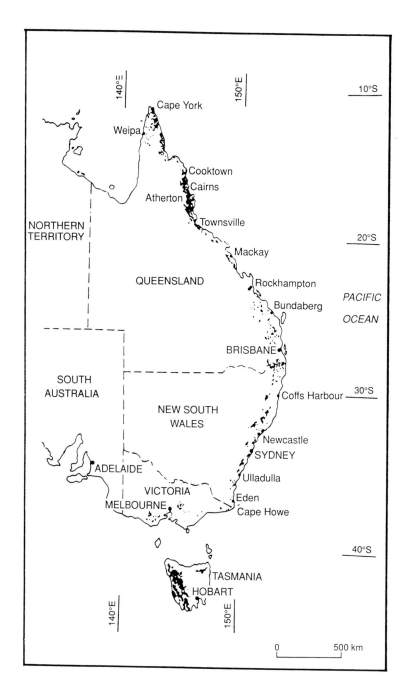

Fig. 3.8 Rainforests in eastern Australia (from Adam 1987*b*, after Werren 1985).

increasing fuel load (as a result of killing trees), fire frequency and/or intensity may be increased. In addition, aggregation of buffalo may cause compaction of the soil.

3.3 NORTHERN CAPE YORK PENINSULA

For present purposes, northern Cape York is considered to be that part of the Peninsula north of 14°S. The southern boundary is formed by the Laura Basin, a low-lying area with very low winter–spring rainfall and moderate summer rainfall which supports very little rainforest.

Most of the Peninsula is of relatively low relief, with the Great Dividing Range close to the east coast. In the south, the McIlwraith–Iron Ranges are rugged mountains of igneous and metamorphic rocks reaching a maximum elevation of 820 m. Further north, extending to the tip of the Peninsula, the Divide scarcely justifies the epithet 'Great', being a low range of sandstone hills up to 100 m elevation. The Peninsula has a strongly seasonal rainfall with a short summer wet season although the McIlwraith and Iron Ranges also receive significant rainfall in the winter (Kikkawa et al. 1981b).

The rainforests in the Peninsula were mapped by F. C. Bell et al. (1987) who estimated that the total area is just over 260 000 ha. The vegetation is reviewed by Lavarack and Godwin (1987), who suggested that the rainforests can be considered as comprising seven broad regions (Fig. 3.9).

The most extensive block of rainforest is the Iron Range–McIlwraith Range region, a region with rugged topography and the most diverse assemblage of rainforest types in the Peninsula. The vegetation is discussed by Lavarack and Godwin (1987) and Lavarack and Puniard (1988). Semi-deciduous mesophyll vine forest occurs on the narrow coastal lowlands and along the alluvial flats of rivers (more extensively in the Iron Range). The forests are tall, with an average canopy of 25 m but with emergents up to 50 m. Plank buttresses are very prominent, as are large vines. Vascular epiphytes are relatively common in the upper canopy. The ground layer is very sparse; possibly a consequence of seasonal flooding. Frequently occurring canopy species include *Buchanania arborescens*, *Tetrameles nudiflora*, *Ficus albipila*, and *F. nodosa*; the major deciduous species are *F. albipila*, *Bombax ceiba*, and *Nauclea orientalis*. Extensive areas of the eastern slopes and plateaux are clothed with notophyll vine forest, normally fairly low (less than 30 m) although on the ridges in the McIlwraith Range there are tall *Araucaria cunninghamii* emergents. There is considerable variation in floristic composition within this type. On the western side of these ranges, there are stands of evergreen notophyll vine forest in which the canopy is virtually pure *Blepharocarya involucrigera*. Further west in drier conditions and often on steep rocky slopes are stands of semi-deciduous notophyll vine forest and deciduous vine thickets. Frequent species include *Cochlospermum gillivraei*, *Bombax ceiba*, *Brachychiton* spp., and *Lagerstroemia archeranum*. Vines are abundant and are often prickly (e.g. *Capparis* spp., and *Ziziphus oenoplia*). Epiphytes include the striking orchid *Dendrobium bigibbum*, the floral emblem of Queensland.

Along some of the watercourses are limited areas of semi-deciduous mesophyll vine forests.

Sand dunes are frequent along the east coast of the Peninsula, varying from narrow bands to extensive dune fields. Rainforest, eucalypt forest/woodland, and heath form a complex, covering much of the dunes. Dense vine thickets tend to form on the exposed frontal dunes, while taller forest occurs in the sheltered lee. Variation is complex and Laverack and Godwin (1987) provide descriptions of some particular regions.

The second-largest block of rainforest on the Cape York Peninsula (Bell et al. 1987) occurs in the Jardine River catchment, occupying the sandstones of the McHenry Uplands. The most extensive rainforest stands are simple notophyll vine forests up to 20 m high, the canopies are evergreen and have numerous vines. Palms are conspicuous. Frequent canopy trees are *Acmena hemilampra*, *Flindersia ifflaiana*, *Podocarpus*

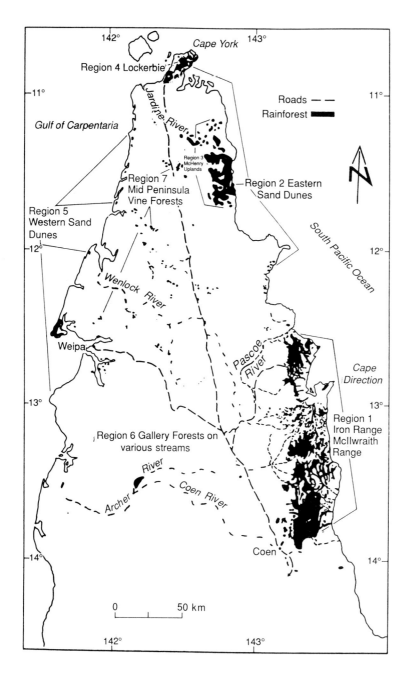

Fig. 3.9 Rainforests of northern Cape York Peninsula (from Lavarack and Godwin 1987).

neriifolius, *Welchiodendron longivalve*, *Syzygium forte*, and *S. fibrosa*. This community merges into a forest type with a lower canopy of mainly *Neofabricia myrtifolia* and *Melaleuca* sp., but with occasional *Callitris columellaris* emergents, and in which vines are not common. These upland forests are more diverse than those found nearby on coastal sand dunes and have strong similarities to heath forests (Heidewald) in Malesia (Whitmore 1984, 1989a), as has been pointed out by Lavarack and Stanton (1977) and Specht (1988). The pattern of variation in the vegetation of this region is related to topography and is probably controlled by the depth of the water table (see Fig. 3.10, Specht 1988).

Immediately south of Cape York, between Bamaga and the Cape is another extensive block of rainforest, known as the Lockerbie rainforests, and covering about 10 000 ha (F. C. Bell *et al.* 1987), a substantial section of this rainforest, *c*.1000 ha, was cleared in 1972 (Winter *et al.* 1987c), little regeneration of the cleared area has occurred. The rainforest occurs on a low tableland of Mesozoic sandstone which is extremely well drained. The vegetation is semi-deciduous mesophyll vine forest with canopy species, including *Alstonia scholaris*, *A. spectabilis*, *Bombax ceiba*, *Calophyllum australianum*, *Canarium australianum*, *Ficus albipila*, *F. virens*, *Myristica muelleri*, and *Terminalia sericocarpa* (Lavarack and Godwin 1987).

The west coast of the Peninsula is fringed with sand dunes. The rainforests on the dunes can be divided into two broad types, those occurring under higher rainfall conditions north from Weipa and those on the drier coast further south.

In the northern section, the rainforests are simple semi-evergreen notophyll vine forests between 10–20 m tall, although in exposed sites reduced to low thickets. Important canopy species include *Dysoxylum oppositiflorum*, *Ganophyllum falcatum*, *Ficus platypoda*, *Mimusops elengi*, and *Pongamia pinnata*.

In the drier southern section, the dune rainforests are low deciduous vine thickets with species such as *Bombax ceiba*, *Lagerstroemia archeriana*, *Cochlospermum gillivraei*, *Gyrocarpus americanus*, and *Terminalia* spp. (Lavarack and Godwin 1987).

Gallery rainforest occurs along all the larger rivers of the Peninsula. Most of these rivers arise in the east coast ranges and flow west and so

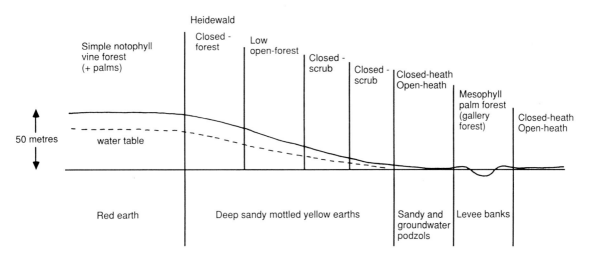

Fig. 3.10 Diagrammatic cross-section of the sequence of sclerophyllous communities in the Jardine River catchment (redrawn from Specht 1988).

provide corridors linking the two coasts. At Archer Bend on the Archer River floodplain there is the largest and most luxuriant example, with a medium to tall semi-deciduous notophyll vine forest.

Scattered through the Peninsula are numerous small patches of notophyll vine forest, mostly less than 50 ha and occurring on deep red earths derived from bauxite and laterite. It is probable that these represent fragments of a once more extensive forest cover; canopy species include *Acacia polystachya*, *Alstonia actinophylla*, *Bombax ceiba*, *Buchanania arborescens*, *Canarium australianum*, *Dysoxylum oppositifolium*, *Ganophyllum falcatum*, *Mimusops elengi*, and *Welchiodendron longivalve* (Lavarack and Godwin 1987).

The tall semi-deciduous mesophyll and notophyll vine forests on the Peninsula, and most of the evergreen notophyll forests contain species which define Webb *et al.*'s B_1 province. The drier deciduous thickets form part of the B_3 province. The summit forests of the McIlwraith Range form an outlier of the B_2 province.

Patterns of variation in the rainforest vegetation of Cape York Peninsula were discussed by Kikkawa *et al.* (1981*b*) who recognized four main environmental gradients (Fig. 3.11). The relationships between the Kimberley and Cape York floras were explored by Clarkson and Kenneally (1988).

To date, there has been relatively little disturbance to most of the Cape York rainforests, although as with the Northern Territory and the Kimberley, fire is an important factor controlling distribution. Except at Bamaga, clearing has been limited. Mining has caused limited local disturbance, but the potential for major disturbance from this source is high (Lavarack and Puniard 1988). Some disturbance by feral animals (pigs and cattle) has occurred.

3.4 THE WET TROPICS OF NORTH-EAST QUEENSLAND

The wet tropics of north-east Queensland contain the largest continuous area of rainforest in Australia. The main extent of rainforest is between Townsville and Cooktown (Fig. 3.12) and the total area in this region is estimated as slightly less than 800 000 ha (Bell *et al.* 1987). In addition, there are a number of outliers extending as far south as the Eungella Range west of Mackay.

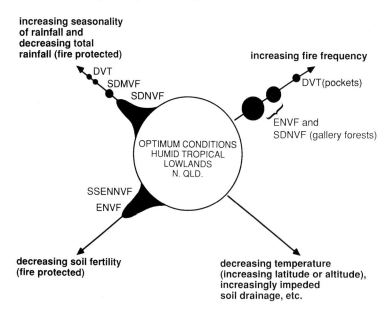

Fig. 3.11 Environmental gradients in Cape York Peninsula rainforests. ● represents rainforests of the structural type indicated, size of the shaded area is proportional to area of rainforest (adapted from Kikkawa *et al.* 1981*b*).

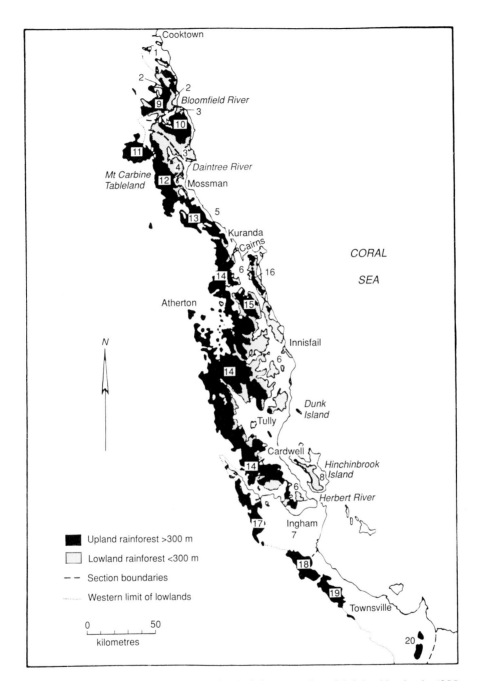

Fig. 3.12 Rainforests of the Queensland wet tropics. Rainforest sections: Mainland lowlands <300 m, 1. Cooktown, 2. Bloomfield–Helenvale, 3. Thornton Peak, 4. Mossman, 5. Macalister foothills, 6. Cairns–Cardwell, 7. Ingham. Offshore islands, 8. Hinchinbrook Island. Mainland Uplands >300 m, 9. Mt. Finnegan, 10. Thornton Peak, 11. Mt. Windsor Tableland, 12. Mt. Carbine Tableland, 13. Black Mountain Corridor, 14. Atherton, 15. Bellenden Ker Range, 16. Malbon Thompson Range, 17. Mt. Lee, 18. Mt. Spec, 19. Mt. Halifax, 20. Mt. Elliot (from F. C. Bell *et al.* 1987).

The rainforests of the Cooktown–Townsville block have been discussed in a number of publications (Breeden and Breeden 1970; Borschman 1984; Keto 1985; Keto and Scott 1986; Russell 1985; DASETT 1987; Figgis 1988a; Stocker and Unwin 1989). The vegetation of the region was mapped by Tracey and Webb (1975) and a monograph on the vegetation has been provided by Tracey (1982).

The region consists of coastal plains, with the rugged slopes of the Great Escarpment to the west and the Tablelands west of the Escarpment. Several steep ranges occur between the main Escarpment and the coast. The highest peak is Mt. Bartle Frere with an elevation of 1622 m. The Atherton Tableland forms an undulating plateau of about 700–800 m elevation. The highest peaks are granitic, while the Atherton Tableland has extensive flows of Cainozoic basalts.

Climatic details for the region are summarized by Tracey (1982). Rainfall regimes vary very widely and in places change considerably over short distances, reflecting local topography. Tracey (1982) recognized six rainfall zones (Table 3.1). Rainfall on the coast declines towards the south and the north of the region. Although rainfall is seasonal, with a summer maximum, the wettest localities have appreciable rain in all months of the year (see Fig. 1.9). Rainfall declines inland from the Escarpment to about 1200–1300 mm per annum at the western limits of rainforest on the Atherton Tableland. Rainfall on the highest peaks is considerable; mean annual rainfall on Mt. Bellenden Ker is in excess of 9000 mm (Keto and Scott 1986), and the rainfall in the 24 hours to 0900 on 4 January 1979 is an Australian record of 1140 mm (Tracey 1982). A consequence of the very high rainfall intensities on the mountain slopes, combined with saturated soil profiles, is that rain cannot be absorbed and widespread overland flow occurs, an unusual feature in tropical rainforests (Keto and Scott 1986). Temperatures are marginally tropical, as mean annual temperatures in the lowlands barely exceed 24°C and the annual temperature range is greater than 5°C (Webb *et al.* 1984). Frost occurs on the Atherton Tableland but very rarely, if ever, on the coastal lowlands (Tracey 1982).

The vegetation of the region has been classified and mapped by Tracey and Webb (1975). The rainforests were classified on the basis of the structural/physiognomical scheme proposed by Webb (1968) (Table 3.2). Floristic data from the region were subjected to numerical analysis by Williams and Tracey (1984). For the most part, this analysis identified units previously recognized intuitively by Tracey and Webb (1975); the only major exception being unit 2a which the numerical analysis suggested to encompass a very wide range of floristic units. Tracey (1982) provided a detailed account and diagnosis of the vegetation types illustrated by profile diagrams; this account provided the basis for the following discussion.

The complex mesophyll vine forest 1a on fertile

Table 3.1 Rainfall zones in the Cooktown/Townsville regions (from Tracey 1982).

Zone	Mean annual rainfall	Mean rainfall May–October	Zone	Altitude
Very wet	>3000 mm	>750 mm	Lowlands	<40 m
Wet	2000–3000 mm	500–750 mm	Foothills	40–400 m
Cloudy wet	2000–3000 mm plus cloud		Uplands	400–800 m
Moist	1600–2000 mm	300–500 mm	Highlands	800–1600 m
Cloudy moist	1600–2000 mm plus cloud			
Dry	1300–1600 mm	200–300 mm		

Table 3.2 Major rainforest types of NE Queensland (from Tracey 1982).

Rainforest type

Complex mesophyll vine forest (CMVF)
    1a. Very wet and wet lowlands and foothills; basalts, basic volcanics, mixed colluvium on footslopes, and riverine alluvia
    1b. Very wet and wet cloudy uplands; basalts
    1c. Moist and dry lowlands; riverine levees (gallery forests)

Mesophyll vine forest (MVF)
    2a. Very wet and wet lowlands and foothills; granites and schists
    2b. Very wet and wet lowlands; beach sands

Mesophyll vine forest with dominant palms (MFPVF)
    3a. Very wet lowlands; feather-leaf palm (*Archontophoenix*) swamps; basaltic and alluvial sands
    3b. Very wet lowlands and lower foothills; fan-leaf palm (*Licuala*), seasonally impeded drainage; schists and granites

Semi-deciduous mesophyll vine forests (SDMVF)
    4a. Moist and dry lowlands and foothills; granites and basalts

Complex notophyll vine forest (CNVF)
    5a. Cloudy wet highlands; very limited areas of basalt and basic rocks
    5b. Moist and dry lowlands, foothills and uplands; basalts

Complex notophyll vine forest with emergent *Agathis robusta* (CMVF + emergent *Agatha robusta*)
    6 Moist foothills and uplands; granites and schists

Notophyll vine forest (rarely without *Acacia* emergents) (NVF + *Acacia* emergents)
    7a. Moist lowlands and foothills along coast including islands; granites and schists
    7b. Moist and dry lowlands; beach sands

Simple notophyll vine forests (often with *Agathis microstachya*) (SNVF + *Agathis microstachya*)
    8. Cloudy wet and moist uplands and highlands; granites, schists, and acid volcanics

Simple microphyll vine–fern forest (often with *Agathis atropurpurea*) (MFF + *Agathis atropurpurea*)
    9. Cloudy wet highlands; granites

Simple microphyll vine-fern thicket (MFT)
    10. Cloudy wet and moist wind-swept top-slopes of uplands and highlands; granites

Deciduous microphyll vine thicket (DVT)
    11. Dry lowlands and foothills; granite boulders

lowlands represents the optimum development of rainforest in Australia. At higher altitudes, but still on fertile basaltic soils, there is reduction in leaf size and changes in species composition (type 1b). However, the average height of these upland forests is greater than in the lowland type 1a. Type 1c is found in drier lowland areas, normally on riverine alluvium. The canopy has an increasing number of deciduous species, and there is a reduction in vascular epiphytes.

On less-fertile soils in the high rainfall lowlands there is a reduction in structural diversity, with fewer species with macrophyll leaves, and fewer plank buttresses, trunk epiphytes, and robust lianas. The analysis by Williams and Tracey (1984) indicated considerable floristic diversity within structural type 2a. Under conditions of impeded drainage, very distinctive palm forest occurs. *Archontophoenix alexandrae* (type 3a) tends to occur on soils of higher nutrient status than the fan palm *Licuala ramsayi* (type 3b). Both types have suffered extensive clearing and only small stands remain.

In the drier lowland areas, on both nutrient-rich (basalt) and nutrient-poor (granite) soils, semi-deciduous mesophyll vine forest occurs. This has deciduous emergents (*Bombax ceiba*, *Gyrocarpus americanus*, *Ficus albipila*, *F. virens*) over a mixed deciduous–evergreen canopy. Plank buttresses are common and, on the emergents, may be very large. In both composition and structure, this type is similar to forests found in favourable lowland sites on Cape York.

The complex notophyll vine forests (type 5) are relatively restricted in extent. In the case of type 5a, this is because the occurrence of basic rocks in the cloudy wet highlands is limited, but in the case of 5b, found at lower elevations on the Atherton Tableland, its rarity reflects extensive clearance.

On less-fertile soils under moist conditions, type 6 complex notophyll vine forest with emergent *Agathis robusta* occurs. This has a much lower canopy than type 5 stands but the emergents may be up to 35 m in height. Some canopy species, such as *Paraserianthes toona* and *Pleiogynium timorense*, are deciduous.

Type 7 notophyll vine forest is restricted to coastal dunes, and on exposed foredunes may be reduced to low dense thickets. Its species composition shows many similarities with other dune thickets around northern Australia.

The most extensive forest type on uplands on granite and other acid rocks is the simple notophyll vine forest type 8. The canopy trees generally have fairly slender trunks of uniform girth (pole-like), plank buttresses, large lianas and deciduous species are rare, athough slender and wiry vines may be common. Tree ferns may be frequent. At higher altitudes microphyll leaves become more frequent.

At higher elevations, type 9, simple microphyll vine–fern forest is found. The canopy height is lower than type 8, although patches of emergent *Agathis atropurpurea* may extend up to 35 m. On exposed ridges or shallow soils, canopy height may be as low as 10 m. As in type 8 most canopy species have pole-like trunks.

On the highest (granitic) mountain tops, simple microphyll vine–fern thickets are found. Similar stands are also found at low altitudes on exposed ridges of isolated mountains. The canopy is wind-pruned with wind-sheared emergents of *Leptospermum wooroonooran*. On very exposed sites, the community may be reduced to prostrate shrubs. Leaves are evergreen, mostly simple and entire, in the leptophyll–microphyll size range with only a few nanophylls. Vines are sparse and slender and wiry, the ground layer is rich in ferns and epiphytic mosses, filmy ferns and small orchids are common.

These mountain top forests are very different in floristic composition from other rainforests in the region and contain an interesting biogeographical mix of species—for example the shrub layer includes *Dracophyllum sayeri* (other species of *Dracophyllum* occur in south-eastern Australia) and *Rhododendron lochae*, the only Australian species of the genus.

This forest type satisfies the definition of tropical upper montane rainforest given by Whitmore (1984). It is interesting that a conspicuous feature of the upper montane forests in South-east Asia described in Whitmore (1984) is the occurrence of *Leptospermum javanicum* (referred to as *L. flavescens* by Whitmore) a species closely related to *L. wooroonooran* (J. Thompson 1989).

Deciduous microphyll vine thickets are found in the northern part of the region under dry climatic conditions among granite boulders. The canopy is low but with taller deciduous emergents; lianas are frequent and are often armed with thorns or prickles. Canopy emergents include *Bombax ceiba*, *Cochlospermum gillivraei*, *Ficus albipila*, and *Gyrocarpus americanus*.

Forest types 12 and 13 are closed forests with sclerophyll emergents (*Acacia* spp. in type 12, *Eucalyptus* spp. in type 13). Most of these forests have dynamic relationships with rainforest (see Chapter 4).

One of the outstanding features of the northeast Queensland rainforests is that it is possible to study variation in structure and floristics along several environmental gradients within contiguous expanses of rainforests (see Fig. 1.13). The total rainforest flora in the region is large, including numerous species of considerable biogeographical and evolutionary interest.

The bulk of the wetter rainforests in the region form the province B_2 of Webb *et al.* (1984). The semi-deciduous mesophyll vine forest type 4 represents an extension of province B_1, while the deciduous microphyll vine thickets of type 11 are part of the province B_3. Bell *et al.* (1987) suggested that the highest mountain peaks, with the simple microphyll vine–fern thickets of type 10 represent an outlier of province A_2. At a generic level, and environmentally, certainly there are strong relationships between type 10 and temperate rainforest, but the overall relationships at species level are with B_2.

To the south, the limits of province B_2 are the stands of complex mesophyll vine forests on slopes and in gorges on the eastern side of the Eungella Range west of Mackay (McDonald and Winter 1986).

3.5 RAINFORESTS OF PROVINCES A_1 AND A_2—SOUTHERN QUEENSLAND AND NEW SOUTH WALES

The subtropical rainforests of the A_1 province have their core area on the Border Ranges and the nearby (now virtually cleared) Big Scrub (see Chapter 9), but outliers extend northwards to near Gladstone and southwards to the south coast of New South Wales. The vegetation of the rainforests has been discussed by Baur (1957, 1962*a,b*), Beadle (1981), McDonald and Elsol (1984), Adam (1987*b*), Helman (1987), Mills

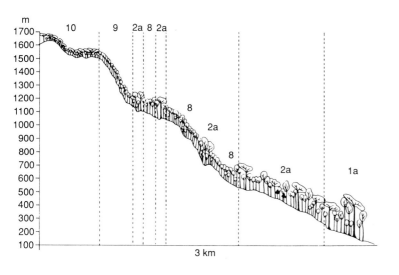

Fig. 3.13 Transect through forest vegetation on Mt. Bellenden Ker. Communities numbered as in Table 3.2 (from Tracey 1982).

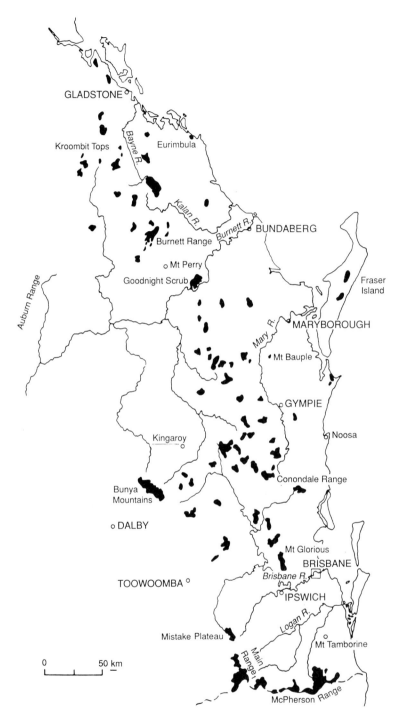

Fig. 3.14 Distribution of rainforest in south-east Queensland (from McDonald and Elsol 1984).

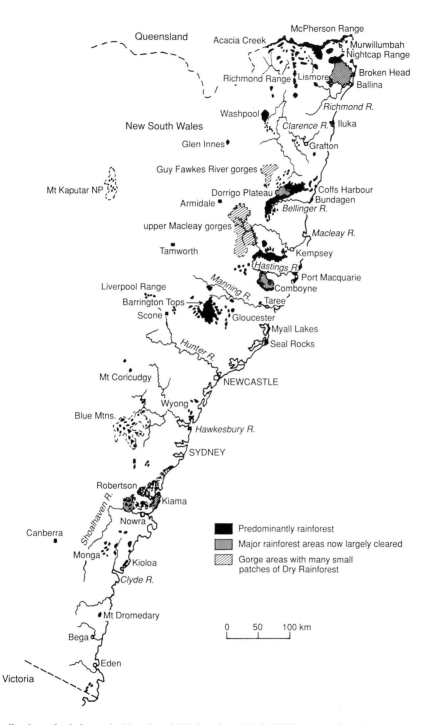

Fig. 3.15 Distribution of rainforest in New South Wales (from W. T. Williams *et al.* 1984).

(1987), Floyd (1987, 1990), Young and McDonald (1987), and Forestry Commission of NSW (1989). Rainforest tree floras are provided by Floyd (1989) and Williams *et al.* (1984). (Figs. 3.14, 3.15).

The highest annual rainfalls in the region exceed 2000 mm. In the north, there is a summer rainfall maximum (but with significant falls in every month), but further south the seasonality in rainfall decreases. On higher peaks there may be persistent mists which contribute to effective precipitation. Temperature declines with increasing latitude.

The major trends of variation in rainforest in northern New South Wales/Southern Queensland are portrayed in Fig. 3.16 and are discussed at greater length by Floyd (1990).

The optimal development of rainforest (in terms of diversity and structural complexity) is in the lowland subtropical complex notophyll vine forests found on soils derived from basalt, generally below 300 m, although extending to 600–900 m in its northern outliers west of Gladstone and Mackay, and experiencing rainfalls above 1500 mm per annum. The species composition is variable, but these lowland forests are characterized by *Argyrodendron* spp.; *A. trifoliolatum* occurs at low altitudes but with gradual replacement by *A. actinophyllum* at greater elevations. (Type 1—Booyong, see Forestry Commission of NSW 1989.) Large lianas, strangling figs, and vascular epiphytes are frequent, and sometimes abundant. Canopy height is generally around 35 m but in some favourable sites may reach 50–55 m (Fig. 3.17).

With increasing altitude, there is a tendency for stands to become less diverse and structurally simpler. Stands tend to be lower, rarely exceeding 30–35 m. Species composition varies considerably between sites but certain species tend to be more frequent in these upland 'cool' subtropical rainforests than in the lowland 'warm' subtropical forests, these include *Sloanea woollsii*, *Dysoxylum fraseranum*, *Caldcluvia paniculosa*, and *Cryptocarya erythroxylon*. (Type 2—Yellow Carabeen, see Forestry Commission of NSW 1989.) Cool subtropical rainforest is a major rainforest type on fertile soils at altitudes above *c*.800 m in northern New South Wales and in southern Queensland, extending northwards to Mt. Glorious and the Conondale Range (McDonald and Elsol 1984). Towards the drier limits of distribution of both type 1 and type 2, *Araucaria cunninghamii* emergents tend to occur.

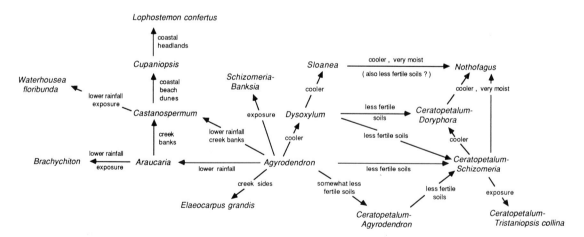

Fig. 3.16 Major environmental gradients reflected in rainforest composition in northern New South Wales and southern Queensland (redrawn from Baur 1962*a*).

Fig. 3.17 Complex notophyll vine forest at Gradys Creek, Border Ranges National Park (note emergent *Araucaria cunninghamii* on skyline, right of centre).

South of the distribution limit of *Argyrodendron* the subtropical rainforest has a less complex structure, in the Forestry Commission of NSW (1989) classification being represented by type 3—Corkwood (*Caldcluvia paniculosa*)—Sassafras (*Doryphora sassafras*)—Crabapple (*Schizomeria ovata*)—Silver Sycamore (*Cryptocarya glaucescens*). Species composition varies within and between stands, but at least three of the species for which the type is named are prominent at any one locality. This type is found on fertile soils under high rainfall (>1200 mm) conditions as far south as the south coast of New South Wales.

Under lower rainfall conditions or at high latitudes, the subtropical rainforest becomes depauperate (the Fig (*Ficus* spp.)—Giant Stinger (*Dendrocnide excelsa*) Type 6 of Forestry Commission of NSW 1989). This type occupies topographically sheltered sites on fertile sites. Species diversity is much lower than in optimally developed subtropical rainforest, with a lower canopy and a less complex structure. The Fig—Giant Stinger forest type is found as far south as the south coast of New South Wales (Helman 1987) (Fig. 3.18). Although at the distributional limit of several component species these southernmost subtropical rainforest stands are notable for the large girth of some of the trees (particularly very large *Dendrocnide excelsa* and *Citronella moorei*) and the abundance of the epiphytic *Asplenium australasicum*. The southernmost example of this type is Mt. Dromedary (Forestry Commission of NSW 1987), where it occurs at low altitudes, along gullies and extending on to the lower slopes on aspects sheltered from hot, dry, westerly winds.

The most inland occurrence of subtropical rainforest is on the Liverpool Range, some 160 km from the sea (Fisher 1985). The range is part of the Great Dividing Range and also forms the northern watershed of the Hunter Valley. Geologically, the range is formed from early Tertiary basalts, weathering to fertile soil. The ridge of the range is oriented east–west, with steep slopes interrupted by occasional cliffs. Rainfall averages *c*.900 mm per year, declining to 800 mm at the western limit of rainforest. There is a pronounced summer maximum in rainfall. Rainfall is probably augmented by fog stripping; the upper slopes of the range (much of the range exceeds 900 m altitude) are often enshrouded in cloud. The rainforest occurs in a series of discrete patches, with sharp boundaries to adjacent eucalypt communities. Most rainforest is on south-facing slopes but in a few places it extends on to the more exposed north-facing aspects.

The majority of rainforest occurs between 650–1000 m elevation although isolated rainforest trees extend down to around 550 m in gullies, and some species are found under eucalypts at higher elevations. At the lower elevations the rainforests are notophyll vine forest, but at higher elevations leaf size declines to microphyll. Fisher (1985) provided details of the floristic composition of the stands, floristic diversity varies between stands but six species occur in virtually all stands (*Daph-*

Fig. 3.18 Complex notophyll vine forest (subtropical rainforest) near Murramurrang, southern NSW, note the palm *Livistona australis* in the background, *Ficus* sp. in the foreground (photo N. C. Wilson).

nandra micrantha, *Alectryon subcinereus*, *Euodia micrococca*, *Hymenosporum flavum*, *Pittosporum undulatum*, *Acmena smithii*). At high elevations *Acmena smithii* predominates. Of the emergents *Ehretia acuminata* and *Toona australis* are deciduous. *Dendrocnide excelsa* is abundant in some stands, but *Doryphora sassafras* is absent; Fisher (1985) suggested that the low rainfall may be responsible for this absence.

One of the most widespread rainforest types in northern New South Wales—south-east Queensland is that designated Type 5 Booyong–Coachwood by Forestry Commission of NSW (1989). Most occurrences are in gullies up to about 400 m elevation; individual stands are small and often show intergradation with eucalypt forest. The topography ensures shelter and relatively moist conditions, while accumulation of alluvium is responsible for the moderate fertility of the soils. Strangling figs are often massive emergents, while the canopy is a mixture of species from the subtropical and warm temperate leagues (including *Argyrodendron* spp., *Gmelina leichhardtii*, *Sloanea woollsii*, *Diploglottis cunninghamii* (the subtropical elements) and *Ceratopetalum apetalum*, *Doryphora sassafras*, *Orites excelsa*, and *Caldcluvia paniculosa* (the warm temperate element)).

On more nutrient-poor soils, the forest structure is simplified to notophyll vine forest or simple notophyll/microphyll evergreen vine forest. Within the region these forests occur in two different situations, on the coast and at higher altitudes on soils derived from acid igneous rocks or relatively nutrient-poor metamorphic and sedimentary rocks.

On the sandmasses of Fraser Island and Cooloola in south-east Queensland, very well-developed rainforest occurs, with a canopy height of up to 55 m (Gillison 1985; Figgis 1988*b*; Webb and Tracey 1975; Young and McDonald 1987). The occurrence of such forests on soils of low nutrient status (Webb and Tracey 1975) is remarkable. Major canopy species include *Syncarpia hillii*, *Lophostemon confertus*, *Agathis robusta*, and *Araucaria cunninghamii*, while *Backhousia myr-*

tifolia is frequent in the understorey. All these species have relatively sclerophyllous foliage. *Schizomeria ovata* is locally common.

In northern New South Wales, well-developed rainforest, with a canopy of up to 30 m, is found on the sandmass at Iluka (Adam 1987*b*); here the dominant species are *Acmena hemilampra* and *Syzygium leuhmannii* (Fig. 3.19). Webb *et al.* (1984) suggested that the flora of rainforest on sand between Iluka and Fraser Island constitutes a recognizable subelement in the A_1 province. These forests of the sandmasses, occurring on giant podzols, are similar to heath forest (*Heidewald*) but of much greater stature than at Cape York (p. 45). Whitmore (1989*a*) has pointed out that heath forest in Malesia, despite its tendency to sclerophylly, is not oligotrophic, at least in a virgin state. Topsoil nutrient levels in these Australian sandmass rainforests, although higher than in adjacent sclerophyll communities, are low, comparable with those in simple notophyll vine forest dominated by *Ceratopetalum* (Webb and Tracey 1975). It would be of interest to compare more closely the nutrient budgets of South-east Asian heath forest and the rainforest on sand on the southern Queensland coast.

Rainforest occurs in a number of situations along the coast, not only on sand but also on rocky headlands. In sites exposed to strong winds and high salt-spray inputs (foredunes and exposed slopes on headlands), the vegetation may be reduced to low microphyll vine thicket (the Tuckeroo and Headland Brush Box forest types of the Forestry Commission of NSW (1989), but

Fig. 3.19 The Iluka rainforest, littoral rainforest in northern NSW. This site is included in the World Heritage List (see Chapter 9).

in more sheltered localities it may be up to 15–20 m tall. On some headlands with more fertile soils, in northern New South Wales, *Araucaria cunninghamii* occurs as an emergent and the floristic composition indicates that the stands are a coastal variant of the lowland Booyong type. These littoral rainforests occupy an environment which is marginal for rainforest development. Rainfall is reliable but is lower than that in the inland uplands. Stands on sand or on steep slopes are often sharply drained. Salt-spray inputs are high and maintenance of the integrity of the canopy is essential for the protection of the stands from salt damage (Fig. 3.20). Nutrient concentrations in the soils, particularly on sands, are low, although nutrient budgets are likely to be augmented by spray inputs.

On poor soils in the uplands, under high-rainfall conditions, the simple notophyll evergreen vine forests have a characteristic and striking structure. The canopy has low floristic diversity and the major species lack buttresses and have untapered pole-like trunks (Fig. 3.21). Tree height is generally about 30 m and emergents are rare (except at localities where *Araucaria cunninghammii* emergents occur, this was probably once more common but many of the *Araucaria*s have been felled). Most vine species are slender or wiry; large robust species are infrequent. At higher altitudes, or on exposed ridges, the general leaf size is reduced to microphyll. The major species in many of these stands is the important timber species, coachwood (*Ceratopetalum apetalum*), which has a distribution extending from the south coast of New South Wales to the Border Ranges with an outlier on the Kroombit Tops west of Gladstone. *C. apetalum* is, however, virtually absent from the Barrington Tops region and in north-east New South Wales has a patchy distribution.

In the Border Ranges, the major species in the absence of *C. apetalum* is *Schizomeria ovata*, which provides a floristic link to the simple notophyll forests on the coastal sandmasses. On exposed ridges, canopy height may be much reduced and *Tristaniopsis collina* becomes an

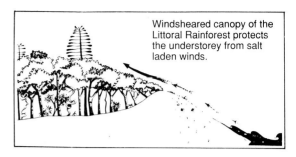

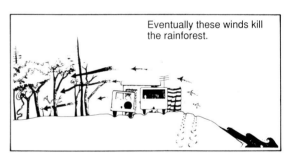

Fig. 3.20 The effect of canopy damage on littoral rainforest (Dept. of Environment and Planning 1988).

important component of the canopy. On some exposed basaltic ridges (between 900–1100 m) in the Border Ranges, with fertile soils, stands with similar simple structure but dominated by *Caldcluvia paniculosa* occur.

Under cooler, or drier conditions, *Doryphora sassafras* becomes more abundant and in some cases *C. apetalum* is virtually absent. This type (Coachwood–Sassafras Type 12 in Forestry Commission of NSW 1989) may occur on more fertile soils than nearby pure *Ceratopetalum* stands.

Fig. 3.21 *Ceratopetalum apetalum*-dominated simple notophyll vine forest, Billilimbra State Forest, New South Wales. The mottling of the bark on *C. apetalum* (left foreground) is due to the presence of lichens

The Forestry Commission of NSW (1989) recognized a number of other rainforest types restricted to special habitats. Stands dominated by palms (either or both of *Livistona australis* and *Archontophoenix cunninghamiana*) occur under conditions of impeded drainage. The occurrence of palm patches in wet sites is not, however, restricted to rainforest, and similar patches may be found in eucalypt forests (particularly *Eucalyptus robusta* and *E. botryoides* communities in coastal lowlands).

A number of authors have recognized gallery rainforests as a distinct type of forest and Forestry Commission of NSW (1989) Type 4 Black Bean

(*Castanospermum australe*) (Fig. 3.22) would be included in this category. However, while there are a number of floristically distinctive assemblages along watercourses, they can be regarded as local, habitat-determined variants of the major local form of rainforest (as in the treatment by Floyd 1990) rather than being treated as a major rainforest subform in their own right. Strips of rainforest along watercourses may extend well beyond major stands into lower rainfall areas and may thus be important in providing continuity of habitat for nomadic fauna.

The Black Bean type occurs on fertile alluvium in north-east New South Wales; but although once widespread it has been extensively cleared. A widespread community, fringing creeks in lower rainfall areas as far south as the Hunter region, is characterized by *Waterhousea floribunda* (formerly *Eugenia ventenatti*) and *Tristaniopsis laurina* (Baur 1975; Floyd 1987).

On the south coast of New South Wales, fringes of *Backhousia myrtifolia* and *Acmena smithii* line many creeks, similar stands are also found further north. *B. myrtifolia* on the south coast of New South Wales is sometimes regarded by foresters as a weed which, at the present time, appears to be invading eucalypt forest.

The A_2 floristic province of Webb *et al.* (1984) encompasses the Cool Temperate Rainforest League of Forestry Commission of NSW (1989), with the addition of certain components of the Warm Temperate League. Structurally, the forests are microphyll fern forests, grading to microphyll fern thickets on exposed sites. Both floristically and structurally these forests are simple, often with a single species dominating. Large lianas are rare or absent, as are palms, but tree ferns may be abundant. Epiphytic bryophytes and lichens are profuse. Although still conforming to the definition of closed forest, light penetration is much greater than in other evergreen rainforests (Lowman 1986).

In the northern part of the range, microphyll fern forest is found at high altitudes (mostly above 900 m), experiencing a microtherm temperature regime with high rainfall (in excess of 2000 mm) which is less seasonal than in the lowlands. In addition, the sites are often enshrouded in cloud, and mist interception may considerably enhance effective precipitation. To the south, these forests

Fig. 3.22 *Castanospermum australe*, northern New South Wales (photo B. S. Wannan).

are found at lower altitudes and experience lower rainfalls (down to 1200 mm), although again this is augmented by the influence of persistent low cloud.

From the Barrington Tops northwards, the major dominant is *Nothofagus moorei* (Figs 3.23, 3.24). On favourable sites, for example Mt. Banda Banda in the Hastings region, canopies may be up to 45–50 m, but in more exposed locations the community may be a denser thicket about 10 m tall. In the taller stands there is often a well-defined lower storey of species such as *Doryphora sassafras*, *Orites excelsa*, *Elaeocarpus holopetalus*, and *Ceratopetalum apetalum*. The small tree *Vesselowskya rubifolia* is characteristic particularly along creeks; the only other species in the genus is restricted to New Caledonia.

Nothofagus moorei is found today as a series of disjunct populations between Barrington Tops and the Border Ranges (Adam 1987b). Although most occurrences are about 900 m, it descends to 500 m along creek lines in the Dorrigo region, probably reflecting cold air drainage. At the lower margins of *Nothofagus* stands there are complex gradations into cool subtropical rainforest (Turner 1976; Webb et al. 1984). There are a number of localities from which it is absent but which, on climatic grounds, would appear to be suitable. The absence from the Bulga and Comboyne Plateaux is explained by clearing early in the twentieth century (see Chapter 9), but absence from the Gibraltar Range and Mt. Hyland is less explicable. On the Gibraltar Range there is a small population of the rufous scrub-bird, a species which is otherwise associated with cool temperate rainforest (Adam 1987b), whose presence may indicate the previous occurrence of *Nothofagus* on the Range. On Mt. Hyland, there is a high-altitude microphyll fern forest dominated by *Doryphora sassafras* and *Cryptocarya nova-anglica*, which is associated with *Nothofagus* at other localities. Other species on the mountain normally occurring in *Nothofagus* stands are *Elaeocarpus holopetalus* (here close to its northern limit), *Trochocarpa* sp. nov. (aff. *T. laurina*), and *Vesselowskya rubifolia*.

The highest altitude *Ceratopetalum* stands (notophyll/microphyll in leaf size) are regarded by Webb et al. (1984) as forming part of the A_2 province. Stands of *Doryphora sassafras* and *Cryptocarya foveolata* (Mountain Walnut—Type 20, Forestry Commission of NSW 1989) occur on the highest peaks of the Dividing Range in southern Queensland (McDonald and Elsol 1984), and on basalt on high peaks as far south as the Barrington Tops (Forestry Commission of NSW 1989) and are also part of the A_2 province.

To the south of Barrington Tops, there is a major disjunction in the distribution of cool temperate rainforest until the Illawarra region (Mills 1987). The core species in cool temperate

Fig. 3.23 Emergent *Nothofagus moorei*, Wrights Lookout, New England National Park

Fig. 3.24 *Nothofagus moorei* canopy below Careys Peak, Barrington Tops National Park, New South Wales

rainforest in southern New South Wales is *Eucryphia moorei*, which is endemic to the State (except for one locality 2 km south of the Victoria border). *Eucryphia* is the only genus in the family Eucryphiaceae; there are two species in Tasmania and a recently discovered, as yet undescribed species found at high altitude in the wet tropics of north-east Queensland. Outside Australia there are two species in Chile. *E. moorei* has compound leaves and can develop into a large tree up to 35 m tall. Frequently it is naturally coppiced and stools can reach considerable dimensions (see Helman 1987). The pattern of variation in *E. moorei* communities is shown in Fig. 3.25.

In the Illawarra region *E. moorei* today has a very limited distribution (Fuller 1980; Mills 1987; Thomas 1990). On the plateau at around 700 m it occurs in the remnants of the Yarrawa Brush, which formerly covered some 2000 ha, and therefore may have been more abundant until recently.

Further south, *E. moorei* is often the sole canopy dominant, although individual stands are mostly small. At higher altitudes *Elaeocarpus holopetalus* and *Atherosperma moschatum* may co-dominate, while at slightly lower elevations *Doryphora sassafras* may be co-dominant (as at Mt. Dromedary: Forestry Commission of NSW 1987). With decreasing altitude *Acmena smithii* becomes an important component. The lowest elevation *E. moorei* stands are found near the coast in Nadgee Nature Reserve.

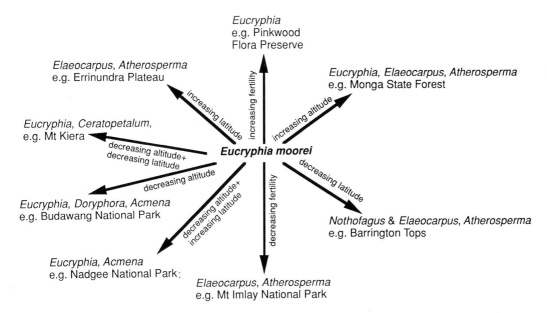

Fig. 3.25 Variation in *Eucryphia moorei* communities (redrawn from Helman 1987).

The *E. moorei* forests form a biogeographical link between the southern *Nothofagus cunninghamii* communities and the northern *N. moorei* forests. *Atherosperma moschatum*, which occurs with *E. moorei*, has a distribution which stretches from Tasmania to Barrington Tops, while *Elaeocarpus holopetalus* has a distribution extending from Victoria to northern New South Wales.

At a number of sites in northern New South Wales, on steep rocky slopes with shallow skeletal soils, there are small stands of closed scrub–low forest forming the *Leptospermum* spp.–*Notelaea venosa*–*Prostanthera ovalifolia* suballiance of Floyd (1987, 1990). Canopy leaf size is in the nanophyll–micropyll range. At similar elevations, but with less steep slopes and deeper soils, taller microphyll/nanophyll forest occurs. While the structure and floristics of these stands is presumed to largely reflect the local environmental conditions, there are, nevertheless, similarities to the montane forests of the peaks of north-east Queensland.

3.6 RAINFORESTS OF VICTORIA

The rainforests of Victoria fall into two broad groupings, those in southern Victoria and on the eastern ranges, falling into Webb *et al.*'s (1984) A_3 province and those in the East Gippsland lowlands, extending to Wilsons Promontory, which form a southern extension of the A_1 province. Rainforest in Victoria occurs in numerous small stands (Fig. 3.26); the total area is estimated at 13 270 ha (Department of Conservation, Forests and Lands 1987), although this figure does not include areas of closed rainforest canopy below emergent eucalypts, which are particularly extensive on the Errinundra Plateau.

Descriptions of Victorian rainforests are provided by Howard (1981), Seddon and Cameron (1985), Cameron (1987), Department of Conservation, Forests and Lands (1987).

The cool temperate rainforests of the A_3 province are nanophyll/microphyll moss forests. In

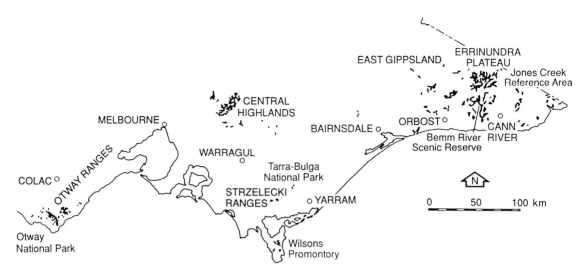

Fig. 3.26 Distribution of rainforest in Victoria (from Department of Conservation, Forests and Lands 1987).

the Otway Ranges, the Central Highlands, the Strzelecki Ranges, and Wilsons Promontory (Fig. 3.26), the dominant (and often only canopy species) is *Nothofagus cunninghamii*. It is generally restricted to sites with annual rainfall (evenly distributed throughout the year) exceeding 1500 mm, but may be found in gullies in areas with as little as 1000 mm annual rainfall. *N. cunninghamii* appears to be indifferent to soil type. Below *c*.600 m altitude, a tall closed forest up to 30 m high is found, but at high altitudes canopy height is reduced.

The other major cool temperate forest type in Victoria is sassafras forest, with a canopy of *Atherosperma moschatum* and/or *Elaeocarpus holopetalus* from which *N. cunninghamii* is absent. The main occurrences are in east Gippsland, but there is an overlap with *N. cunninghamii* forest in the Central Highlands, Strzelecki Ranges, and Wilsons Promontory (Seddon and Cameron 1985). *Acacia melanoxylon* is frequently found as a component of the canopy or as an emergent. The largest stand of this type, and the largest single stand of rainforest in Victoria, is in the Errinundra Flora Reserve (Seddon and Cameron 1985). This stand is at about 1000 m altitude and the annual rainfall is about 1050 mm, although this relatively low figure is augmented by cloud stripping. Epiphytic bryophyte ferns and lichens are abundant and the tree fern *Dicksonia antarctica* is a prominent component of the vegetation (Figs. 3.27, 3.28).

At Goonmirk Rocks in the Errinundra Flora Reserve, at a higher elevation (1200 m) and with a high rainfall (*c*.1200 mm a year) there is an unique stand of microphyll–nanophyll moss thicket characterized by *Podocarpus lawrencei*, an endemic species (Cameron 1987), which, as well as being a major component of the canopy, forms emergents up to *c*.8 m. In addition to *Podocarpus*, the canopy is composed of *Pittosporum bicolor*, *Prostanthera lasianthos*, *Telopea oreades*, *Lomatia fraseri*, *Persoonia silvatica*, *Leucopogon maccraei*, *Tasmannia lanceolata*, and *T*. sp. aff. *xerophila* (Seddon and Cameron 1985). This last species has low, branching, almost horizontal stems, makes the stand almost impenetrable, and creates a resemblance to the horizontal scrubs of *Anodopetalum biglandulosum* in Tasmania. Species diversity of ferns is low,

Fig. 3.27 Mottled trunks of *Acacia dealbata*, Errinundra Plateau (photo N. C. Wilson).

Fig. 3.28 *Atherosperma moschatum* in a gully, Errinundra Plateau (photo N. C. Wilson).

although *Polystichum proliferum* is abundant. There is very abundant growth of epiphytic bryophytes and lichens. Seddon and Cameron (1985) refer to this stand as elfin forest, and in terms of structure it has a striking similarity with some cloud forests at much higher altitudes on tropical mountains.

The lowland rainforests of East Gippsland are best described as simple notophyll–microphyll vine–fern forest (Cameron 1987). They are found along gullies and riversides within 50 km of the coast and up to 500 m altitude. The westernmost stand is on Mt. Moornapa; there is then a gap of about 150 km to the isolated small stands on Wilsons Promontory (Ashton and Frankenberg 1976). Annual rainfall varies from less than 800 mm to *c*.1400 mm. Stands have an uneven canopy between 20 and 30 m tall and are frequently dominated by *Acmena smithii*; large vines are conspicuous (including *Cissus hypoglauca* and *Pandorea pandorana*). On poorly drained sites, *Rapanea howittiana* is frequent, while on drier sites, *Pittosporum undulatum* and *Elaeocarpus reticulatus* are more frequent. On creek sides *Tristaniopsis laurina* is common. At two sites, Brodribb River Flora and Fauna Reserve and Cabbage Tree Creek Flora Reserve, the palm *Livistona australis* has its only Victorian localities. Seddon and Cameron (1985) provide accounts of the vegetation of the more accessible stands.

The *Acmena smithii* stands on Wilsons Promontory have been described in detail by Ashton and Frankenberg (1976). They occur along streams and on slopes protected from the prevailing north-west to south-west winds. At higher elevations, *A. smithii* is replaced by *Nothofagus cunninghamii* and *Atherosperma moschatum*, but at Sealers Creek and Paradise Gully, mixtures of *Acmena* and *Nothofagus* are found virtually at sea-level (Ashton and Frankenberg 1976)). At sites exposed to salt spray, *A. smithii* may form low wind-pruned thicket less than 20 m high; similar patches are frequent on exposed headlands in New South Wales. The streamside stands have a much simpler structure than those in East Gippsland, and the large robust vines are absent, although various wiry species do occur. Ferns are both diverse and numerous. *A. smithii* is the major canopy species, although overtopped by emergent *Acacia melanoxylon* and eucalypts.

3.7 TASMANIAN RAINFORESTS

Tasmania has about a third of the rainforest which remains in Australia. Vines are absent or very rare but epiphytic ferns, bryophytes, and lichens are often extremely abundant, while epiphytic angiosperms are rare (a marked contrast to cool-temperate *Nothofagus moorei* forests in New South Wales, where a number of epiphytic orchids occur). All the woody plants are evergreen except for the winter deciduous *N. gunnii*.

Tasmanian rainforests can be subdivided into four major groups, on the basis of structure and floristics (Jarman *et al.* 1984, 1987). The features of these groups are listed in Table 3.3 and are illustrated in Boyer and Hickey (1987).

The term 'callidendrous' is derived from the Greek *kalos* (beautiful) and *dendron* (tree), a reference to the well-formed trees (mostly *Nothofagus cunninghamii*) forming the canopy. The understorey is open.

'Thamnic' is derived from the Greek *thamnos* (shrub), referring to the well-developed shrub layer in thamnic rainforest. Implicate rainforest (from the Latin *implicatus*—tangled) has a dense understorey of interwoven stems which makes access extremely difficult. The canopy is normally of low stature and often uneven.

Open montane forests are, as the name suggests, restricted to high altitudes and are associated with boulder fields or rock outcrops. They are dominated by low, widely spaced *Athrotaxis*. Ferns are rare. Although open montane forest shares some species with the other rainforest groups, its open canopy structure is closer to woodland (*sensu* Specht 1981*a*) than to forest.

In addition, Jarman *et al.* (1984, 1987) recognized a gallery rainforest dominated by *Leptospermum riparium* which forms a narrow band along some of the major rivers. A number of floristically defined communities, with structural characteristics intermediate between callidendrous and thamnic rainforest, have also been documented.

Callidendrous rainforest has a sharply different distribution pattern from the other three major groups with a boundary which divides Tasmania diagonally from north-west to south-east (see Fig. 3.29). Callidendrous rainforest is found to the east of this line but also occurs in Victoria. The other three major groups (endemic rainforest in Jarman *et al.* 1987) are floristically defined by species restricted to the south-west. However, many of the species in callidendrous rainforest, including the canopy dominants, are found in rainforests throughout Tasmania, so that the absence of callidendrous rainforests as a distinctive forest type from the south-west is not due to the absence of its characteristic species.

3.8 THE RAINFORESTS OF PROVINCES C_1 AND C_2

The broad overview of rainforest vegetation in Australia has concentrated so far on types largely

Table 3.3 Characteristics of the major rainforest types in Tasmania (from Jarman *et al.* 1987).

Rainforests group	No. of communities	Characteristics	Typical species	
Callidendrous	3	Trees tall, well-formed (lowland) Understorey park-like Angiosperm diversity low Fern diversity high (lowland)	Dominants Shrubs Ferns	*Nothofagus cunninghamii, Atherosperma* *Pittosporum, Aristotelia, Coprosma quadrifida,* *Telopea truncata, Tasmannia* *Dicksonia, Polystichum*, epiphytic species
Thamnic	6	Trees of moderate height Understorey shrubby but relatively open Angiosperm diversity moderate Fern diversity moderate	Dominants Shrubs Ferns	*Nothofagus cunninghamii, Eucryphia lucida,* *Lagarostrobos, Athrotaxis selaginoides,* *Phyllocladus* *Anodopetalum, Anopterus, Cenarrhenes,* *Richea pandanifolia, Trochocarpa, Archeria* *Blechnum wattsii*, sporadic
Implicate	5	Trees low, forest canopy open Understorey tangled Angiosperm diversity high Fern diversity low	Dominants Shrubs Ferns	*Nothofagus* spp., *Lagarostrobos, Phyllocladus,* *Athrotaxis selaginoides, Leptospermum* spp., *Eucryphia* Dominants above + *Richea pandanifolia,* *Anodopetalum, Agastachys* *Blechnum wattsii* sporadic
Open montane	3	Trees low, canopy open Understorey dense—shrubs 1–8 metres Angiosperm diversity moderate Fern diversity low	Dominants Shrubs Ferns	*Athrotaxis cupressoides* *Nothofagus* spp., *Diselma, Microstrobos,* *Podocarpus, Orites* spp., *Richea* spp., *Baeckea gunniana, Leptospermum rupestre* None consistent

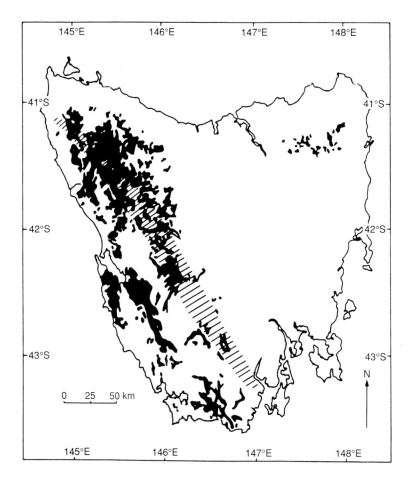

Fig. 3.29 Distribution of rainforest in Tasmania (based on a figure in Boyer and Hickey 1987). Thamnic and implicate rainforests occur on the west coast, callidendrous rainforests in the north-east and all three types occur in the cross-hatched area

(but not entirely) experiencing annual rainfalls in excess of 1200 mm (the lower cut-off suggested by Schimper 1903). It is now necessary to backtrack in order to consider those vegetation types frequently discussed by Australian ecologists under the heading 'rainforest', but which are found in drier regions.

The core area of province C_1 is the coastal lowlands and adjacent ranges of southern Queensland from just north of Brisbane to Mackay, but with extensions northwards and to the south. The major structural type is araucarian notophyll/microphyll vine forest and is found on soils of high to moderate fertility. The rainfall is between 1000 and 1500 mm a year; under higher rainfall conditions the araucarian notophyll/microphyll vine forest is replaced by forests of the A_1 province, while under lower rainfall elements of the C_2 province appear. Towards the lower limits of rainfall (c.1000 mm per annum) the structural type becomes a low microphyll vine forest with *Araucaria* emergents, which are much lower than those under more humid conditions. The most widespread emergent is *Araucaria cunninghamii*, *A. bidwillii* being more local in occurrence.

These araucarian forests were once very widespread (Watson 1989) but substantial areas have been cleared. This province is marked by the importance of a number of families (Sapindaceae, Euphorbiaceae, Celastraceae, Rutaceae, Rubiaceae, and Ebenaceae) which, although well represented in other Australian rainforests, are particularly prominent in the C_1 province. Other widespread families are poorly represented, particularly in the northern part of C_1 (Proteaceae, Meliaceae, Monimiaceae, Escalloniaceae) (Young and McDonald 1987).

In northern New South Wales, on fertile soils at sites experiencing 900–1250 mm annual rainfall, notophyll–microphyll vine forest with emergent *Araucaria cunninghamii* is widespread. The dense canopy ranges from 8 to 30 m, above which is a scattered emergent layer of *Araucaria* (up to 45 m) and a number of deciduous or semideciduous species, principally *Flindersia* spp. and *Brachychiton discolor*. (Type 21—Hoop Pine of Forestry Commission of NSW 1989.) Other locally important emergents include *Ficus* spp., *Ehretia acuminata*, and *Grevillea robusta*. In the main canopy, Sapindaceae and Euphorbiaceae are well represented. The transition between the rainforest types assigned to province A_1 and those of C_1 is a gradual one and *Araucaria* emergents are a feature of the stands at the drier end of the range of A_1.

At drier, or otherwise more unfavourable sites (for example steep, rocky slopes), and to the south of the distribution of *Araucaria* the main canopy is usually lower (10–15 m), but with a scattered deciduous overstorey in which *Brachychiton discolor* is often prominent (Yellow Tulipwood, Type 22 Forestry Commission of NSW 1989). This type contains few species of economic value and is floristically heterogeneous. A number of floristic communities within the type have been recognized (Floyd 1987, 1990).

Under still drier conditions the rainforest community is reduced to a low closed forest–vine thicket. Numerous stands of this type, with a distinctive flora (characterized by such species as *Backhousia sciadophora*, *Alectryon forsythii*, and *Notelaea microcarpa*), occur on the steep, rocky slopes of gorges on the upper reaches of rivers below the Northern Tablelands (Fig. 3.30). In some of these stands, rainfall may be augmented by fog interception. A striking feature of some

Fig. 3.30 Apsley Gorge, below the Great Escarpment. Small patches of rainforest occur on the steep, rocky slopes

stands is the abundance of epiphytic orchids, bryophytes, and lichens (Adam 1987b). Structurally similar stands are found further south. In the Shoalhaven Gorge on the South Coast are patches of semi-evergreen notophyll vine thickets with emergent deciduous *Toona australis* and *Melia azedarach* (Mills 1987). Stands of *Backhousia myrtifolia*, widespread on the south coast, may also be found in relatively dry sites (Floyd 1990).

Elements of the C_1 province extend into drier parts of northern Queensland where, as Kahn and Lawrie (1987) suggested, there is a tendency for species characteristic of several provinces to co-occur in many of the patches.

The C_2 province is characterized by thickets, although sometimes with large emergents. Microphyll vine thicket is found on calcareous sand dunes north from Bundaberg (Young and McDonald 1987), but shares many species with the drier, semi-evergreen to deciduous vine thickets of more inland localities, which normally experience annual rainfall below 800 mm. Although semi-evergreen and deciduous thickets can be distinguished structurally and also have characteristic species assemblages, they also share many species and Kahn and Lawrie (1987) regarded them simply as convenient reference points along a continuum of variation.

In central Queensland, where the Dividing Range is low, there is a merging of the C_1 and C_2 provinces, under drier conditions occurring with increasingly lower canopies and stunted *Araucaria* emergents. To the south, where the Range is higher, there is an attenuation of the C_1 province westwards, with lower, more species-poor stands being increasingly restricted to sites protected from fire and an intermingling of C_1 and C_2 species. However, the main part of the distribution of the C_2 province is distinct, being centred on the western slopes of the Dividing Range, in areas receiving between 800 and 500 mm of rain per year. In northern New South Wales and southern and central Queensland, vine thickets within the brigalow (*Acacia harpophylla*) belt (Gasteen 1987; Johnson 1984) belong to the C_2 province. These vine thickets are mostly small and scattered, their number having been considerably reduced by recent clearing (see Chapter 9). Many of them occur on more fertile soils derived from basalt, but this is not always the case, and there are examples on a range of substrates including poor sandstone country.

Descriptions of the vegetation of these vine thickets were provided by Beadle (1981), Johnson (1984), and Gillison (1985), and for the more northerly examples by Kahn and Lawrie (1987). Most stands have a dense, closed canopy about 5–10 m high with widespread species including *Geijera parviflora*, *Croton insularis*, *Canthium* spp., *Ehretia membranifolia*, *Planchonella cotinifolia*, and *Cassine australe* (Johnson 1984), with the shrub layer containing such species as *Carissa ovata*, *Acalypha eremorum*, *Citriobatus spinescens*, and *Alectryon diversifolius*. Vines are often abundant, including a number of members of the Asclepiadaceae, *Cissus opaca*, *Jasminum didymum* subsp. *racemosum*, and *Parsonsia* spp. Above the canopy is almost invariably a sparse layer of taller deciduous emergents which include *Brachychiton rupestris*, *B. australis*, *Flindersia* spp., and *Cadellia pentastylis*. The emergents may be 30 m+ tall. The *Brachychiton* spp. are bottle trees with swollen trunks and in growth form are similar to *Adansonia gregorii* in north-west Australia. In a relatively limited region of the central brigalow belt, on deeply weathered basalt, a distinctive variant of these thickets occurs, in which the canopy is dominated by *Macropteranthes leichhardtii* (Beadle 1981; Johnson 1984).

In less-fertile sites *Eucalyptus* spp. may occur as emergents. There are also stands of brigalow in which vine thicket species form an understorey and *Brachychiton* spp., *Ventilago viminalis*, and *Flindersia collina* may be components of the canopy (Johnson 1984; Gasteen 1987).

3.9 RAINFORESTS OF THE GREAT BARRIER REEF ISLANDS

Two different types of islands are associated with the Great Barrier Reef, continental islands which are detached pieces of the mainland, and coral cays. Continental islands support a range of communities, including rainforests which are similar to those on the mainland (Brennan 1988). The coral cays, however, support very distinctive communities. The cays provide an extreme environment with limited water availability, salt spray, and extremely alkaline soils. The vegetation is reviewed by Cribb and Cribb (1985). The cays on the northern part of the Reef may have stands of mixed closed forest in which *Diospyros ferrea* and *Planchonella obovata* are important elements. In the central and southern sections of the reef, many of the cays support one of the most distinctive forest types in Australia; that dominated by *Pisonia grandis* (Dale *et al*. 1984; Hulsman *et al*. 1984). *Pisonia* may grow up to 20 m tall, although on the exposed windward side of the cays it may be stunted and wind pruned. The leaves of *P. grandis* are large and cast a dense shade, although in prolonged dry periods there may be a considerable loss of foliage. The understorey under the densest canopy is lacking (Fig. 3.31) but the bare sand is frequently riddled with nest burrows of wedge-tailed shearwaters (*Puffinus pacificus*). The canopy may provide nest sites for vast colonies of black noddies (*Anous minutus*) (Fig. 3.32). The birds are responsible for substantial inputs of nutrients, particularly phosphorus, to the cays (Allaway and Ashford 1984). Phosphates in the rain of guano are dissolved in the acid humus layer but precipitate in a hardpan in the underlying alkaline sand (Cribb and Cribb 1985).

P. grandis has a very wide distribution in the Indo-Pacific, extending from the Seychelles to Pitcairn Island, mostly on small islands and associated with large seabird colonies. The fruits exude a sticky resinous compound which is responsible for the death of many birds. If a bird touches a developing cluster of fruit, the sticky exudate attaches to the plumage and the individual fruit or whole infructescence breaks away. Birds may be crippled by this sticky burden, and, unable to fly, starve to death.

Fig. 3.31 *Pisonia grandis* forest on Heron Island. Note virtual lack of ground cover (photo A. E. Ashford).

3.10 NORFOLK ISLAND

Norfolk Island is a small island of volcanic origin lying between New Zealand and New Caledonia, approximately 1400 km east of Australia (29°02′S, 167°57′E). The island covers an area of 3450 ha; and off its southern coast are two smaller islands, Nepean Island and Philip Island.

Between 1788 and 1814, and again between 1825 and 1856, the island was maintained as a convict settlement. After 1856, the descendants of the mutineers from Captain Bligh's *Bounty*, then living on Pitcairn Island, were moved to

Fig. 3.32 *Anous minutus*, Heron Island (photo A. Pitman).

settlement of the island. Common species in the tree canopy included *Baloghia lucida*, *Celtis paniculata*, *Dysoxylum patersonianum*, *Elaeodendron curtipendulum*, *Lagunaria patersonia*, *Nestigis apetala*, *Streblus pendulinus*, and *Zanthoxylum pinnatum*. In gullies, dense groves of palms (*Rhopalostylis baueri*) and tree ferns (*Cyathea* spp.) occurred (Schodde *et al.* 1983). Today, forest vegetation is reduced to small patches, most of which are infested with the exotics *Lantana camara* and *Psidium guajava*.

Takhtajan (1986) treated Norfolk Island as a province within the Neozeylandic Region of the Holoantarctic Kingdom, and thus phytogeographically distinct from the Australian Kingdom, but although prominent Australian elements, such as *Eucalyptus* and *Acacia*, are absent, only five taxa are shared with New Zealand.

3.11 LORD HOWE ISLAND

Lord Howe Island is situated some 700 km north east of Sydney (31°30′S, 159°05′E) and, administratively, is part of New South Wales. It is the eroded remnants of a shield volcano which first erupted 6.9 million years ago.

The island covers an area of about 1500 ha and is one of the smallest natural areas included upon the World Heritage List. Inclusion on the List acknowledges the spectacular scenery (culminating in the two major peaks at the south of the Island—Mt. Lidgbird: 765 m and Mt. Gower: 866 m, Fig. 3.33) and the very high endemicity in both flora and fauna.

The temperature variations on the island are small, reflecting its oceanic position. The annual rainfall exceeds 1600 mm; with a winter maximum and a late summer (February) low. Wind speeds are high throughout the year.

Accounts of the natural history of the island are provided by Hutton (1986) and Mosley (1988) and a detailed vegetation description was given by Pickard (1983).

Although first colonized in 1834, clearing has been restricted and much of the area is still

Norfolk Island. The vegetation of the island has been extensively cleared and substantially modified by European settlement.

Norfolk Island was formed about 3 million years ago, but, despite its young age, had a high degree of endemism in both flora and avifauna (of the native vascular flora of 174 species, some 52 are endemic—Takhtajan 1986).

Two hundred years ago, large parts of the islands were forested with a closed-canopy community with emergent *Araucaria heterophylla* (the endemic Norfolk Island pine). In structure these forest stands were similar to rainforest with emergent *A. cunninghamii* in eastern Australia. *A. heterophylla* was seen as a valuable source of timber for naval purposes (particularly the provision of spars), and was the main stimulus for the

Fig. 3.33 Lord Howe Island, Mt. Lidgbird on the left, Mt. Gower on the right. Norfolk Island pines (*Araucaria heterophylla*) in the foreground have been planted and are not native to the island (photo K. Gillett, courtesy of National Parks and Wildlife Service).

occupied by rainforest (Figs. 3.34, 3.35). Although the main peaks are—in absolute terms—not high, they demonstrate very well the change in forest structure with altitude, (Fig. 3.35) from canopies 20 m high in sheltered lowland sites to low gnarled thickets, covered with epiphytic bryophytes on the summits. In addition, there is a reduction in average leaf size and the summit communities meet the definition of montane forest.

The native angiosperm flora is 180 species, of which 57 are endemic, including all the palm species which are a conspicuous component of the vegetation. In the lowlands are two *Howea* spp. (*H. forsteriana* and *H. belmoreana*) in the uplands *Hedyscepe canterburyana* occurs above 350 m, while *Lepoidorrhacis mooreana* is restricted to above 700 m. Hutton (1986) illustrated the majority of the native flora.

Takhtajan (1986) included Lord Howe Island as a province within the Neozeylandic region. However, while Australian elements like *Eucalyptus*, *Acacia*, and *Casuarina*, and the Proteaceae are absent, the flora has more genera in common with eastern Australia (129) than New Zealand (75 genera). The occurrence of some species remains difficult to explain in terms of known dispersal mechanisms, one of the most enigmatic species being *Dietes robinsoniana* of the Iridaceae, as the genus is otherwise restricted to South Africa.

3.12 WHAT IS RAINFOREST?—THE PROBLEM REVISITED

Following this overview of those vegetation types which, in Australia, are regarded as rainforest,

Fig. 3.34 Rainforest vegetation on Lord Howe Island (*Ficus columnaris* and *Howea forsteriana*) (photo K. Gillett, courtesy of NPWS).

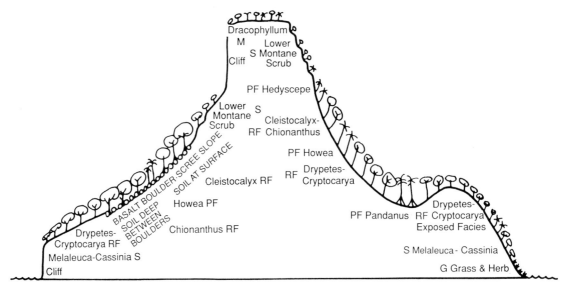

Fig. 3.35 Transect across Lord Howe Island (redrawn from Pickard 1983).

some comment is required as to whether these vegetation types can be regarded as rainforest within a wider context. Despite the very broad definition proposed by Schimper (1903), rainforest is all too often regarded as synonymous with tropical rainforest, a category which in Schimper's treatment is only one subset of the total range of rainforest types.

Relationships between rainforest types recognized in Australia and those elsewhere in the world are indicated in Table 3.4. Very little vegetation in Australia meets the strict definition of tropical ombrophilous forest (tropical rainforest). Even in north-east Queensland, the average temperatures and seasonality in rainfall make categorization of the region as strictly tropical marginal. Nevertheless, there is vegetation in Webb et al.'s B_2 province which can be classified as lowland, sub-montane, and (on some peaks) montane tropical ombrophilous forest. Lowlands and upper lowlands in the B_2 province also support tropical evergreen seasonal forest. All of these forest types are tropical rainforest as defined by Schimper (1903) (see also table 12.1 in Whitmore 1984).

The communities of Webb et al.'s ecofloristic region A are subtropical and temperate in their distribution, but most of them experience more than 1200 mm annual rainfall, and although rainfall may be seasonal, significant falls (>60 mm) occur in every month of the year in most years. These communities therefore fall within the broader definition of rainforest suggested by Schimper (1903). Some stands are found under lower rainfall conditions, but their similarities, both structurally and floristically, preclude exclusion from the rainforest category. In the case of subtropical rainforests (provinces A_1 and A_2) there are strong similarities with the rainforests for province B_2 in terms of patterns of floristic and structural diversity. Richards (1952) recognized that the similarities between tropical and subtropical rainforests were far greater than the differences, and that no clear boundaries can be drawn between the two. The majority of forests included within province A_3 (the temperate rainforests) meet Schimper's definition of rainforest under both climatic and structural criteria, although the inclusion of some open montane communities within the definition of Tasmanian rainforest extends the concept of rainforest to its limits.

The wettest stands included in the C_1 province are rainforest in the strict sense, being a subtropical equivalent of the tropical evergreen seasonal forest (Table 3.4) but the drier parts of C_1 and the whole of C_2 fall outside Schimper's definition. The tall semi-deciduous mesophyll vine forests of province B_1 and the deciduous vine thickets of province B_3 are also outside the conventional definition of rainforest. The forests of the B_1 province are representative of Schimper's monsoon forest, and the drier parts of C_1 could be regarded as a subtropical equivalent of this type. The drier forests also have many obvious links to vegetation discussed by Schimper (1903) under the broad title of periodically dry forest and woodland.

What justification is there for including these drier forests within the heading 'rainforest' in Australia? The major reason is the high degree of similarity, both structural and floristic, between these communities and 'true rainforest', and the very sharp differences between them and eucalypt communities. Despite their fragmented geographical distribution, there is a continuum of variation between all the so-called rainforest communities and a very sharp break between the continuum and other vegetation types. Webb et al. (1984) demonstrated that there is a substantial pool of shared genera between ecofloristic regions A, B, and C. In view of the degree of similarity at the present day and the evidence for the differentiation of rainforest over geological time (Chapter 6) it is reasonable to treat all the communities under the one heading of rainforest.

3.12.1 What is warm temperate rainforest?

Authors from Schimper onwards have distinguished between tropical, subtropical, and temperate

Table 3.4 Relationships between Australian and overseas rainforests (Derived from Kershaw and Whiffen 1989).

UNESCO (1973) formation groups	Webb (1978a) structural types	Webb et al. (1984) provinces	Forest formation (Whitmore 1984)
1.A. Mainly evergreen forest			
1.A.1. tropical ombrophilous forest			
1.A.1.a. lowland forest	Complex mesophyll vine forest—CMVF	B_2	1. Tropical lowland evergreen rain forest
	Mesophyll vine forest—MVF	B_2	
*upper lowland or upland	Complex notophyll vine forest—CNVF	B_2	2. Tropical lower montane rain forest
	Simple notophyll evergreen vine forest—SNEVF	B_2	
1.A.1.b. submontane forest	Simple notophyll evergreen vine forest—SNEVF	B_2	
	Microphyll vine/fern forest—MV/FF	B_2	
1.A.1.c. montane forest	Microphyll vine/fern thicket—MV/FT	B_2	3. Tropical upper montane rain forest
1.A.1.e. cloud forest			
1.A.1.f. alluvial forest	Complex mesophyll vine forest—CMVF	B_2	1. Tropical lowland evergreen rain forest
1.A.1.g. swamp forest	Mesophyll feather-palm vine forest—MFPVF	B_2	12. Swamp forest
	Mesophyll fan-palm vine forest—MFAPVF	B_2	
1.A.2. tropical evergreen seasonal forest			
1.A.2.a. lowland forest	Complex mesophyll vine forest—CMVF	B_2	13. Tropical semi-evergreen rain forest
	Mesophyll vine forest—MVF	B_2	
*upper lowland or upland	Complex notophyll vine forest—CNVF	C_{1+}	
	Notophyll vine forest—NVF	C_{1+}	
	Araucarian notophyll vine forest—ANVF	C_{1+}	
1.A.2.b. submontane forest			
1.A.2.c. montane forest			
1.A.3. tropical semi-deciduous forest			
1.A.3.a. lowland forest	Semi-deciduous mesophyll vine forest—SDMVF	A_1	14. Tropical moist deciduous forest

	Semi-deciduous notophyll vine forest—SDNVF	A_1
	Semi-evergreen vine thicket—SEVT+	C_{2+}
1.A.3.b. montane or cloud forest		15. Monsoon forest
1.A.4. subtropical ombrophilous forest		
1.A.4.a. (warm) subtropical lowland forest *cool subtropical lowland and upland forest	Complex notophyll vine forest—CNVF Complex notophyll vine forest—CNVF	A_1 A_1
	Araucarian notophyll vine forest—ANVF	A_1
1.A.4.b. lower montane (submontane) subtropical forest	Simple notophyll vine forest—SNVF	A_2
	Microphyll fern forest—MFF	A_2
1.A.6. temperate evergreen ombrophilous forest		
1.A.6.a. (1) warm temperate without conifers	Simple notophyll vine forest—SNVF	A_2
1.A.6.a. (2) Warm and cool temperate lowland and upland with evergreen conifers in places	Microphyll fern forest—MFF	A_3
	Nanophyll fern forest—NFF	A_3
	Nanophyll mossy forest—NMF	A_3
1.A.7. temperate evergreen seasonal broad-leaved forest	Complex/simple notophyll/microphyll vine forest—C/SN/M VF	A_2
	Araucarian notophyll/microphyll vine forest—AN/M VF	A_2
1.A.9. tropical evergreen needle-leaved forest		
1.a.9.a. lowland and submontane forest		
1.a.9.b. montane and subalpine forest		
1.B. Mainly deciduous forest		
1.B.1. tropical drought deciduous forest		
1.B.1.a. lowland and submontane forest	Deciduous vine thicket—DVT	B_3
1.B.1.b. montane and cloud forest		15. Monsoon forests

UNESCO formation groups taken from Mueller-Dombois and Ellenberg (1974), those marked * are additional terms introduced by Webb et al. (1984). The UNESCO classification has not been widely used, but provides a scheme which encompasses temperate and subtropical as well as tropical rainforests. For the Webb (1978) structural types see the key on p. 26. The ecofloristic provinces (Webb et al. 1984) listed are those in which the relevant group or structural type has its major representation (+ indicates that the province listed is the 'core' area, but that the vegetation type occurs more widely). For tropical vegetation types, synonymy with the formations from table 12.1 in Whitmore (1984) is given.

rainforest. In Australia there is also a tradition of recognizing warm temperate rainforest.

The term 'warm temperate rainforest' is widely used, particularly in New South Wales where it is recognized as one of the major subforms of rainforest (Floyd 1987, 1990; Forestry Commission of NSW 1989; Fig. 2.1 this volume). The vegetation included within the grouping is characterized structurally by the absence of buttressing and the uniform pole-like appearance of the trunks, and the relatively low species diversity of the canopy (in many cases the dominants belong to the Cunoniaceae and Monimiaceae). Most of the vegetation assigned to the warm temperate grouping falls into Webb *et al.*'s (1984) A_1 province, although some would be included in A_2. Webb *et al.* (1984) argued that the name 'warm temperate' is inappropriate, as in northern New South Wales the subtropical rainforest (complex notophyll vine forest) and warm temperate rainforest (simple notophyll (or notophyll/microphyll) vine forest) may be found in close proximity, but occur on different soil types, with the subtropical rainforest being restricted to more-fertile soils.

The distinction between subtropical rainforest on basalt and 'warm temperate' rainforest on acid igneous rocks is often striking. Acid soils are generally relatively rich in aluminium and a number of rainforest species are notable for their accumulation of soluble aluminium in tissues (Webb 1954). Interestingly this trait was found by Chenery and Sporne (1976) to be correlated with a number of 'primitive' morphological characteristics. Aluminium accumulation is a marked feature of a number of species characteristic of New South Wales 'warm temperate rainforest'—*Ceratopetalum apetalum*, *Schizomeria ovata*, and *Orites excelsa* (Webb 1954; Turner *et al.* 1981). The litter fall from such species increases the aluminium content in the upper soil profile, to the extent that it may inhibit the growth of aluminium-sensitive species (Turner *et al.* 1981). *Ceratopetalum apetalum*-dominated, and *C. apetalum–Schizomeria* stands may therefore be regarded as an edaphic climax on infertile acid soils. However, characteristic 'warm temperate' species are not restricted to nutrient-poor soils, but are also found as components of subtropical rainforest on more nutrient-rich sites, and although the relative abundance of species may be related to a fertility axis, it is also influenced by elevation (and thus, by inference, climate) (see Fig. 3.16). At higher latitudes 'warm temperate rainforest' is found at lower altitudes. *Doryphora sassafras*, which also dominates stands of 'warm temperate' structure, but is not an aluminium accumulator (Webb 1954), occurs on infertile soils but is also found on basalt. It is interesting that in the northern Blue Mountains, west of Sydney, rainforests on basalt caps are dominated by *D. sassafras*, while *Ceratopetalum* occurs in deep sandstone gorges. However, interpretation of this distribution, in terms of edaphic differentiation, is complicated by the different susceptibility of the species to fire, the thin-barked *C. apetalum* being much more sensitive than the thicker-barked *D. sassafras*.

In northern Queensland, simple notophyll vine forests (Type 8 in Tracey 1982) are the major wet forest of uplands on acid soils. These Queensland forests are similarly composed of canopy species with relatively slender pole-like trunks, although canopy species richness is considerably greater than in New South Wales. There are floristic similarities at the generic level with canopy species in Queensland, including *Ceratopetalum succirubrum*, *Schizomeria whitei*, and *Doryphora aromatica*.

North of the Hunter Valley, the 'warm temperate' rainforest types, which were the principal source of veneer logs in post-war years, are best regarded as an edaphically determined derivative of the subtropical complex notophyll vine forest, with distinctive structural and floristic composition. South of the Hunter River, complex notophyll vine forest is increasingly restricted to more coastal sites. Simple notophyll vine forest, although similar in floristic composition to that found on acid, infertile soils further north, is found on fertile upland soils and can be regarded as warm temperate rainforest in a climatic sense.

What can also be considered as climatic warm temperate rainforests are the simple notophyll–microphyll vine forest stands, frequently dominated by *Acmena smithii*, such as occur on the highest parts of the Liverpool Range (Fisher 1985), the south coast of New South Wales, and in Victoria (Patton 1930).

3.13 TRENDS OF VARIATION IN AUSTRALIAN RAINFOREST

Australian rainforest occurs as numerous discrete patches, each, in various ways, different from any other. It is possible to recognize various broad groupings of these patches, as in Webb *et al.*'s (1984) analysis. Nevertheless, there are identifiable trends in variation which transgress the boundaries between groups. Webb (1968) showed that at the continental scale there were trends in structural/physiognomic features which correlate with environmental gradients. At this scale, three major environmental factors assume importance: temperature, rainfall (both annual total and seasonal distribution), and soil nutrient status (unfortunately, there are still relatively few reported soil analyses for Australian rainforests and soil nutrient status is inferred from the geology) (see Figs. 3.36, 3.37, 3.38). The climatic and topographic factors tend to override any effect of soil nutrient availability at the climatic extremes of rainforest distribution (tropical monsoonal and cool temperate), but in the wet tropics and subtropics, soil fertility is an important correlate of physiognomic forest type (Webb 1968, 1978*a*). Pictograms representing some of the major trends of variation are presented in Fig. 3.37.

Stocker and Unwin (1989) implied that Webb (1968, 1969) may have overly stressed the importance of soil chemistry in differentiating between rainforest types. While there is acceptance that, in general, sclerophyll forest can be differentiated from rainforest on the basis of soil chemistry (although, as Webb (1968), emphasized, the distinction is sharpened by the effects of fire, rainforest being more resilient on more nutrient-rich sites), discussion of variation within rainforest is hampered by the paucity of measured, as distinct from inferred, data. However,

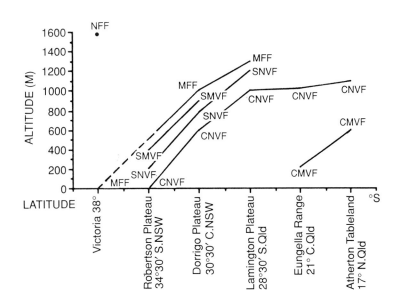

Fig. 3.36 Altitudinal zonation of structural types of rainforest under conditions of high rainfall and soil fertility (from Webb and Tracey 1981*a*).

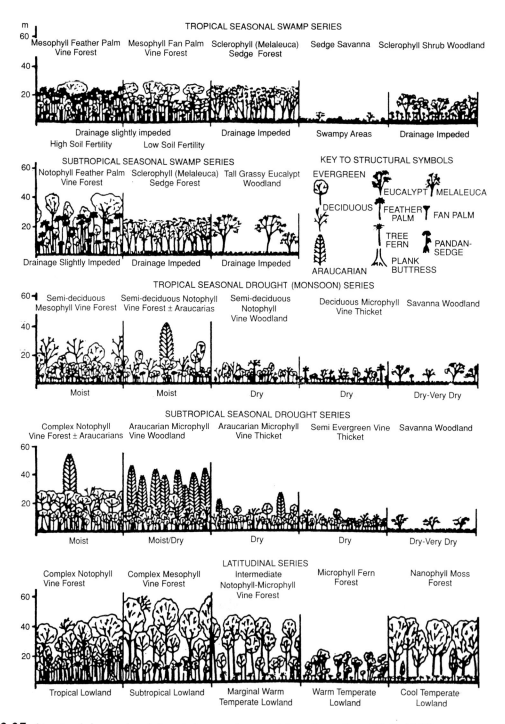

Fig. 3.37 Structural changes in rainforest along environmental gradients (from Webb 1968).

AN OVERVIEW OF AUSTRALIAN RAINFORESTS

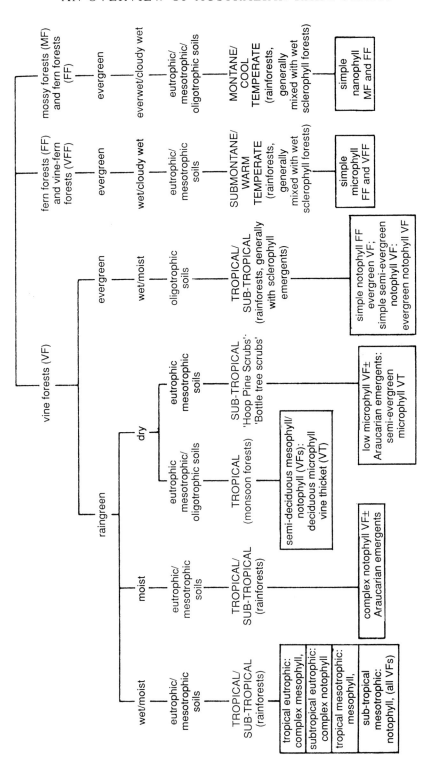

Fig. 3.38 Inferred soil nutrient status of major structural rainforest types (from Webb 1978a after Webb 1968).

the chemical distinctiveness of the soils in simple notophyll vine forest dominated by *Ceratopetalum apetalum* in northern New South Wales, is well established (see above); in addition to higher aluminium, the soils in these forests have lower nitrogen, phosphorus, and calcium levels than complex notophyll vine forest (Turner and Kelly 1981), and it is plausible that the relationships suggested by Webb (1968—see Fig. 3.38) will be confirmed by soil analysis. The role of soil physical factors is less clear (Tracey 1969), but the distribution of some distinctive rainforest types is determined by physical factors (for example palm domination is usually associated with poorly drained sites—Turner and Kelly 1981; Tracey 1982).

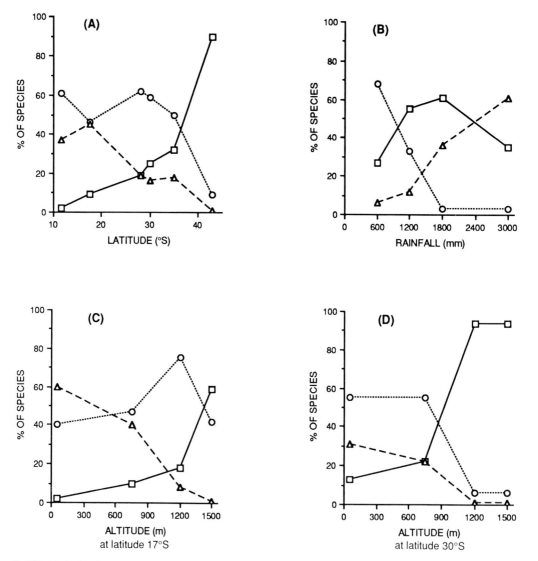

Fig. 3.39 Variation in representation of different leaf-size classes along different environmental gradients ○ = notophyll, □ = microphyll, △ = mesophyll (redrawn from Webb 1968).

Although Webb (1968) showed that at the climatic extremes for rainforest distribution there was little effect of soil type on occurrence of rainforest, this is not to argue that the distribution of individual species is not strongly influenced by soil factors. Jackson (1968, 1983) has pointed out that the floristic composition of rainforest stands in Tasmania is related to soil fertility, with *Nothofagus cunninghamii* dominant on richer soils and *Phyllocladus aspleniifolius* dominant on the poorest sites.

As well as influencing the structural and floristic composition of rainforest, soil fertility is a major determinant of productivity (see Vanclay 1989).

The trends in the distribution of leaf size (Fig. 3.39) are well marked and are strongly related to climate. The trends in Australia are similar to those reported from the northern hemisphere by Wolfe (1979) and encourage the belief that leaf-size spectra in fossil deposits may provide an accurate indication of palaeoclimates. As well as variations in leaf size, there are also parallel trends in other leaf features. At low latitudes, leaf margins tend to be smooth and entire; lobing and various forms of serration increase with latitude. The incidence of compound leaves declines with increasing latitude, but, even in Tasmania, there is a small number of species with compound leaves. Wolfe (1979) illustrated the very high correlation between the proportion of entire-margin leaved species in forest floras and mean annual temperature, but suggested that in the southern hemisphere there is, in general, a higher proportion of entire-margined species. Bailey and Sinnott (1916) attributed the difference to the low diversity of broad-leaved deciduous species in the southern hemisphere. Nevertheless, among evergreen species in Australia, there is a considerable increase in non-entire-margin leaved species with increasing latitude and/or altitude.

One interesting structural trend which cannot, as yet, be explained, is an increase in stand basal areas at higher altitudes. This is seen in north-east Queensland (Stocker and Unwin 1989), where, in general, stand basal areas are high compared with those in other tropical rainforests and in New South Wales, where the highest basal areas are in high-altitude cool temperate *Nothofagus moorei* forest (Baur 1962b). Stocker and Unwin (1989) suggested that the trend reflected the greater impacts of cyclones on lowland rather than upland forests, but while this may be plausible in northern Queensland, it may be less so in northern New South Wales, which is south of the cyclone belt, and storm damage, although locally serious, is more restricted both spatially and temporally.

3.14 HOW MUCH RAINFOREST?

Australian rainforest stands are part of a great crescentic archipelago of habitat islands (endpapers), but the total area is small (Fig. 3.40, Table 3.5), approximately 2 million hectares (Webb and Tracey 1981a,b, Winter et al. 1987b). Even before European settlement the area was,

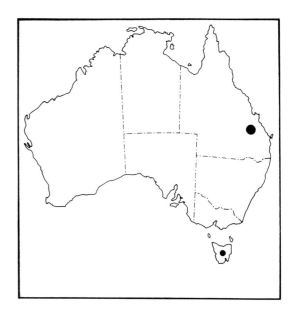

Fig. 3.40 The total area of rainforest remaining in Australia, represented by two circles, one 70.6 km in diameter and the other 38.1 km in diameter (from Winter et al. 1987b).

Table 3.5 Occurrence of rainforest in different Australian states (from Baur 1989).

	Area of rainforest (ha)	% of State under rainforest	% of Australian rainforest
Queensland	1 237 200	0.72	54.6
New South Wales	226 100	0.33	11.7
Victoria	13 300	0.06	0.6
Tasmania	711 500	10.50	31.4
Northern Territory	38 000	0.03	1.7
Western Australia	2000	<0.01	<0.1
Australia	2 266 000	0.3	100.0

compared with the size of the continent, not large—perhaps 8 million hectares. The biotic diversity of this rainforest is vastly in excess of what might be expected for its limited area. Estimates that half the total terrestrial biota of Australia occurs in rainforest (Bell 1981) are not unreasonable.

3.15 AUSTRALIAN RAINFOREST IN WORLD CONTEXT

As is shown in Table 3.4, Australian rainforest vegetation types can be related to categories recognized elsewhere. However, where are the analogues of Australian communities to be found?

The range of communities seen in ecofloristic province B_2 is very similar to that in South-east Asia, with the patterns of variation being related to similar environmental gradients (Whitmore 1984).

The more strongly seasonal communities of provinces B_1 and B_3 have very clear links, both structural and floristic, with monsoonal forests in Asia (Webb *et al.* 1985; Whitmore 1984).

The temperate rainforests of South-east Australia have similarities in floristics and structure with New Zealand and the southern Andes, and also with montane communities in New Guinea. Structurally, there are also similarities to montane rainforests in the tropics, and in the northern hemisphere, there are similarities to communities in southern Japan.

The subtropical drier communities (ecofloristic region C) have general similarities with communities in equivalent climatic regions in other continents. However, the sharp floristic distinction between open sclerophyll (*Eucalyptus* and/ or *Acacia*) communities and the closed rainforest communities is not paralleled elsewhere.

Subtropical rainforest (provinces A_1 and A_2) has not been well documented outside Australia, and the broad reviews in Schimper (1903) and Richards (1952) still provide the best overview. Subtropical rainforest, similar in structure and complexity to that in Australia, was described by Schimper (1903) from Mexico, south-east Brazil, and northern Argentina. Webb (1966) compared Australian forest communities with those in south-east Brazil and found that equivalent communities occurred in both continents, although the edaphic differentiation of communities in Australia was less obvious in Brazil. Webb (1978*b*) also showed that rainforest in the North Island of New Zealand could be regarded as belonging to the subtropical rainforest formation similar to that in New South Wales. In the northern hemisphere there is subtropical rainforest in northern Laos, Vietnam, and southern China, but detailed ecological studies of these areas have yet to be published.

Littoral rainforest, which occurs on sand dunes and headlands, is found throughout the Indo-Pacific region, where it has a similar structure and widespread floristic elements. Forest that is structurally and physiognomically similar to Malesian heath forest, is found in Australia in both ecofloristic regions A and B. It is interesting that the Australian floristic elements are strongly represented in Malesian heath forest (Richards 1943; Specht and Womersley 1979; Whitmore 1984).

4 RAINFOREST BOUNDARIES AND THE PROBLEM OF MIXED FORESTS

The total area of rainforest in Australia is small, and is composed of numerous discrete stands. Within the broad geographical regions which contain rainforest there appear to be many sites with climatic and soil conditions appropriate for rainforest growth but which support other natural communities.

The apparent failure of rainforest to achieve its potential may reflect slow migration and recovery from some earlier period of adverse climatic conditions. While a lag between a change in climate and the response of trees would be expected, this explanation would not hold for local absences within regions of extensive rainforest. For a more general explanation, we must seek a currently active ecological factor which could control the position of rainforest boundaries.

Rainforest boundaries may be external, where rainforest stands are surrounded by non-rainforest communities, or internal, where pockets of non-rainforest vegetation are surrounded by rainforest. The non-rainforest community may be sclerophyll forest or woodland, or in some cases grassland. The boundary may be extremely sharp (Figs. 4.1, 4.2) or there may be a very broad transition zone).

In general, boundaries in northern Australia are sharp, while further south there are extensive transitions with closed rainforest canopies below tall eucalypts (Ash 1988). These transitions have been called mixed forest, particularly in Victoria and Tasmania.

There is no single factor which explains the position of all rainforest boundaries, but a number of major factors are clearly involved—rainfall, substrate, topography, and fire (Ash 1988). The interactions between these factors are likely to be complex and site-specific in their outcomes, leading to the diversity of boundary types.

4.1 NORTH-EAST QUEENSLAND

Ash (1988) has provided a detailed analysis of rainforest boundaries in two areas in north Queensland, the Atherton Tableland, and the McBride region which includes the vine thickets on the Kinrara Basalt Flow (Kahn and Lawrie 1987, see page 70).

In the Atherton Tableland area 38 per cent of the boundaries coincided with a geological boundary. Rainforest stands centred on basalt rarely extend on to other substrates, but stands on granite and metamorphic rocks frequently cross geological boundaries. The soils on different substrates vary in their physical and chemical properties, but there is also an interaction with topography, the granites and metamorphic rocks being associated with more rugged topography than basalts. In the McBride area the vine thickets are associated with basalt, but not all basalt outcrops support rainforest.

There is a clear interaction between rainforest distribution, soil type, and rainfall (Fig. 4.3); on more fertile soils rainforest can extend into drier regions.

Topography also influences rainforest distribution. Where both rainforest and other kinds of vegetation are found on the same hillslope rainforest is generally found downslope. This may reflect soil differences, soils on the lower slopes generally being deeper, more nutrient-rich, and frequently with better water supply. However, slope also has considerable influence on fire behaviour; fires tend to burn more rapidly and with higher intensity upslope than downslope, although the behaviour of any particular fire will be influenced by wind speed and direction.

There are exceptions to these generalizations, where rainforest occurs on summits above slopes of non-rainforest vegetation. In these circumstan-

Fig. 4.1 A sharp boundary between rainforest (complex notophyll vine forest) and eucalypt forest—Cambridge Plateau, northern New South Wales

ces, environmental conditions on the summits are especially favourable to rainforest growth—higher rainfall, greater cloudiness, or improved soil fertility.

Towards the drier climatic limits of rainforest, there is increasing restriction to more fertile soils, gallery stands along rivers, and sites topographically sheltered from fire. The driest rainforests are often found on broken rocky slopes, where the patchy cover limits the spread of fire.

The absence of rainforest from fertile sites experiencing high rainfall can be related to topography, the slope and aspect of these sites being such that they would be exposed to frequent high-intensity fires.

Although fire may maintain pockets of vegetation surrounded by rainforest, the origin of such patches is still uncertain. Ash (1988) suggested that they are either remnants of large areas engulfed by expanding rainforest or alternatively that they result from local fires within rainforest.

4.1.1 Stability of boundaries

The analysis by Ash (1988) indicated that fire, acting alone or interacting with other factors exerts considerable control on rainforest boundaries. Variations in fire regimes would be expected to have considerable impact on the stability of boundaries. Unwin et al. (1985) suggested that, in the short term, the sharp rainforest boundaries in north Queensland are stable,

88 RAINFOREST BOUNDARIES AND MIXED FORESTS

Fig. 4.2 Sharp edge of rainforest, Cedar Brush Nature Reserve, Liverpool Ranges, New South Wales

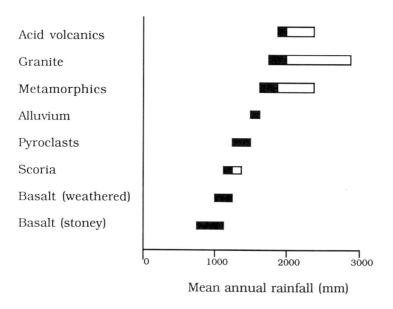

Fig. 4.3 The location of rainforest–pyrophytic vegetation boundaries in relation to mean annual rainfall and substrate in north-eastern Queensland. Boundaries on rugged topography are indicated in white (from Ash 1988).

but that readjustment may have taken place over longer periods. Low to moderate intensity fires burning into rainforest are rapidly extinguished. The shrubs and small trees characteristic of the boundary recover well after fire, principally by coppicing, so that the boundary position is maintained. Higher-intensity fires are likely to advance further into rainforest. In much of north Queensland, fires are sufficiently frequent to limit fuel accumulation, thus restricting possible intensities. Unwin et al. (1985) noted that the alien shrub *Lantana camara*, which often invades rainforest margins, burns more fiercely under hot, dry conditions than the native species of the habitat, so that greater damage to the rainforest may result. On more fertile soils the re-establishment after fire of rainforest species may be favoured over non-rainforest species (Webb 1968; Unwin et al. 1985).

Monitoring of boundaries on the Herberton Highland (on the western edge of the Atherton Tableland, see Fig. 4.4) by Unwin (1989) has demonstrated that even very abrupt transitions between rainforest and sclerophyll communities may be highly dynamic. Transects (Fig. 4.5) showed that rainforest canopy height decreased towards the boundary (although species diversity was maintained). At the boundary, there was commonly a narrow belt of 'transition forest' with a closed canopy of rainforest species below tall *Eucalyptus grandis* (flooded gum). This was fringed by tall, open forest, in a narrow band 10–50 m wide, dominated by *E. grandis* up to 52 m in height, over a grassy understorey with scattered shrubs and vines. The open forest beyond was lower (10–25 m), with *E. intermedia* as the main canopy species. Observations over twelve years showed that the rainforest species were growing vigorously in the 'transition forest' and expanding their distribution in places. Seedling regeneration of *E. grandis* advanced ahead of the rainforest margin.

The average rate of expansion of the rainforest was 1.2 m per year. Unwin (1989) indicated that rainforest expansion was currently widespread in north Queensland (although near settlements and

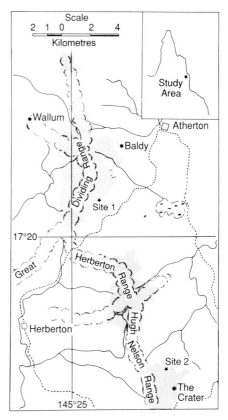

Fig. 4.4 Rainforest patches (shaded) on the Herberton Highland, northern Queensland. The locations of the transects in Fig. 4.5 are indicated as Sites 1 and 2 (from Unwin 1989).

adjacent to sugar farms on the coastal plain attrition from fire was frequent). Whether the expansion of rainforest is a continuing process, or part of a long-term cycle of expansion and contraction, remains to be determined (Unwin 1989). Various perturbations, such as fire, drought, or cyclone damage, could slow or reverse the current expansion.

Narrow bands of tall open forest, similar to those described by Unwin (1989) are of considerable extent in north Queensland; for example, along the western boundary of the main rainforest massif, there is a narrow band of tall *Eucalyptus*

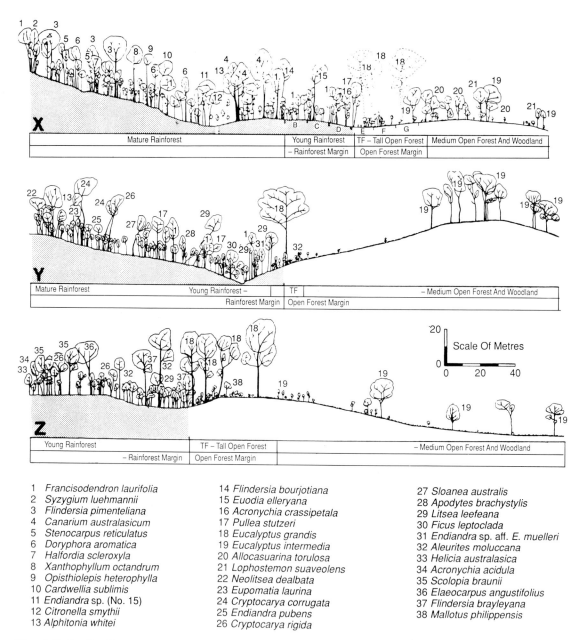

Fig. 4.5 Profile diagrams across rainforest boundaries on the Herberton Tableland. Trees less than 6 m tall omitted. Profile X, site 1 and Profiles Y and Z, site 2 (from Unwin 1989).

grandis and *E. resinifera* extending 360 km from Mount Halifax north to the Mount Windsor Tableland (Bell *et al.* 1987). The ground cover is predominantly blady grass, *Imperata cylindrica*.

4.2 SOUTH-EAST AUSTRALIA

Further south, there are many examples of very sharply bounded rainforest stands, and the boundary positions demonstrate relationships to ecological factors like those which occur in the north. However, there are also extensive areas where there is a broad transition between rainforest and non-rainforest vegetation.

This transition is characterized by tall eucalypts above a continuous rainforest canopy (Fig. 4.6). The number of eucalypt species is relatively small, perhaps forty in total (Ashton 1981*a*), and these species in general have extensive geographical distributions. This group of species dominates the vegetation type defined by the Specht (1981*a*) scheme (Table 2.1) as tall open forest, which is found in a discontinuous area on the east coast from Tasmania to southern Queensland, and also in the south-west of Western Australia (Fig. 4.7). These forests are visually striking, containing the tallest flowering plants in the world, and also yield some of the commercially most valuable hardwoods in Australia.

Tall open forest is also widely known as wet sclerophyll forest. Although found under high rainfall conditions 'wet' in this case refers to the mesic nature of the understorey (often including abundant ferns and tree ferns) in contrast to the sclerophyllous understorey of dry sclerophyll

Fig. 4.6 Tall *Eucalyptus pilularis* (blackbutt) over complex notophyll vine forest. Black Scrub region of the New England National Park

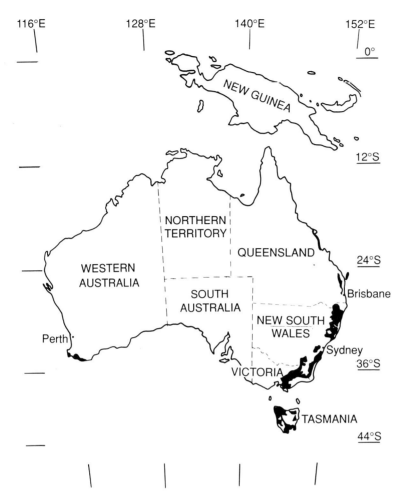

Fig. 4.7 Distribution of tall open forests in Australia (from Specht 1981b).

forest. Wet and dry sclerophyll forest can be found in close proximity under the same regional climate, with segregation on soil fertility or topographic shelter (dry sclerophyll on ridges and wet sclerophyll in gullies).

The ecology of tall open forests was reviewed by Ashton (1981a,b). The optimum climatic conditions are annual rainfall of 1500–2000 mm or more, with more than 50 mm rainfall in the driest month, although on fertile soils, tall open forests may be found under rainfalls as low as 1000 mm. With frequent low-intensity fires on fertile soils, the understorey may be open and grassy, but more frequently there is an understorey of tall shrubs (Fig. 4.8), while adjacent to rainforest there is a forest within a forest with eucalypts above a closed rainforest which may be up to 40 m tall.

These transitions are frequently interpreted as rainforest advancing at the expense of eucalypts, as seedlings of *Eucalyptus* are light-demanding and cannot establish under the dense shade of the rainforest. (Several of the eucalypts only regenerate from seed although others, may, at least in some populations, vegetatively regenerate from lignotubers.)

Fig. 4.8 *Eucalyptus deanei* over a tall, shrubby understorey. Wheeny Creek, northern Blue Mountains, New South Wales.

Studies on the regeneration of tall open forest (Ashton 1981 *a,b*) have permitted a better understanding of the interrelationships of eucalypts and rainforest (see also Gilbert 1959; Cremer 1960). This forest presents the paradox of a vegetation type which produces very high fuel loads and so generates extremely high-intensity fires, but which is dominated by fire-sensitive species that, nevertheless, require fire during their seed-bearing lives for regeneration. The tall open forest eucalypts are unusual in that very little viable seed is stored in the soil (Ashton 1981*b*) (although many understorey species are represented there). The major seed store is in capsules in the canopy. Although these capsules are small they confer sufficient protection to permit seed survival during fierce crown fires. After a fire, seeds are shed and seedling establishment is enhanced by an 'ash bed' effect variously attributed to enhanced nitrogen or phosphorus availability, destruction of soil micro-organisms or destruction of various inhibitory compounds in the soil. The outstanding feature of tall open forest eucalypts is the very rapid early height growth, up to 2 m a year during the first decade (Ashton 1981*a*). In consequence of this reproductive behaviour, stands are even-aged.

If the understorey is invaded by rainforest then, unless fire occurs, the eucalypts will senesce (after perhaps 400 years) without regeneration, and the stand will become a pure rainforest. Under most circumstances, rainforest does not carry fire, so that subsequent opportunities for the re-establishment of eucalypts will be very limited. The occurrence in some rainforest stands of scattered emergent-standing dead eucalypts indicates the expansion of rainforest into tall open forest. While eucalypts are still alive above the rainforest, the fire risk is greater, because of the higher inflammability of the eucalypt foliage, the amount of flammable litter (eucalypt litter decays more slowly than that of rainforest species), and the propensity of eucalypts to promote spot fires due to the presence of loose strips of bark in the crown (Jackson 1983). In addition, Jackson (1983) argued that eucalypt root systems are more effective in water uptake than those of rainforest, so that in dry periods the eucalypt overstorey may place the rainforest under water stress and this limits rainforest distribution towards the dry edge of its range.

If fire burns tall open forest with a closed rainforest understorey, there will be rapid regeneration of the eucalypt layer but it may take a century for the rainforest layer to become continuous again. Repeated fires at intervals of less than 100 years would lead to a loss of the rainforest element. Very frequent fires would prevent the eucalypts reaching reproductive maturity and promote replacement by bracken (Ashton 1981b).

The re-establishment of rainforest below eucalypts is demonstrated in Table 4.1, albeit in somewhat unusual circumstances. The site concerned, near Coffs Harbour, was formerly complex notophyll vine forest which was destroyed by fire in 1951. After the fire the site was planted with *Eucalyptus grandis* (flooded gum) tube stock but with no other site preparation, fertilizing, or weed control. After 27 years the stand

Table 4.1 Composition and average number of stems per hectare in three height classes in the understorey of a *Eucalyptus grandis* stand near Coffs Harbour (from Turner and Lambert 1983).

Species	Size class (height)		
	<1.3 m	1.3–10 m	10–20 m
Rainforest			
Daphnandra micrantha	2118	1354	—
Rhodamnia trinervia	1848	812	9
Cryptocarya obovata	1098	783	5
Elaeocarpus obovatus	352	478	6
Ficus stenocarpa	22	75	—
Rhodamnia argentea	166	94	—
Quintinia sieberi	160	7	4
Synoum glandulosum	129	48	—
Ceratopetalum apetalum	120	145	—
Diploglottis australis	13	16	—
Schizomeria ovata	—	434	—
Toona australis	—	—	10
Tristaniopsis laurina	—	23	—
Total rainforest	6233	4269	34
Sclerophyll			
Eucalyptus gummifera	388	150	26
E. propinqua	—	23	12
E. microcorys	—	80	4
E. pilularis	—	15	118
Lophostemon confertus	—	54	—
Casuarina spp.	244	—	—
Acacia spp.	—	48	13
Total sclerophyll species	632	370	173
Ferns and palms			
Tree fern	—	48	—
Archontophoenix cunninghamiana	189	18	—
Total ferns and palms	189	66	—

averaged 790 stems of *E. grandis* per hectare with a basal area of 30.4 m^2 ha^{-1} (Turner and Lambert 1983). Other species regenerated without planting, and the composition of the understorey recorded in Table 4.1 indicates that eventual reversion to rainforest is likely.

In many circumstances, therefore, tall open forest can be viewed as a successional stage leading to rainforest. Whether or not the succession proceeds to rainforest will be governed by fire frequency which, in turn, will be governed by climate, topography, and ignition sources which for at least 40 000 years on the mainland have included humans. It is an unusual forest succession, as the rainforest climax is much shorter in stature than the seral eucalypt stage.

In southern Australia under present climatic regimes, conditions conducive to fire in tall open forest, are relatively infrequent. In the interval between fires, the fuel build-up is considerable, so that when fire does occur, it is fierce; major fires in tall open forest in Victoria and Tasmania are probably the most intense in the world. Widespread fierce fires, such as those in Victoria on 13 January 1939, covering 20 300 km^2 (W. S. Noble 1977) and Ash Wednesday 1983, have been responsible for major property damage and loss of life but in terms of the open forest ecosystem should not be regarded as exceptional.

The broad transition zones between rainforest and sclerophyll forest in southern Australia are an indication of low fire frequencies. However, when fire does occur, the high flammable loads close to rainforest increase the risk that the core rainforest itself will be damaged so that, over an extended time period and large areas, the distribution of rainforest may exhibit instability. By contrast, the high frequency but low intensity of fire in north Queensland may result in rainforest boundaries which are relatively stable. If changes to fire regimes in north Queensland have indeed resulted in widespread expansion of rainforest, as indicated by Unwin (1989), it will be interesting to see whether, in time, much broader transitional forests develop.

4.3 CAN MIXED FOREST BE STABLE?

The common view that it is fire alone which prevents the spread of rainforest into tall open forest is not unchallenged. Florence (1963, 1964) has argued that the distribution of tall open forest and rainforest is determined largely by gradients in soil physical and chemical properties (Fig. 4.9), although recognizing (Florence 1964) that fire may introduce sharp discontinuities in vegetation gradients. The sclerophyll canopy composition varies with environmental conditions, although some canopy species such as *Eucalyptus pilularis* and *E. acmenoides* are found over a wide range of soil conditions, under lower nutrient conditions the understorey tends to dry sclerophyll, and rainforest elements rarely occur. Under more fertile conditions there may be no difference in soils under eucalypts and rainforest, and long-term maintenance of eucalypts is dependent on fire or other disturbance. (Turner and Kelly (1981) showed that soils under flooded gum (*E. grandis*) in northern New South Wales were not nutritionally distinguishable from those under complex notophyll vine forest.) In the absence of fire or other disturbance, eucalypts would not regenerate under a closed rainforest canopy but, nevertheless, Florence (1964) doubted that, under lower-fertility conditions, the resultant rainforest would be stable. He suggested 'that a rainforest-element stratum which is not capable of self-perpetuation following senescence of the dominants' [tall open forest eucalypts and other sclerophyll elements] 'would itself decline in time, leading directly to regeneration of the dominants, or creating the vegetational condition conducive to its own destruction by fire'. Both the fire-climax and Florence's model predict that rainforest boundaries are unstable, the difference being that the fire-climax model allows for the possibility of permanent extension of rainforest, while, under Florence's model, expansion of rainforest beyond a core area can only be a transient, one-generation, phenomenon.

The possibility that the advancing rainforest causes a sufficient amelioration of soil conditions

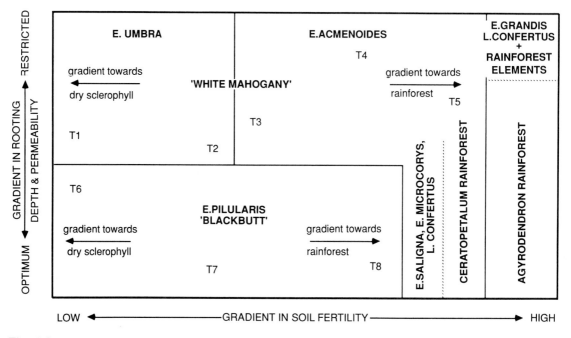

Fig. 4.9 Relationships between rainforest, sclerophyll forest, and soil types (after Florence 1964).

so as to permit its long-term survival has been proposed (see Ash 1988). Rainforest litter has different chemical and physical properties from that of sclerophyll forest (Bowman and Wilson 1988) and the closed canopy creates a very different microclimate. In consequence, soil organic matter content and moisture retention are generally higher beneath rainforest than in adjacent eucalypt forest. In addition, Beadle (1954, 1962) found evidence that nutrient levels, most particularly phosphorus, were higher in rainforest foliage than in sclerophyll species, and as the rainforest margin advanced, the shedding of leaves enriched the soil. Whether either mechanism produces significant changes in soil properties cannot be determined on the present evidence.

There are insufficient data on historic fire frequencies in rainforest ecotones to permit a testing of these various hypotheses, although the study of the Terania Creek basin reported by Turner (1984) does provide some information for that site. In the basin, pure rainforest is found on fertile soils derived from basalt. On the slopes of the basin, there is tall open forest over a rainforest understorey on less-fertile soils derived from rhyolite. On the more fertile lower slopes, the canopy dominant is *Lophostemon confertus*, while at higher levels *Eucalyptus pilularis* dominates (Fig. 4.10). If fires were completely excluded, the fire climax model would predict eventual permanent replacement of all the tall open forest by rainforest. Florence's model would suggest that, on the poorer soils, rainforest would fail to regenerate even if it temporarily achieved dominance, although on the more fertile lower slopes *L. confertus* could potentially be replaced by rainforest. Turner investigated the soil profiles in the forests and consistently detected eight discrete layers of charcoal under *Lophostemon*, and more than twelve layers under *Eucalyptus*. It is suggested that these layers represent major

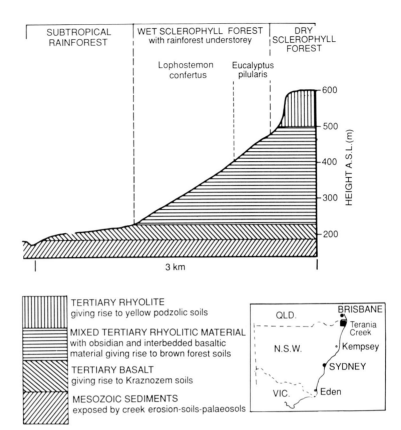

Fig. 4.10 Distribution of forest types at Terania Creek (from Adam 1987b, after Turner 1984).

fires. Radio carbon dating of charcoal samples indicated a return fire frequency of 300–400 years in the *Lophostemon* area and 200–300 years in the *Eucalyptus*. Unfortunately, there is no information on the nature of the previous vegetation, but fires of this frequency would permit the regeneration of the tall open forest species before senescence and replacement by rainforest. These data illustrate the importance of fire in the dynamics of the rainforest ecotone, and suggest that with the fire frequencies prevailing in recent history, mixed forest may have persisted (and could then be regarded as stable) over large areas even on fertile soils over thousands of years, but do not permit a prediction of what might occur if fire were excluded.

The question of the stability or otherwise of tall open forest–rainforest boundaries has also been the subject of debate in Tasmania. Gilbert (1959) in one of the earliest detailed studies of the problem argued that, if undisturbed, mixed forest would be a stage towards the development of rainforest, except in the case of very infertile soils under high rainfall when mixed forest could form the climax. Cremer (1960) suggested that under such conditions the rainforest understorey would be so open as to scarcely merit description as rainforest, and so this sort of tall open forest would not be a mixed forest.

Jackson (1968, 1978) argued that in the high-rainfall areas of Tasmania the theoretical climax in the absence of fire would be rainforest which is difficult to burn because of the low flammability of foliage, the humid microclimate, and the low

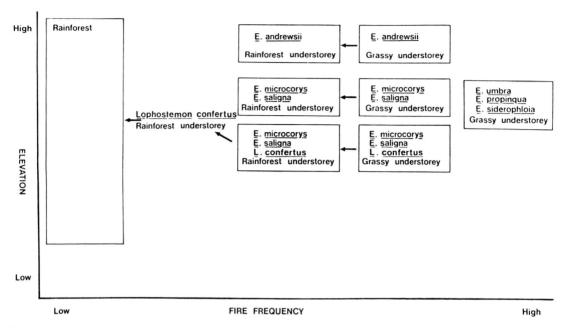

Fig. 4.11 Inferred relationships between eucalypt dominants, rainforest, and fire frequency in Washpool National Park (Fox 1983).

inflammable litter load. In consequence, Jackson suggested that rainforest species are poorly adapted to withstand fire and regenerate poorly after fire. Nevertheless, under some circumstances rainforest is burnt in fires (see, for example, Hill 1982). The frequency of fire acts as a selective force; frequent burning promotes more inflammable vegetation which, in itself, increases fire frequency. In addition, Jackson argued that substantial nutrient losses occur in fires, so that repeated fires cause a drop in soil fertility, which creates a further feedback because the communities on infertile soils tend to be open, slow growing, and fire prone. Rainforest is restricted to areas where the topography affords protection from high fire frequency. With high fire frequencies, all trees are eliminated and button-grass (*Gymnoschoenus sphaerocephalus*) sedgeland results. Jackson (1978) suggested 'that chance processes in the distribution of fires lead to a process of "Ecological Drift" in which any given area moves gradually towards climax rainforest or towards sedgeland'.

Mount (1979) proposed a feedback mechanism under which fire frequency is determined by the pattern of fuel accumulation. Under a particular set of climatic conditions, the model would predict a stable pattern of vegetation types. However, fire frequency depends not only upon fuel and vegetation features, but also upon the behaviour of ignition sources. While lightning may have been the major initiator of fires in the past, Jackson and Bowman (1982) argued that the higher and more variable fire frequencies since the arrival of Aborigines will have had considerable impacts on vegetation distribution. Jackson and Bowman (1982) agreed with Mount (1979) that each vegetation type has its own average burning cycle, but suggested that this cycle is not strongly dependent on fuel. Mount (1979) predicted a low variance in fire frequency, but Jackson and Bowman (1982) maintained that the variance is very large and that the extremes in the distribution of inter-fire interval can promote vegetation change.

Although there has been a vigorous debate between the proponents of the different models (see Mount 1982; Jackson and Bowman 1982; Bell 1983), the evidence appears to be more in favour of Jackson's (1968, 1978) ecological drift model. One important postulate of Mount's (1979) model is relative stability of vegetation boundaries, but a number of studies have demonstrated dynamic changes in rainforest boundaries and successional shifts in the directions suggested by the ecological drift model (see, for example, Brown and Podger 1982; Bell 1983; Ellis 1985).

4.3.1 Conclusions

It seems reasonable to suggest that there are many areas where soil and climatic conditions would be suitable for rainforest establishment were it not for the occurrence of fire. There are also some areas of tall open forest, with rainforest elements in the understorey, occurring on soils which, under present climatic conditions, do not support pure rainforest even in sites sheltered from fire. In these sites, the long-term replacement of tall open forest by rainforest appears unlikely.

The alteration of fire frequencies in eastern Australia by Aborigines over the last 40 000 years (see Chapter 9) may have limited the spread of rainforest in response to improved climatic conditions after the end of the last glacial period. Changes to fire regimes since European occupation have resulted in recent extension of rainforest species into tall open forest.

4.4 EVOLUTION OF TALL OPEN FOREST

The tall open forests with relatively short-lived canopy species which regenerate from seed after fire and in the absence of fire revert to rainforest, pose an interesting problem. Have these canopy species evolved since the arrival of humans when fire frequencies increased, or were pre-human fire frequencies sufficient to promote tall open forest?

Smith and Guyer (1983) suggested that it was not necessary to postulate a long coevolutionary history with fire, but that some apparently fire-adapted features may have developed in fire-free environments. They classified the canopy species of tall open forests into the category of ecological nomad (pioneer species in the terminology of Swaine and Whitmore 1988—see Chapter 5) as defined by van Steenis (1958b). Nomads are species which occupy successional sites in rainforest, such as landslip sites or in gaps caused by tree fall. Characteristics of nomads are shade-intolerance in seedlings, rapid growth in early stages and reproduction at a relatively early age, production of numerous, well-dispersed seeds, and relative indifference to edaphic factors (J. M. B. Smith and Guyer 1983). These characteristics are valuable in a fire-prone environment but may be pre-adaptations to fire having evolved in response to other forms of disturbance.

The fossil record is, as yet, insufficient to indicate the antiquity of tall open forest eucalypts. Little information is available on the long-distance dispersal characteristics of the species; the seeds, although small, do not have any special dispersal features. In the absence of fire, the disturbed areas would probably have been discrete and widely separated; was dispersal sufficient to colonize these sites?

Although tall open forest and rainforests have obvious interrelationships in eastern Australia, this is not the case in south-west Australia where a number of eucalypts (Ashton 1981b), dominate tall open forests with no rainforest elements. These western forests are separated from eastern Australia by several thousand kilometres. If tall open forest eucalypts evolved as nomads in rainforest, why has only one element of the rainforest (the nomads) survived in the west?

4.5 THE DEFINITION OF RAINFOREST REVISITED

The occurrence of eucalypt overstoreys over rainforest adds to the problem of defining rainforest. Where along the continuum is the line to be drawn between rainforest and eucalypt forest? The matter has been the subject of considerable controversy in Tasmania, Victoria, and New South Wales.

In Tasmania there is a very extensive area of mixed forest, and similarly on the Errinundra Plateau in Victoria, rainforest elements are well represented under eucalypts. In Tasmania where a moratorium on logging pure rainforest is in place, rainforest species are still harvested from mixed forest. Pure rainforest is defined as having less than 5 per cent eucalypt canopy, but this figure is arbitrary and there is no evidence of any marked difference in stand ecology above or below this level.

However, if it is accepted that rainforest is the climax in much of the wetter forests, then there is a case for including all mixed forest with rainforest, since the mixed forest is but a transient seral stage. The definition of rainforest proposed by Dale *et al.* (1980) specifically included seral communities with sclerophyll overstoreys within the category of rainforest.

In New South Wales, the tall open forest/rainforest ecotone often includes two other species of capsular Myrtaceae in addition to *Eucalyptus* spp. namely *Lophostemon confertus* (brushbox) and *Syncarpia glomulifera* (turpentine), and these are conventionally regarded as sclerophylls. *Lophostemon*, however, has relatively large, flat, dark-green leaves which, while possessing some scleromorphic features, are much more mesic than the leaves of either *Eucalyptus* or *Syncarpia*; indeed in overall features *Lophostemon* leaves differ little from those of several *Ficus* spp. The canopy of *Lophostemon* is much denser than that of most eucalypts (Fig. 4.12).

Whether stands with *Lophostemon* and eucalypt emergents should be regarded as rainforest was of considerable significance in the debate over rainforest conservation in New South Wales, and the issue was addressed at some length in the inquiry into logging at Terania Creek (Isaacs 1982—see Chapter 9 this volume). The Forestry Commission of NSW contended that the brushbox stands at Terania Creek were neither rainforest nor seral to rainforest, arguing that soil

Fig. 4.12 Canopy of tall emergent *Lophostemon confertus*, Mt. Boss State Forest, New South Wales

fertility and fire frequency would prevent the development of pure rainforest (see Fig. 4.10), although conceding that in more-fertile situations successional development of rainforest might occur. The Inquiry's analysis of the various arguments embodied some novel views of successional processes, and provided a strong argument against using judicial (or quasi-judicial) methods to resolve environmental issues.

A further view is that it is proper to regard the Brush Box as Rainforest because it is part and parcel of a vibrant or growing vegetation which is undergoing transformation by what is called 'seral succession'.

Shortly put, this theory of seral succession involves the concept that when in the course of time the Brush Box trees become extinct they will be overtaken and the area occupied by the Coachwood–Crabapple type Rainforest and thus there would be complete Coachwood–Crabapple Rainforest in due course.

Before proceeding to examine this theory of seral succession it is pertinent to point out that if extinction of the Brush Box trees has to be achieved before you get a complete Rainforest of Coachwood–Crabapple type vegetation, this in itself, in my view, destroys the concept or view that Brush Box is Rainforest. If it is Rainforest why do you have to wait for Rainforest Brush Box to be extinct before the area becomes a full Rainforest? This seems to me to add only another nail in the coffin to which the arguments against so classifying or typing Brush Box have already been consigned.

When the theory was first advanced, I ventured the view that by logging the Brush Box, Forestry was helping conservationists and ecologists in the realisation of their ambition because it was said that this transition involving extinction of Brush Box would take something between 300 to 400 years to achieve. By removing the trees now so as to enable the Rainforest of the Coachwood–Crabapple type to cover the area entirely would accelerate this objective of seral succession and the Forestry was really, on this view, doing ecologists a favour and not causing any detriment to the environment. Isaacs (1982).

Given that there is a general agreement that the closed understorey is, in some sense, rainforest, there is a strong case for treating the whole mixed forest complex as rainforest, albeit one with special features. In terms of biomass the sclerophyll element is predominant, reflecting the substantial bulk of the overstorey trees, but in terms of numbers of individuals the rainforest element is by far the larger (Table 4.2).

Earlier authors were less concerned to narrow the definition of rainforest. P. W. Richards (1952) accepted that *Eucalyptus saligna* and *Syncarpia glomulifera* formed an open overstorey over part of the Barrington Tops rainforest described by Fraser and Vickery (1938) without questioning that the vegetation concerned was correctly categorized as rainforest. Schimper (1903), who coined the term rainforest, referred to tall open forest in fern gullies in Victoria as temperate rainforest, widening the concept of rainforest farther even than current Australian usage.

Table 4.2 Composition of three 30 × 30 m quadrats near Wilson River, Mount Boss State Forest, New South Wales. In each there is a sparse emergent canopy of very tall trees regarded as sclerophylls by the Forestry Commission of NSW, these trees contribute more than 50 per cent of the total basal area, although rainforest trees are more numerous

Species	1		2		3	
	No. of stems	Basal area (m^2)	No. of stems	Basal area (m^2)	No. of stems	Basal area (m^2)
Rainforest						
Archontophoenix cunninghamiana	1	0.03	14	0.29	11	0.25
Argyrodendron actinophyllum			3	0.07	5	0.44
Caldcluvia paniculosa	7	1.30	7	0.74	6	0.79
Ceratopetalum apetalum	2	0.20	1	0.03		
Cryptocarya glaucescens	5	0.77	2	0.13	1	0.10
Daphnandra micrantha			1	0.08		
Doryphora sassafras			1	0.02	6	0.10
Euroschinus falcata			1	0.01		
Orites excelsa			1	0.01		
Polyscias elegans			3	0.07	1	0.01
Sloanea australis	1	0.10	1	0.02	5	0.88
Sloanea woollsii	3	0.44	4	0.27	1	0.20
Schizomeria ovata	1	0.40	9	0.76	3	0.13
Synoum glandulosum			1	0.04	3	0.05
Trochocarpa laurina	1	0.02	1	0.04		
Total basal area of rainforest species		3.26		2.55		2.98
% of total stand basal area as rainforest spp.		42.60		36.80		47.50
Sclerophyll						
Eucalyptus microcorys			1	0.13		
Eucalyptus saligna	3	2.80	1	1.50	1	1.70
Lophostemon confertus	1	1.60	1	1.70	1	1.60
Syncarpia glomulifera			1	1.04		
Total basal area sclerophyll species		4.40		4.37		3.30
% of total stand basal area as sclerophyll spp.		57.40		63.20		52.50

5 REGENERATION AND RESPONSES TO DISTURBANCE

Rainforest is complex, both structurally and floristically. How is this complexity maintained? The answers to this question are of considerable interest in building up our understanding of the ecology of rainforest, and are of vital importance to those who wish to manage rainforest, whether for some form of harvest or for long-term preservation.

In this chapter a general model for forest regeneration is described. Various aspects of the reproductive and regenerative biology of Australian rainforest species are then discussed in the light of this model.

5.1 THE GAP-FILLING MODEL OF FOREST REGENERATION

Forest regeneration cycles are driven by disturbance which creates gaps in the canopy. These gaps are filled and the continuity of the canopy restored. At any one time a forest is a mosaic of patches at different stages of maturity (see Hopkins *et al.* 1977; Whitmore 1978, 1982, 1984, 1988, 1989*b*, Swaine and Whitmore 1988; Baur 1989). The nature of the stages in the cycle at any particular site will depend on the size of the gaps and the environmental conditions within them, which in turn will be related to the intensity, scale, and frequency of disturbance (Hopkins 1990).

The smallest gaps in the canopy are formed by the falling of individual branches (which can be brought about by several agencies) or the slow death and *in situ* decay of a single tree. These gaps are rapidly filled by the expansion of neighbouring tree crowns or by a growth spurt from previously suppressed trees below the gap (Hopkins *et al.* 1977). The continuous regeneration of shade-tolerant species under a nearly continuous canopy has been termed *diffuse* regeneration by van Steenis (1958*a*).

Except in species-poor forest such regeneration is likely to result in slow, kaleidoscopic changes in canopy floristic composition, as it is unlikely that the distribution and composition of the understorey would match that of the canopy.

In larger gaps the process of regeneration back to the mature canopy phase is slower (the rate depending on gap size), and in its early stages involves light-demanding species. At the smaller end of the spectrum of large gaps, the disturbance factor is most frequently lightning strikes or wind damage. Often, when a single tree falls, the tangle of vines linking its canopy to its neighbours drags down large parts of the canopies of adjacent trees, so creating a much larger gap, which cannot be filled by lateral expansion of neighbours. Frequently the gap-creating tree-fall badly damages the lower strata, so that growth of suppressed stems is no longer possible. Many rainforest trees have shallow root systems which are ripped out of the ground when a tree falls (rather than the trunk splitting above the ground); in this way hummock and hollow microrelief may be created. The effect of a big tree falling is to produce a vacant space, often with disturbed soil and, depending on the size of the gap, a markedly different microclimate then prevails under a closed canopy. Regeneration in these larger canopy gaps was termed *spotwise* by van Steenis (1958*a*).

There is a continuum between gaps created by the fall of single trees to large openings, covering hectares. Recovery of rainforest over larger areas is usually treated as an example of secondary succession.

The size of a gap determines the extent to which environmental conditions within it are different from those under a closed canopy. The response of species to gaps is related to their requirement for (or tolerance of) the conditions prevailing in the gap. Various categorizations of species into response groups have been proposed, but Swaine and Whitmore (1988) have argued that there are two qualitatively distinct groups—pioneer and climax (or non-pioneer) species. Pioneer species have seeds which only germinate in gaps in the canopy large enough for full sunlight to reach the ground for at least part of the day, and require high irradiance levels for seedling establishment and growth. Seedlings and young plants of these species are never found under a closed canopy (Swaine and Whitmore 1988).

Climax species have seeds which can germinate under shade. The seedlings can establish in shade; many species can persist in shade for prolonged periods, although some require alleviation of shade conditions if they are to survive (Swaine and Whitmore 1988). Young plants of climax species are commonly found below a closed canopy, but they are not necessarily restricted to such a habitat, the seedlings establishing in a gap may include both pioneer and some climax species (Swaine and Whitmore 1988).

Within both groups there is variation in the response of species to a number of environmental factors. While subdivision of each group is possible, the variation is continuous and the boundaries of the subgroups are arbitrary. Pioneer and climax species are distinguished absolutely on the basis of germination and seedling characteristics, but Swaine and Whitmore (1988) document other characteristics often associated with either pioneer or climax species.

Secondary succession in Australian rainforests has been described by Hopkins (Hopkins *et al.* 1977; Hopkins 1981) who recognized four broad groupings of species, groups A and B are subdivisions of the pioneer category of Swaine and Whitmore (1988) while group D corresponds with the climax group (see Fig. 5.1). The characteristics of the groups were documented by Hopkins *et al.* (1977). Group A is divided into two subgroups. A_1 consists of herbaceous annuals or short-lived perennials which grow up to 1–2 m tall. Several of the herbs which are now widespread in the early stages of secondary succession are aliens. A_2 species are short-lived, normally soft-wooded small trees or shrubs growing up to 4–8 m. One of the most widespread species in this group is the introduced *Solanum mauritianum*. All species of group A produce large numbers of well-dispersed seeds (spread either by wind in the case of various herbs in the Asteraceae—*Senecio* spp., *Ageratum* spp., or *Eupatorium* spp.—or by birds as in *S. mauritianum*), with long-viability.

Group B species were termed 'Early Secondary Species' by Hopkins *et al.* (1977) and correspond to the nomads of van Steenis (1958*b*). They are fast-growing trees growing, under good conditions, to 10–25 m and living 15–50 years. They are shade intolerant and produce large numbers of seed at frequent intervals. Seed dispersal is good and viability in the soil is long enough to span the period between seed years. Group B species are most likely to constitute the early regrowth in small to medium-sized gaps, although if a seed is available later stage species may also commence growth at the same time. As well as occurring in gaps, these species are also found in other well-lit open habitat, such as river banks. Road construction, associated with logging operations, has created a network of habitats for group B species which now occur in ribbons through many rainforest stands.

The number of group B species is small compared with the size of the total rainforest tree flora (in north-east Queensland about 20 species out of a total of 700—Thompson *et al.* 1988). Most of these species have very wide geographical distributions. The low number of pioneer tree species is a general feature of tropical rainforest (Whitmore 1989*b*).

The group C species become canopy components some way into the regeneration cycle (stage VI in Fig. 5.1) and were referred to as late secondary species by Hopkins *et al.* (1977). Swaine and Whitmore (1988) and Whitmore (1989*b*) have argued that this terminology should be abandoned, as there is not a physiologically distinct grouping of species between pioneers and

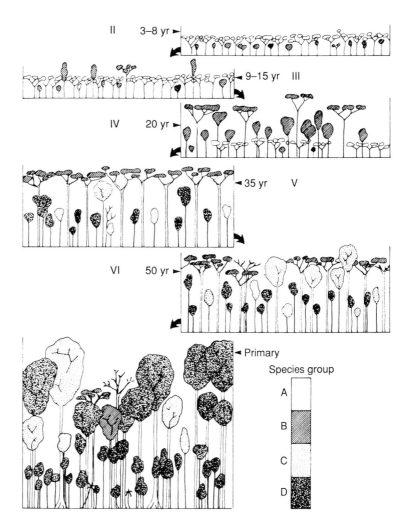

Fig. 5.1 Schematic representation of regeneration in complex notophyll vine forest (from Hopkins 1981).

climax species. In fact group C may contain both long-lived, large-growing pioneers, and rapid-growing light-tolerant climax species which may colonize large gaps simultaneously with pioneers, or establish and grow rapidly as the initial pioneer canopy begins to develop gaps (Queensland maple, *Flindersia brayleyana*, is an example of a species with wide shade tolerance—Thompson *et al.* 1988).

Group D, climax species, are long-lived and slow-growing. Fruiting is often irregular, and dispersal is normally limited, seed viability is short (often only a few weeks). The seedlings are tolerant of shade. An important consequence of the generally short viability of the seeds of climax species is that they will not be represented in the soil seed bank and re-establishment of these species will be dependent on dispersal from nearby primary or older secondary forests (Hopkins 1990). Recolonization will be dependent on the size of the gap and the dispersal characteristics of individual species. The larger the size of the gap the greater the likelihood that at least some species may be lost from the canopy (i.e. local extinction) (Hopkins 1990). The dependency of the regenerating forest on an outside seed source will continue until all climax species have established large enough populations of reproductive trees to maintain themselves. If the seed source is itself disturbed or destroyed there is a possibility that the original climax canopy (or something approximating to it) will never be restored (Hopkins 1990). Examples of this have not been reported from Australia but are referred to as 'stagnant' secondary growth elsewhere in the tropics (Lovejoy 1985).

Although climax species may become prominent members of the community after about 50 years (Fig. 5.1), they are not reproductive at this time. Gaps developing in the canopy at this stage are likely to be filled by spotwise regeneration of group B species (depending on gap size), although over time diffuse regeneration becomes more important unless disturbance creates a large gap and resets the cycle.

The model outlined above and illustrated in Fig. 5.1, suggests that over a sufficiently long period secondary succession can lead to restoration of a forest similar in structure (but, except fortuitously, not identical in composition) to the pre-disturbance primary forest. This is not always the case, and Hopkins (1981) has outlined a number of situations where the succession is either arrested or culminates in a forest type unlike the original (see Fig. 5.2). Hopkins (1981) suggested that these changes to the typical sequence may arise from changes to the physicochemical environment and/or changes in the biota.

The physicochemical environment can be changed in many ways, but the impacts and reversibility/irreversibility of most changes are poorly known, although Hopkins (1990) suggests that rainforests on more nutrient-poor soils are more vulnerable to nutrient loss following disturbance than those on nutrient-rich soils, and that this can result in long-term changes to the vegetation. Repeated clearing and burning of rainforest on deep nutrient-poor sand in northern Australia has resulted in increases of species characteristic of beach dunes, while similar disturbance to rainforest on basaltic alluvium has not resulted in such a floristic alteration (Hopkins 1990).

The secondary succession may be altered if species creating conditions unsuitable for later-phase species become established. Cyclone-damaged areas may be smothered by vines, which may prevent re-establishment of a tree canopy (Webb 1958). The sprawling shrub *Lantana camara*, an introduced South American species which is a group A species, can block succession at an early stage in some circumstances (Williams *et al.* 1969*a*; Webb *et al.* 1972). The establishment of some species may lead to changes in the disturbance regime, so as to prevent rainforest re-establishment. Eucalypts are more inflammable than rainforest, as is the grass *Imperata cylindrica*. Once these species enter an area, there is a probability of increased fire frequency maintaining non-rainforest vegetation.

Forest regeneration processes occur over a time-scale longer than a human lifetime. Predictions about the outcome of the processes involve extrapolation from short-term observations on the basis of understanding of the growth processes and ecology of the species concerned; this understanding is either expressed verbally in generalized models (such as that of Hopkins 1981) or incorporated in formal mathematical models.

There have only been a few detailed studies on rainforest regeneration in Australia, most of them addressing the recovery of forests following logging operations. Although they concentrate upon merchantable species rather than the total flora, the observations reported are not at variance with the picture already outlined. (The effects of logging were reviewed by Horne and Hickey (1991)—see also Chapter 9.)

One way in which secondary succession following logging and some forms of natural disturbance may differ from that on former agricultural land, is that regeneration by coppice growth or root suckering may occur. Vegetative growth would permit an early representation of species charac-

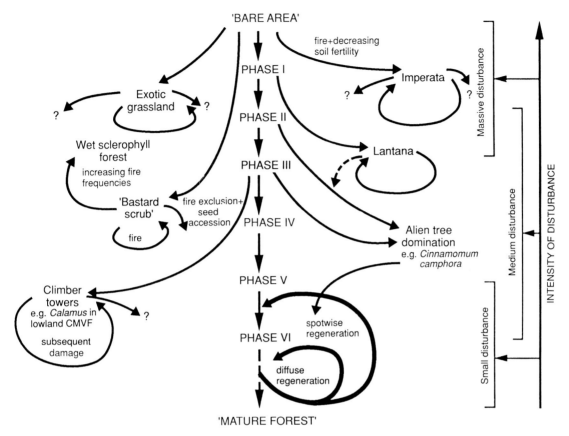

Fig. 5.2 Schematic representation of the pathway of secondary succession or natural regeneration in vine forest in Australia, showing some of the variations that may occur in response to disturbance of various intensities (phases II–IV correspond with those illustrated in Fig. 5.1) (from Hopkins 1981).

teristic of the late stages of the regeneration cycle in the reforming canopy and an acceleration of the recovery process. In simple notophyll vine forest (dominated by *Ceratopetalum apetalum*) in New South Wales (King and Chapman 1983), and in complex notophyll vine forest in southern Queensland (Williams *et al.* 1969a) coppice regeneration, although significant, was not the major component in regeneration. In north Queensland, Stocker (1981) reported a predominance of coppice or root-sucker regeneration in complex notophyll vine forest on the Atherton Tableland (see also Department of Forestry, Queensland 1983), although the early height growth of coppice stems was less than that of seedling regeneration (in southern Queensland, Webb *et al.* 1972 found height growth of coppice to be greater than that of seedlings). The treatment of the study plot was more severe than in normal logging, because following harvesting of timber, all remaining trees were felled and the debris burnt. Growth of stinging trees, *Dendrocnide moroides*, was controlled by several sprayings with the herbicide 2-4D, while *Solanum mauritianum* was subject to control by use of 2-4-5T (Stocker 1981). The concentration of 2-4D used is thought to have been too low to have adversely affected seedling regeneration by other species.

Individuals of species regenerating from seed show a tendency to be clumped rather than evenly distributed (Burgess *et al.* 1975; King and Chapman 1983; Department of Forestry, Queensland 1983), a consequence of their dispersal characteristics.

Regeneration of complex notophyll vine forest in Wiangaree State Forest in northern New South Wales has been modelled by Horne (1981), Horne and Gwalter (1982), Shugart *et al.* (1980), and Shugart (1984) and early experimental observations were reported by Burgess *et al.* (1975). In terms of basal area (Horne 1981) and species composition (Shugart *et al.* 1980), the models suggest that recovery from heavy logging treatment would take about 200 years. Recovery from lighter logging would be much quicker. The compositional changes predicted by the KIAMBRAM model used by Shugart *et al.* 1980) (a shift toward greater representation of pioneer species) are similar to those proposed by Hopkins *et al.* (1977). Horne (1981) emphasized that his modelling referred to the effect of a single logging cycle; there are no data to predict the behaviour of these forests after multiple logging cycles. Even extrapolation to two hundred years hence must be regarded with caution, as changes in environmental factors over an extended period may deflect the theoretical course of succession.

The only study which has reported on the recovery of the total flora of a stand following logging is that of G. C. King and Chapman (1983) at Banjo Creek in the Hastings region of New South Wales. This study is unusual in that, fortuitously, there had been a comprehensive inventory of the site prepared some years prior to logging (Burges and Johnston 1953). The forest was simple notophyll vine forest dominated by *Ceratopetalum apetalum*. It was logged in 1955-56, with 90 per cent of the canopy removed on flat ground and lower slope, and 35 per cent removed on the upper slope. A log dump was established on flat ground near the creek. Recording of recovery took place in 1981, 25 years after the end of logging.

All vascular species reported in 1949 by Burges and Johnston (1953) were still present in 1981. In addition, a large number of pioneer species (including woody herbs) were recorded in 1981. Many more bryophyte species were recorded in 1981 than in 1949, but this probably indicates more comprehensive collecting and identification rather than a genuine increase. G. C. King and Chapman (1983) stressed the importance of the residual canopy trees in providing habitat for vascular epiphytes, and suggested that in their absence the recovery of the epiphytic component in the flora would have been much slower. Based on the first 25 years regrowth, G. C. King and Chapman (1983) estimate that full restoration of the original forest structure would occur in 140-190 years. This evidence for recovery of floristic composition after a major disturbance is impressive, but it would be unwise to extrapolate the results to more complex, floristically diverse rainforests (see also Chapter 9).

There have been few detailed studies of the effects of logging or other disturbance on fauna, although opposition to rainforest logging has often highlighted potentially adverse impacts. During logging there is obviously disturbance or destruction of faunal habitat, while roads constructed in the course of logging may permanently fragment a larger region. Logging also alters the extent and spatial distribution of patches of forest in different stages of the regeneration cycle; for species associated with particular stages the populations may be reduced to levels which are not sustainable in the long term. In the absence of appropriate studies, both before and after logging, claims of little impact on the fauna can be neither supported nor refuted. There is also an absence of data on the association between fauna and the stages in the natural regeneration cycle. Crome (1990) suggests that a few species of birds and mammals may be characteristic of early regrowth stages, although confirmatory data are required. One of the few relevant studies was conducted by Burgess *et al.* (1975) on the avifauna in the early stages of post-logging recovery of complex notophyll vine forest in northern New South Wales. Although there were changes in diversity compared with pre-logging, there were no changes in the overall composition. In recently logged areas the number of insectivorous birds increased and high densities of insectivores were associated with dense undergrowth. Crome (1990) similarly found that in north Queensland there was rapid recovery of the avifauna (less than 30 years) following logging, but suggested that if secondary rainforest becomes isolated from primary (or older secondary) forest the recovery may be slower and less certain.

The distribution and ecology of insectivorous bats in Australia are still poorly known but it is clear that rainforests provide an important habitat. Crome and Richards (1988) have shown that, in rainforest on the Windsor Tableland in northeast Queensland, vegetation structure is an important determinant of bat community composition, with different species assemblages in closed canopies and in gaps. The species in each group had different wing morphologies. The species in gaps were not rainforest gap specialists but were also found in open (eucalypt) woodland. (However, the members of the closed canopy group have all been recorded outside rainforest, but they do not forage in rainforest gaps. Their foraging behaviour outside rainforest is not known but they may utilize other dense microhabitats (Crome and Richards 1988). Changes in the proportion of gap to canopy area might result in changes in the abundance of different species of bat (Horne and Hickey 1991).)

Logging and extensive natural disturbance also alter the physical environment. Log landings and snig trails alter physical and chemical properties in the soil (Gillman *et al.* 1985). Some of the patchiness in the regeneration of particular species may be associated with these features. Local climate is also affected. High isolation may result in drought stress, and desiccation may damage species (particularly bryophytes and epiphytes) in the newly exposed edge habitat adjacent to the cleared area. At higher altitudes, the incidence of ground frosts may increase in clearings. Baur (1968) illustrated the effect of clearings on microclimate (Fig. 5.3) and suggested that frosts could

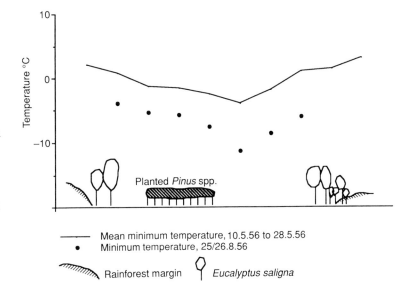

Fig. 5.3 Cross-section through a grassland area in complex notophyll vine forest in Clouds Creek State Forest, northern New South Wales, at 650 m altitude showing mean minimum temperature from 10 June 1956 to 29 August 1956 (–), and minimum temperatures on the nights 25/26 August 1956 (redrawn from Baur 1968).

delay, or even prevent rainforest regeneration. King and Chapman (1983) suggested that in their study area delayed regeneration in the largest clearings reflected frost damage, and not as originally suspected, soil compaction on the log dump site and tracks.

5.1.1 Regeneration in rainforests with species-poor canopies

The models and evidence discussed above refer to rainforests with a relatively large number of canopy (or potential canopy) species. These species have different physiological responses to environmental conditions (particularly light quality and quantity of nutrients—Thompson *et al.* 1988) and thus gaps of different sizes are recolonized by different suites of species. However, not all the rainforests in Australia have species-rich canopies. In floristically simpler forests is there still a distinction between pioneer and climax species or does the regeneration cycle differ?

Simple notophyll vine forest in New South Wales is floristically simple and frequently dominated by a single species, *Ceratopetalum apetalum*, which falls into Swaine and Whitmore's (1988) climax species category. Germination can occur in heavy shade and suppressed seedlings can survive long periods in shade, until released following gap creation and exposure to higher levels of irradiance (McGee 1990). However it is versatile and can successfully establish in large open gaps. It also fits the climax syndrome as seed viability is maintained for only a few days (Baur 1989; Floyd 1990). In other respects *C. apetalum* is perhaps an atypical climax species as seed production is copious, with individual seeds being very small with limited reserves for seedling establishment (Floyd 1990). Successful establishment of seedlings in shade may be dependent on mycorrhizal associates which permit transfer of organic compounds from other species to the seedlings (McGee 1990).

It would thus seem that *C. apetalum*-dominated forest could be maintained in the absence of major disturbance through recruitment to small gaps, but that, provided that a seed source is available, regeneration directly in large gaps would also be possible. However, fruiting in *C. apetalum* is restricted to January and February (Floyd 1990). While a large gap created during this time could be immediately recolonized, gaps at other times of the year will be initially occupied by other species, although because of its shade tolerance *C. apetalum* could establish itself under the developing canopy during its next (and subsequent) fruiting seasons. (*C. apetalum* may also regenerate vegetatively.)

Schizomeria ovata is a simple notophyll vine forest species with rather different behaviour from *C. apetalum*, with which it frequently occurs, although it is also locally a canopy dominant (see Chapter 3). Fruiting occurs over a long period from summer to winter and seed viability is retained in the soil for a considerable time. *S. ovata* is therefore able to respond to disturbance at any time of the year (Floyd 1990). These are frequently attributes of pioneer species. However, the seeds are much larger than those of *C. apetalum*, and seedling establishment draws upon these reserves. In this respect *S. ovata* has some attributes of climax species.

Cool temperate rainforests (microphyll moss/fern forests) in Australia are also characterized by low tree species diversity. The major canopy species show a tendency, even in apparently undisturbed sites, for vegetative reproduction by formation of multistemmed stools. The seeds of the upper canopy species are small and wind- or gravity-dispersed (with the exception of the rainforest gymnosperms with fleshy fruit which may be bird dispersed). In this respect, Australian temperate rainforests differ from those in southern Chile, which have more woody species and a much higher incidence of biotic dispersal except among emergents where anemochory predominates (Armesto and Rozzi 1989). Emergents in Australian cool temperate rainforest are infrequent and are simply larger specimens of the main canopy species.

Dispersal distances of wind-dispersed canopy species in Tasmania are small (Hickey *et al.*

1983). Seed production varies from year to year (markedly so in *Nothofagus cunninghamii*—Hickey *et al.* 1983; Howard 1973*a*). Germinability of *N. cunninghamii* was much higher in years of heavy seed fall (Hickey *et al.* 1983) than in years of low production. Germination of *Eucryphia lucida* was more consistent and generally higher than in *N. Cunninghamii*; *Atherosperma moschatum* gave low germination rates (Hickey *et al.* 1983).

Under what circumstances does germination and seedling establishment of cool temperate rainforest species occur?

Secondary succession in Tasmanian cool temperate rainforests has been studied by Read and Hill (1983), whose data indicate significant differences from the model described above for tropical and subtropical complex vine forests.

In complex vine forests, the fast-growing group B species form a relatively high canopy early in the succession. In microphyll moss/fern forest the number of potential forest-canopy-forming species is small, essentially limited to those which form the primary forest.

In the old field sites, studied by Read and Hill (1983), there was a pioneer shrub stage (corresponding with group A species). The species were mostly bird-dispersed. Commonest was *Tasmannia lanceolata*; other species were *Pittosporum bicolor*, *Coprosma quadrifida*, *Cenarrhenes nitida*, and *Cyathodes* spp. The invasion of shrubs was patchy, determined by available perch sites (old logs, fence posts, tree ferns, etc.). *Nothofagus cunninghamii* and *Atherosperma moschatum* (the climax big tree species) regenerated at the boundary between the field and undisturbed forest, but little invasion into the field had occurred. The low germination of *Atherosperma* and the poor dispersal of both these species, plus the requirement for competition-free establishment sites must have limited their spread. However, Read and Hill (1983) suggested that as the few successfully invading trees reach reproductive maturity, there will be increased inoculum potential and subsequent more rapid spread. Thus the regeneration of the forest, although slow, appears to involve fewer suites of species and fewer stages than that of more complex tropical and subtropical rainforest.

Regeneration of cool temperate rainforest in the absence of major disturbance has been extensively studied in *N. cunninghamii* forests, and to a lesser extent in *N. moorei* forests (Howard 1981; Read 1988; Read and Hill 1985*a*,*b*). *Atherosperma moschatum* is a frequent component of *N. cunninghamii* forests. Gilbert (1959) suggested that because of its ability to regenerate under a closed canopy, *A. moschatum* was more shade tolerant than *N. cunninghamii*. He described stands in the Florentine Valley (Tasmania), which showed little evidence of *Nothofagus* regeneration, and argued that in the absence of disturbance there would be progressive replacement by *Atherosperma*. Unfortunately, the area has been cleared so this hypothesis cannot be confirmed (Read and Hill 1985*b*). Noble and Slatyer (1980, 1981) accepted that *Atherosperma* had greater shade-tolerance, and their succession model predicted that, in the absence of disturbance, Tasmanian rainforest would be dominated by *Atherosperma*. They noted, however, that most stands were in fact a mixture of *N. cunninghamiii* and *A. moschatum*, and suggested that some disturbance must be occurring in order to explain the persistence of *Nothofagus* (Noble and Slatyer 1979). Other authors (Howard 1981; Read and Hill 1985*a*) have described *N. cunninghamii* forests as self-regenerating through vegetative regrowth and seedling establishment in small gaps (Ellis 1985). This persistence of the supposedly shade-intolerant *Nothofagus* without major disturbance is anomalous when compared with the behaviour of species in tropical and subtropical rainforest.

Read (1985) confirmed that *Atherosperma* seedlings were more shade-tolerant than those of either *Nothofagus* or *Eucryphia lucida*. Both *N. cunninghamii* and *E. lucida* showed highest rates of photosynthesis under full sunlight, although *N. cunninghamii* showed relatively better growth under intermediate shade than any of the other species examined by Read (1985). Seedling regeneration of both species is predominantly clumped in canopy gaps (Read 1985; Read and Hill 1985a; Ellis 1985).

The dependence of *N. cunninghamii* on gaps for seedling regeneration and the superior growth of seedlings of *Atherosperma* under canopy shade would suggest that the model advanced by Noble and Slatyer (1979, 1980, 1981), predicting long-term replacement of *N. cunninghamii* by *A. moschatum* in the absence of disturbance, is plausible. Nevertheless, field observations provide little indication that the process is occurring. Read (1985) suggested that the failure of *A. moschatum* to replace *Nothofagus* can be explained by its low seed viability (Hickey *et al.* 1983) and poor seedling establishment (Read and Hill 1985a). *A. moschatum* largely generates vegetatively by growth of basal sprouts from stem bases, which permits its persistence, but limits its ability to spread through the forest.

N. moorei seedlings have very similar photosynthetic characteristics to *N. cunninghamii* (Read and Hill 1985b), but it is suggested that the regeneration characteristics of the two species show some differences (Howard 1981; Read and Hill 1985a,b).

Seedling establishment beneath *N. moorei* is rare, except at the most north-western site for the species, Mount Nothofagus on the New South Wales–Queensland border (Howard 1981; Adam 1987b). At some sites the size-class distribution of stems indicates continuous regeneration by vegetative regrowth (Read and Hill 1985a). At other sites, where there is a well-developed undercanopy of species such as *Ceratopetalum apetalum*, *Doryphora sassafras*, *Orites excelsa* and, at Mount Banda Banda, *Sloanea woollsii* (Fig. 5.4), Read and Hill (1985a) suggested that *N. moorei* will not regenerate but will be replaced by the undercanopy species. Read and Hill (1985b) showed that *C. apetalum* and *D. sassafras* seedlings were shade-tolerant and would be able to establish under *N. moorei* canopies.

These data suggest that relatively pure stands of *N. moorei* will be maintained by vegetative growth, but if invasion by other species occurs then *N. moorei* will be replaced. Compared with *A. moschatum*, it would appear that species such as *C. apetalum* and *D. sassafras* exhibit greater success in seedling establishment. The apparent failure of *N. moorei* seedlings to establish in gaps within *N. moorei* stands is not easily explained (Read and Hill 1985a), although seedling regeneration is vigorous in disturbed areas, such as track

Fig. 5.4 *Sloanea woollsii* (note buttresses) in *Nothofagus moorei* forest at Mount Banda Banda

sides. Turner (1976) reported seedling invasion of *N. moorei* into eucalypt woodland at high altitude on the Barrington Tops.

If the size class and species composition data from some *N. moorei* stands are correctly interpreted as indicating that *Nothofagus* is being replaced, then questions arise as to whether this is of recent origin, possibly in response to climatic change as suggested by Turner (1976), or whether it is part of a recurrent cycle in which major disturbance promotes the regeneration of *Nothofagus*.

Nothofagus spp. are important components of temperate rainforests in New Zealand and Chile. In New Zealand they are frequent at high altitudes, on disturbed sites and infertile soils but are usually not competitive in lowland fertile sites (Wardle 1983). In Chile *Nothofagus* is dominant at high altitudes and latitudes where, in the absence of more shade-tolerant species, it regenerates continuously (Veblen *et al.* 1981) (this is a similar situation to *N. cunninghamii* in Australia). On lowland and mid-altitude sites *Nothofagus* is gradually replaced by more shade-tolerant species. However, parts of Chile are tectonically active and landslides and volcanism are major disturbances which continually provide open sites for the establishment of shade-intolerant *Nothofagus*. The long-term survival of *Nothofagus* in lowland and mid-altitude sites in Chile depends on these infrequent catastrophic disturbances (Veblen and Ashton 1978; Veblen *et al.* 1981). In lowland forests, seedling regeneration of *Nothofagus* does not occur in the absence of major disturbance, even in the relatively large gaps created by the fall of large, old trees (Veblen *et al.* 1980; Veblen *et al.* 1981). Comparative studies on the photosynthetic characteristics of Chilean and Australian species of *Nothofagus* (Read and Hill 1985*b*) suggested that Chilean species are not markedly different from those in Australia, but in Tasmania *N. cunninghamii* would readily regenerate in such gaps. The differences in behaviour appear to be related to the properties of the associated flora. The most shade-tolerant species in Tasmania, *Atherosperma*, regenerates poorly and so rarely replaces *Nothofagus*. There is a relatively poorly developed understorey in most Australian stands of *Nothofagus*: when a small gap appears in a canopy any seedling *Nothofagus* establishing in the higher irradiance now experienced on the forest floor is likely to be free from competition (Read *et al.* 1990*a*). In Chile the associated flora is richer and any gaps are rapidly filled by previously suppressed seedlings of shade-tolerant species, or at higher elevations, bamboos (*Chusquea* spp.) (Veblen *et al.* 1981). Regenerating *Nothofagus* in New Guinea also faces competition from a rich understorey and lower canopy flora (Read *et al.* 1990*a*). In Chile, *Nothofagus* spp. regenerate following large-scale disturbance. The response of Australian species to disturbance is less certain. The major natural disturbance likely to affect temperate rainforest in Australia is fire. While *N. cunninghamii* might be maintained after fire, if the fire interval is sufficiently long and a seed source is available (Gilbert 1959), there is much evidence that fire has been responsible for the regression of rainforest (Jackson 1968, 1978, see Chapter 4).

Temperate rainforest in Australia is thus unusual in that, even in the absence of major disturbance, a canopy of species with pioneer characteristics can be maintained. Seedlings are shade intolerant, but absence of understorey and climax species permits establishment in small gaps, while a propensity for coppicing also assists persistence of the canopy species. Where true climax species establish under *Nothofagus*, the prognosis is that they will eventually form a canopy and replace beech (Read and Hill 1985*a*).

Although there are differences in photosynthetic physiology between *Nothofagus* spp., relative shade intolerance seems to be a general feature of the genus, and disturbance, in some form, a requirement for successful regeneration. *Nothofagus* has a very long history in Australia (see Chapter 6). If the fossil record is to be interpreted as indicating that *Nothofagus* was a component of mixed forest (rather than being restricted to some specialized habitat islands within a more widespread diverse forest), then the question arises as to whether its persistence indicates regular extensive disturbance (Hill 1987). If disturbance was necessary to prevent the suppression of *Nothofagus* regeneration, then the nature of the disturbing factors operating throughout the Tertiary is unknown.

Despite the apparent anomaly of the maintenance of *Nothofagus cunninghamii*, the importance of seedling response to irradiance as the key to regeneration is again confirmed, as it is also by the behaviour of the Tasmanian rainforest conifer *Athrotaxis selaginoides*. *A. selaginoides* is dependent on disturbance creating gaps for regeneration in evergreen forests (i.e. it is a pioneer). However, in forests dominated by the deciduous *Nothofagus gunnii*, with greater light penetration to the forest floor, *A. selaginoides* can regenerate continuously without any requirement for disturbance (Cullen 1987), similar behaviour is also shown by other Tasmanian rainforest conifers (Read 1988).

5.1.2 Nutrient distribution following disturbance

After disturbance there may be a greater potential for erosion and loss of nutrients. There has been a number of studies which have looked at nutrient concentrations in foliage of subtropical rainforest in northern New South Wales (Webb *et al.* 1969; Turner and Kelly 1981; Lambert *et al.* 1983; Lambert and Turner 1986). The reported concentrations and distribution of mineral nutrients within plants vary between species. However, there are interesting differences between climax rainforest species and early successional (pioneer) species which have important implications for rainforest management (Lambert and Turner 1986).

Lambert and Turner (1986) studied an area in Wiangaree State Forest which had also been investigated in studies of recovery after logging (Burgess *et al.* 1975; Shugart *et al.* 1980; Horne and Gwalter 1982) and in earlier nutrient studies (Lambert *et al.* 1983). The shrub stratum developed following logging had the highest foliar nitrogen and phosphorus concentrations. The remaining canopy trees (the logging treatment involved 50 per cent canopy retention) had higher nitrogen and phosphorus levels than the canopy of undisturbed stands. This suggests that the disturbance, and changed microclimate, enhance mineralization and increase nutrient availability.

The colonizing shrub layer had generally high foliage nutrient concentrations. Lambert and Turner (1986) suggested that the high accumulation of nutrients is a mechanism for nutrient conservation after disturbance, a mechanism which would be particularly significant if the disturbance led to increased availability of leachable nutrients.

Little is known about nutrient uptake in pioneer species. Swaine and Whitmore (1988) suggested that pioneers might, in general, be shallow rooted. However, Hopkins (1990) reported that some pioneers in Australia (*Polyscias elegans*, *Alphitonia petriei*) have deep tap roots which may permit them to obtain nutrients and water not available to shallow-rooted species. *Acacia* spp. which may be important early colonizers, are nitrogen-fixers, the importance of their contribution to soil fertility has not been evaluated.

If Lambert and Turner's (1986) findings can be extrapolated to other rainforest types, they would suggest that the resilience of rainforest in the face of disturbance may depend on the rapid establishment of nutrient accumulating pioneer species. Where rainforest is cleared for agricultural or other purposes, the planted species may not be able to conserve nutrients and the prospects of being able, at some future date, to re-establish rainforest may be reduced.

5.2 ASPECTS OF THE LIFE CYCLE OF RAINFOREST PLANTS

The model of the regeneration cycle discussed above postulates that the process of filling gaps is determined by the seed and seedling properties of individual species. If the future course of vegetation development in any particular gap is to be predicted, then information will be required on the availability of seed, germination requirements, growth responses of seedlings, and factors affecting seedling mortality.

5.2.1 Flowering, pollination, and fruit-set

The availability of seed at a particular site at a particular time will be determined by the complex interaction of many factors. However, a fundamental factor is successful fruiting of the species concerned and this, itself, is the end point of a series of events commencing with the initiation of flower primordia and including pollination.

There have been few studies of flowering and seed-set in Australian rainforest trees, although Baur (1989) and Floyd (1990) have provided important reviews for the south-east part of the continent.

A few species flower and fruit annually but irregularly through the year. Most species, however, have definable flowering and fruiting seasons, although their length varies considerably between species. In northern New South Wales there is a peak of flowering in late spring/early summer and of fruiting in autumn and winter (Baur 1989; Floyd 1990), but with some flowering and fruiting throughout the year.

Although flowering and fruiting seasons are predictable, seed-set may vary considerably from year to year. Some pioneer species produce regular annual crops (for example *Omalanthus populifolius*, *Trema aspera*, *Polyscias* spp.—Floyd 1990), while other species, including *Elaeocarpus reticulatus*, *Glochidion ferdinandi*, and *Neolitsea dealbata* fruit annually but alternate heavy and light crops (Floyd 1990). For climax species heavy fruiting occurs at longer intervals, every three to five years in the case of *Argyrodendron trifoliolatum* and *Doryphora sassafras*, every six years or more in *Dysoxylum fraseranum* and *Sloanea woollsii* (Floyd 1990). A 'good' seed year may result in an abundance of seed on the forest floor, with sufficient escaping predation to permit regeneration (Floyd 1990). The temporal variation in availability of the recalcitrant seeds of many climax species suggests long-term instability in canopy floristic composition at any particular site.

Although the major canopy species in cool-temperate rainforest have pioneer properties, there is substantial variation between years in seed-fall (Hickey *et al.* 1983).

A prerequisite to fruit set is pollination. The pattern of pollination will also have important consequences for the genetic structure and evolution of species. Fedorov (1966) argued that self-pollination was the predominant mode in rainforest, and that successive generations of inbreeding resulted in genetic drift and non-adaptive segregation of small populations, leading to the widespread occurrence of closely related sympatric species, which can be observed in many stands. The hypothesis that tropical rainforest trees would be chiefly inbreeding was first suggested by Corner (1954).

A diametrically opposed hypothesis was advanced by P. S. Ashton (1969). He proposed that outbreeding predominated, accompanied by strong natural selection and narrow niche differentiation.

Assessment of selection pressures operating on long-lived trees and measurement of niche breadth are difficult tasks, but determination of breeding systems would allow differentiation between the two hypotheses.

Studies of breeding systems in Malayan rainforest are reviewed by Whitmore (1984), while those in Costa Rica are discussed by Bawa (1979). A range of breeding systems has been found in both areas, with many species being outbreeding. Whitmore (1984) concluded that breeding systems were diverse and complex, but in themselves did not provide an adequate explanation of rainforest species diversity.

The breeding systems of many Australian rainforest trees remain to be described, but the spectrum from apomixis to obligate outbreeding is represented. (It is an interesting sideline of botanical history that apomixis was first described from an Australian rainforest species, *Alchornea ilicifolia* (Euphorbiaceae), by Smith in

1841, on the basis of observations of specimens in cultivation at Kew (see Briggs and Walters 1984). Unlike the situation in apomictic northern hemisphere herbs with a myriad of described microspecies, *A. ilicifolia* has not attracted taxonomic splitting.) Among the obligate outbreeders, there are a number of dioecious species (see House 1989).

For outbreeding species, the question arises as to what are the methods of pollination. Although there is a long history of studies of pollination biology in Australia (Armstrong 1979), there are few investigations of rainforest species (Hopper 1980; Crome and Irvine 1986; Irvine and Armstrong 1988; House 1989; Sands and House 1990).

A brief overview of pollination in rainforests is provided by Whitmore (1984). One of the problems with pollination studies is that recording the frequency and length of time of visits to flowers by potential pollinators does not provide an indication of the relative importance of different vectors in accomplishing successful pollination. This point was well made by Crome and Irvine's (1986) study on the pollination of *Syzygium cormiflorum*. It was concluded that bats were the most effective pollinators (accounting for 55.4 per cent of successful pollinations), despite the much greater frequency of visits by birds and insects. The floral biology of *S. cormiflorum* and its production of large volumes of weak nectar conform to the chiropterophilous syndrome reported from elsewhere in the tropics, but nevertheless, birds and insects also effected successful pollination, and self-pollination was also possible (Crome and Irvine 1986).

Bird pollination is widespread in the Australian flora, particularly among sclerophyll elements (Ford *et al*. 1979). Over a hundred species of birds have been recorded as flower visitors. The honey eaters (Meliphagidae) are the largest family of passerine birds in Australia with over 70 species and probably all of them are involved in pollination to varying degrees. However, only one species, Lewin's honey eater (*Meliphaga lewinii*), is widespread in rainforest (from northern Queensland to Victoria), although there are a number of species in rainforest in north-east Queensland and Cape York (including *Xanthotis flaviventer*, *X. macleayana*, *Meliphaga notata*, *M. gracilis*, and *Lichenostomus frenatus*). The abundance of honey eaters in heathland and sclerophyll woodland and their relative scarcity in many rainforests may indicate that bird pollination is of lesser importance in rainforest, but this remains to be determined.

For most rainforest trees, flowering is discontinuous in both space and time. In view of this, it might be suggested that pollination mechanisms are likely to be generalist, and this is borne out by the few studies conducted to-date. However, there are exceptions, and one of considerable interest is provided by *Eupomatia laurina*, which has many extremely primitive features. *E. laurina* has a very wide geographical range from Victoria to northern Queensland, but throughout this range appears to be pollinated by a single vector, the weevil *Elleschodes hamiltoni* (Hamilton 1897; Irvine and Armstrong 1988). Pollination by beetles—cantharophily—is widespread in the Australian flora, and is a feature of many rainforest species, either as part of flexible generalist systems or in more specialist associations as in *Eupomatia* (Irvine and Armstrong 1988; House 1989).

Irvine and Armstrong (1988) suggested that the specialized pollination of *Eupomatia* renders the species vulnerable to habitat disturbance. Isolated patches of rainforest may not support the pollinator, or may be vulnerable to pesticide drift from adjacent agricultural land. Whitmore (1984) also argued that there was a strong case for conservation of inviolate patches of rainforest to protect interactions between plants and pollinators.

Both Hopper (1980) and House (1989) have reported pollen collections from rainforest trees by the introduced honey-bee, *Apis mellifera*, but there is doubt that it is an effective pollinator (House 1989). Honey-bees, both from hives and feral colonies, are widespread in rainforest (and in Tasmanian rainforest production of leatherwood (*Eucryphia*) honey is of considerable economic importance, see Chapter 9). For rainforest, nothing is known about interactions between honey-bees and native pollinators or about effects of honey-bees on pollination rates and seed-set in the flora, although in sclerophyll communities there are indications of competitive displacement of native species, nectar robbing and reduced seed set (see Pyke 1990).

5.2.2 Seed dispersal

Seedling establishment in gaps requires a source of viable propagules, which can take one of two forms, immediate dispersal from undisturbed areas, or a seed bank in the soil which reflects the accumulated result of dispersal.

There are two principal modes of dispersal for pioneer rainforest species: wind and animals. Whitmore (1984) reviewed evidence for dispersal in South-east Asia, and concluded that animal vectors were more effective in dispersal than wind. Pioneer tree species in Australia are mainly animal dispersed (Hopkins 1990).

Examination of the diaspores of Australian rainforest species shows that in many rainforest types there is a predominance of fleshy fruit suggesting dispersal by animals (Webb and Tracey 1981*a*; Russell-Smith and Dunlop 1987; Russell-Smith 1988; Hopkins *et al.* 1990*a*; Crome 1990; Floyd 1990), although there have been few detailed studies of the effectiveness of dispersal. Although many species have fleshy fruit, Baur (1989) has stressed that in the upper canopy of complex notophyll vine forest there is a high proportion of trees with wind-dispersed seeds.

Of the rainforest avifauna, a significant proportion is composed of either obligate or opportunistic frugivores. The most important species for seed dispersal are probably pigeons, for which Australia is a major centre of diversity (Frith 1982). The frugivorous rainforest species are nomadic, moving both latitudinally and altitudinally and taking advantage of locally available fruit (Floyd 1990). Seed is excreted; Frith (1982) provides evidence of the range of fruit taken by different species but there are few data on the viability of excreted seed. In northern Australia, the Torres Strait pigeon (*Ducula spilorrhoa*) is regarded as probably the major dispersal vector for rainforest species (Russell-Smith and Dunlop 1987). The pigeons migrate in flocks between northern Australia and New Guinea, and in north Queensland travel daily between the mainland and offshore islands where they roost and nest (Frith 1982). Identification of seeds excreted at nesting sites revealed a diversity of rainforest species (Frith 1982), similar to that on which it has been observed foraging on the mainland (Crome 1975).

The role of fruit bats as dispersal agents has been poorly studied although the perception that these species are a major threat to the livelihood of fruit farmers has frequently led to calls for reduction in their numbers (Ratcliffe 1938). Recent studies of *Pteropus poliocephalus* in northern New South Wales (Eby, pers. comm.) have demonstrated that this species can disperse a number of rainforest species, possibly over very long distances (dispersal ability assessed by the successful germination of seeds collected from faeces). Concerns had been expressed about the possibility of *P. poliocephalus* introducing invasive weeds into rainforest, but preliminary results suggest that this is not a serious problem. Radio-tracking of individuals suggests that there are not permanent discrete local populations but rather a single diffuse east coast population, which at various times is concentrated on a small number of sites. While it will be difficult to quantify the relative significance of *Pteropus* compared with other dispersal agents, these data indicate that a policy of managing fruit bats as pests could have serious implications for the long-term maintenance of both isolated rainforest remnants and larger stands.

There remains a significant portion of the canopy species in complex mesophyll and notophyll vine forests with large-seeded large fruit (Webb and Tracey 1981*a*; Webb *et al.* 1986), which are unlikely candidates for dispersal by any present indigene of the canopy. These fruit belong to the gravity-dispersal category of Webb and Tracey (1981a), and fall to the forest floor close to the parent. Stocker and Irvine (1983) have shown that, in north-east Queensland, large fruit on the forest floor may be eaten by cassowaries. Seeds of more than 70 rainforest trees were recovered from cassowary droppings, and the majority showed some germination capacity. Stocker and Irvine (1983) and Jones and Crome (1990) have suggested that cassowaries are the main dispersal agents for many large-fruited trees. The decline in cassowary numbers as a result of habitat disturbance and fragmentation (Crome and Moore 1990) may have important implications for future patterns of rainforest regeneration in north Queensland.

At the present time, cassowaries have a restricted distribution in Australia, being limited to north-east Queensland, but large-fruited species are more widespread (Fig. 5.5). Does the distribution of these trees reflect an earlier and wider distribution of cassowaries or other large birds? One of the species dispersed by cassowaries in north-east Queensland is *Elaeocarpus angustifolius*. In northern New South Wales, *E. angustifolius* is characteristic of creek sides and is rarely found in forest away from water. Baur (1989) has suggested that this restricted distribution is a consequence of the absence of dispersal by fauna; the only available dispersal mechanisms are gravity and water. Another characteristic gallery species in northern New South Wales is *Castanospermum australe*, which is also large-seeded— the composition of gallery communities (and the present absence of gallery species throughout rainforest) may be the function of the changing availability of dispersal vectors.

Another large flightless bird, the emu, has a distribution today centred on drier vegetation types, but at some localities may be found at the edge of rainforest and perhaps historically played a role in the distribution of rainforest propagules. At Iluka, on the north coast of New South Wales, there is an isolated relic population of emus and Floyd (1977, 1990) reported stimulation of the germination of the seeds of *Syzygium leuhmanii*, a

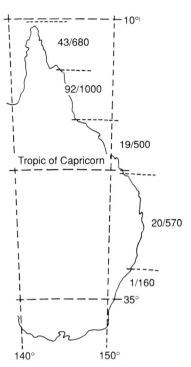

Fig. 5.5 Present distribution of larger-fruited tree species in the woody flora of rainforests in different regions of eastern Australia (i.e. in the northernmost region 43 out of 680 species have large fruit). Cassowaries today are found only in the two northern regions (from Stocker and Irvine 1983).

major species in the littoral rainforest, after passage through an emu.

This speculation on the past role of cassowaries and other large birds as dispersal agents draws attention to the fact that trees, with their long generation times, may survive long after the animal species they were once associated with have disappeared. The late Tertiary rainforest vertebrate fauna was very different from that of today (see Chapter 7), but at least some rainforest tree species have survived, apparently unchanged, from that time to the present. Some tree species may provide a link to long extinct fauna. One of the botanical specialities of the north-east Queensland rainforest is *Idiospermum australiensis*, which is regarded as an extremely primitive angiosperm. This produces very large, and, to mammals, highly toxic seeds, which today simply drop to the ground and have no known dispersal agent. Crome (1990) has speculated that 'if they were once swallowed whole (biting and crunching them up would kill them) by an animal now extinct, and we were to scale them up on the basis of what horses can swallow, the most feasible extinct disperser I can conjure up is a small dinosaur, about the size of a five-ton truck'.

Russell-Smith and Dunlop (1987) and Russell-Smith (1988) indicated that feral pigs may distribute the seeds of large-fruited rainforest species in the Northern Territory. Given the difficulty of controlling feral animals, pigs may become important dispersal agents in many Australian rainforests in the future.

Native rodents may disperse seeds of smaller fruit falling on the forest floor (Stocker and Irvine 1983), but little is known about dispersal distance. Rodents are probably the major consumers of fallen seed (Willson 1988). On Lord Howe Island the introduced *Rattus rattus* is an important seed predator: unless controlled it may limit the potential regeneration of the palm *Howea fosteriana* (Pickard 1982).

One of the striking features of the Australian sclerophyll flora is the very high proportion of ant-dispersed species. Little is known about the importance of ants in dispersing rainforest seeds, although Howard (1974) has reported that seed of *Tasmannia*, one of the components of the flora regarded as being primitive, is ant-dispersed, and Floyd (1990) suggested that *Omalanthus populifolius* and *Codonocarpus attenuatus* are also myrmecochores.

5.2.3 The seed bank

The presence of viable seed banks in the soil plays an important role in rainforest regeneration, both in secondary succession following disturbance and in spotwise regeneration in smaller gaps.

There have been a number of studies of the seed bank in Australian rainforest soils which confirm findings from elsewhere in the world.

Hopkins and Graham (1983) investigated the composition of the soil seed bank in four lowland rainforest sites in north-east Queensland, by promoting germination in soil samples. The majority of germinations were of pioneer or early secondary species (groups A and B species), while climax rainforest species (group D species) constituted less than 1 per cent of the seedlings. Both the number of species and individuals germinating declined with depth in the soil, with relatively few germinations from depths greater than 50 mm. The numbers of individuals germinating (taken as an index of the size of the soil seed bank) were comparable with those reported from Sabah by Liew (1973), but were lower than those reported from temperate agricultural soils.

There was a substantial similarity in the species representation in the seed bank at the different sites, although when a number of unlogged sites was compared, they could be distinguished on the basis of seed-bank composition, which reflected the present forest type (Graham and Hopkins 1990). Nevertheless, Hopkins and Graham (1983) pointed out that there is greater variation in the regrowth in gaps and clearings than might be predicted from the composition of the seedbank. The heterogeneity of the composition of regrowth within and between sites is presumably a reflection of environmental patterns and variation between species in the requirements for germination and establishment. Surprisingly, a number of common group B species were apparently absent from the seed bank. The sampling strategy was inadequate to determine whether this was a real absence, or a consequence of patchy distribution or particular germination requirements (Hopkins and Graham 1983). An interesting feature of the results was the presence of large numbers of seeds of figs which normally establish as epiphytes (a result confirmed by Hopkins *et al.* (1990*a*) and Graham and Hopkins (1990)).

The seed bank from unlogged sites within a large block of unlogged forest showed an absence of weed species and a smaller suite of group B pioneers than earlier studies (Graham and Hopkins 1990). Graham and Hopkins (1990) suggested that the 'missing' pioneer species would, in virgin forest, be restricted to permanent natural gaps such as stream sides and steep unstable slopes. Logging and clearance may provide an opportunity for such species, and other initially rare components of regeneration phases, to increase in abundance and spread to become more generalist pioneers. Continued broad-scale disturbance may therefore permanently change the composition of the rainforest (Graham and Hopkins 1990).

Hopkins and Graham (1984b) investigated the seed bank in eight north Queensland sites, all believed to have been under rainforest prior to European settlement, but now, with the exception of one undisturbed site, supporting vegetation which reflects different disturbance regimes. Even in the undisturbed site, the seed bank was dominated by pioneer species (species groups A and B), although the number of seeds was much smaller than under later secondary rainforest. Under sites experiencing a high frequency of fire (*Imperata* grasslands and eucalypt forest) the seed bank was very small. Hopkins and Graham (1984b) carried out preliminary experiments on the effect of heating on the viability of the seed bank. Half of the species and individuals were killed by exposure to 60 °C heat for one hour, and all of the seeds were destroyed at 100 °C. This suggests that the composition of both the seed bank and of regeneration could be affected by temperatures experienced in fires or by exposure of the soil surface to full sunlight.

Hopkins and Graham (1987b) extended their study by investigating the behaviour of seeds buried in rainforest soil and recovered at various intervals up to two years.

A high proportion of the seeds of pioneer group A and B species retained viability after two years burial, and germinated rapidly when exposed to greenhouse conditions after exhumation. In some of the species, dormancy must have been enforced by burial, as fresh seed was found to give high germination percentages. In other cases, there was an initial period of innate dormancy due to the presence of a hard impermeable seed coat. Earlier studies (Hopkins and Graham 1983, 1984b) had shown that these pioneer species were major components of the soil seed bank. The retention of viability suggests that the seed bank may accumulate over time, so that the composition of subsequent regeneration integrates fruiting success over a period, rather than merely reflecting the immediate seed rain. The presence of a large seed bank of groups A and B species under 'primary' rainforest does not, however, preclude a significant role for the contemporary seed rain in recovery from disturbance.

The climax species (species groups C and D) remained viable for less than one year when buried, and in most cases did not survive six weeks burial (i.e. would be classified as recalcitrant). Some species were obligate immediate germinators regardless of whether they were buried or not, although other species, with some form of tough endocarp protecting the seed, retained viability for long periods before germination or decay. Fresh samples of these species showed a pattern of extended slow germination; there was no evidence for induced or enforced dormancy.

For many pioneer species, germination in the field appears to be dependent on disturbance of the habitat. An interesting feature of the results obtained by Hopkins and Graham (1984a) was that an appreciable proportion of the soil seed bank could germinate, following disturbance of the soil, under dense shade (Fig. 5.6). The triggers for germination following disturbance are frequently assumed to be related to canopy opening—changes in soil temperatures and irradiance. For at least some species, other factors must also be capable of initiating germination. (Webb and Tracey (1967) report A. Floyd's observations that localized disturbance of the litter layer by brush turkeys may permit germination but not establishment of *Araucaria cunninghamii*.)

The rapid germination of the seed bank after

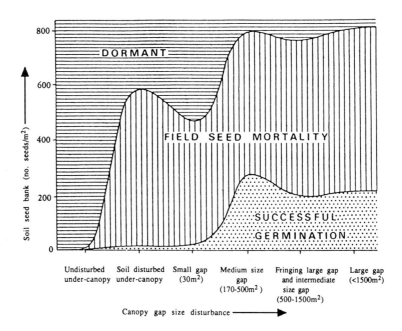

Fig. 5.6 The effect of canopy gap size on germination and establishment of seeds from the soil seed-bank in north-east Queensland rainforests. Note that disturbance in small gaps under the canopy results in substantial field seed mortality (this includes seeds which germinated but for which the seedlings did not survive long enough to be recorded as successfully germinating, as well as seeds suffering loss of viability or being lost by other means from the soil seed bank); successful germination requires a much larger disturbance (redrawn from Hopkins and Graham 1984a).

release from enforced dormancy suggests that, after major disturbance, the seed bank would be completely depleted. The large, diverse seed bank recorded from secondary forest developing after a single disturbance (Hopkins and Graham 1984b) reflects the high rate of accession from the local seed rain once the recolonizing species reach reproductive maturity. The depauperate seed banks in sites subject to repeated disturbance (Hopkins and Graham 1984b) indicate the importance of local seed sources, rather than acquisition by long-distance dispersal, in replenishing the stored seed.

The effects of disturbance on the composition of the seed bank in complex notophyll vine forest (subtropical rainforest) were examined by Abdulhadi and Lamb (1988). The initial effect was a reduction in the number of species in the seed bank. Maximum species richness was shown in a site 50 years post-disturbance, after which diversity declined. The number of seeds declined immediately after disturbance, but highest densities ($c.4\times$ those in undisturbed forest) occurred 20 years after disturbance. An interesting feature, even in the undisturbed forest, was the high incidence of exotic species, which reflects the dispersal powers of many weed species. In Abdulhadi and Lamb's study site, the Lamington National Park in southern Queensland, exotic species (mainly herbs) are likely to be a feature of the early stages of any recovery from disturbance. Abdulhadi and Lamb (1988) suggested that a single episode of disturbance will be reflected in the soil seed bank for at least 60 years.

The absence of 'primary' species from the seed bank suggests that, unless there is a vegetative regeneration, group D species must be dispersed to disturbed sites. In the case of small disturbed areas within rainforest, the seed rain from surrounding undamaged forest may ensure rapid establishment of these species, but if large areas are disturbed, the spread of 'primary' species may be slow. Olsen and Lamb (1988) suggested that any surviving canopy trees which provide perches for birds may be important foci for regeneration. Where the disturbed area becomes remote from undisturbed forest, dispersal of propagules of group D species to the site may become unlikely.

The seeds of group D species are generally large, with substantial food reserves, which support the seedling in its early development. Large seeds have a large moisture requirement for germination, are frequently poorly dispersed, are produced in relatively small numbers, and are very susceptible to predation (Harper et al. 1970). It is unlikely that large seeds are buried in large numbers (Hopkins and Graham 1983). Although some large-seeded species are obligately immediate germinators (recalcitrant species), others may survive for extended periods on the soil surface if not subject to predation (Hopkins and Graham 1983). However, in north Queensland, native rodents effectively dispose of most seeds on the surface (Hopkins and Graham 1983, 1987b). Hopkins and Graham (1987b) noted that some species can sustain massive damage from rodent or insect predation and still germinate. The burial studies of Hopkins and Graham (1987b) showed that while some large seeds germinated while buried, the seedlings could survive prolonged periods of total darkness and grew successfully following exhumation. In the case of *Castanospermum australe*, some seedlings survived burial for two years. The physiological basis of this behaviour remains to be elucidated. Hopkins and Graham (1987b) suggested that group D species store seedlings—in contrast to seeds—in the soil. The size of the seedling bank is likely to be very small compared with that of seeds, given the low probability of the large seeds being buried. While such seedlings could be significant in diffuse and spotwise regeneration, it is unlikely that they are of importance following major disturbance.

Hopkins et al. (1990a) have investigated the seed bank in a seasonally dry semi-evergreen vine thicket in northern Queensland. The number of seeds was much greater than in either complex mesophyll or complex notophyll vine forest, but the majority were weedy herbs rather than rainforest species. Three explanations were suggested for these differences. With increased deciduousness the rainforest would be subjected to invasion by seeds of rapidly maturing herbs. In addition, the study site was exposed to continued disturbance by cattle and feral pigs, which might disperse seeds into the forest and create microsites suitable for the germination and establishment of herbs. The deciduous nature of the canopy means that annually the forest floor would be exposed to light and temperature conditions similar to those experienced in disturbed, closed evergreen forest. Hopkins et al. (1990a) suggested that the consequences of this would be that the long-lived seeds of rainforest species would be stimulated to germinate and so would not accumulate. If seedlings from this germination survive, then regeneration in semi-evergreen vine thickets may be less dependent on disturbance than evergreen rainforest, and regeneration may be continuous rather than episodic. These possibilities remain to be investigated.

5.2.4 Germination

The specific triggers for germination of most rainforest species are unknown. Although under field conditions many species respond to disturbance, many factors which could promote germination will change in disturbed sites (irradiance, temperature, and nutrient availability). Baur (1989) and Floyd (1976, 1990) provided reviews on germination studies on New South Wales rainforest species. Floyd (1990) suggested that one generalization which can be made is that for fleshy-fruited species the pulp must be removed from around seeds before germination will occur.

5.2.5 Seedling establishment and mortality

Seedling densities of rainforest species may be high (for example up to 4×10^6 ha^{-1} for *Toona australis* on the Atherton Tableland, Department of Forestry, Queensland 1981). There is frequently heavy mortality among seedlings (Hopkins and Graham 1984a), but high densities can sometimes persist into the sapling stage (Webb et al. 1972). Nevertheless, complex vine forests (mesophyll and notophyll) are characterized by high species diversity and, generally, little gregariousness among canopy species. At some stage, therefore, there must be considerable mortality, although when this occurs in different situations is still poorly documented, as are the mechanisms of mortality (predation, disease and competition). The maintenance of diversity in species-rich rainforest has attracted much speculation. In order to explain why some species in any stand of rainforest are 'rare' and others more common, two contrasting schools of thought have developed (Connell *et al.* 1984). One assumption is that each species has an equilibrium population size to which it returns after perturbation. Thus common species remain common, while rare species remain rare: rarity and commonness are features of species. The opposing hypothesis suggests that, at any one site, currently rarer species are favoured over more common ones. As a species becomes commoner its rates of recruitment, growth, or survival will be reduced, with the opposite trend in species becoming rarer. This hypothesis implies that rarity or commonness are not intrinsic properties of species but are descriptions that apply at a particular site at a particular time. Processes which would permit these changes in relative abundance of species have been called 'compensatory mechanisms'.

Connell *et al.* (1984) conducted long-term studies on recruitment in rainforests in both northern and southern Queensland. The data permitted the testing of two specific hypotheses: firstly that at a scale of 1–2 hectares, commoner species have lower rates of recruitment and growth and higher mortality than do rarer ones, and secondly that the close proximity of other individuals is more deleterious (i.e. causes slower growth or higher mortality) if they are conspecific than if they are not. Because of the difficulties of sampling sufficient individuals of rare species, this second hypothesis could only be tested for the more common species.

Connell *et al.* (1984) rejected the first hypothesis for overstorey species, but the data did not permit rejection in the case of recruitment of subcanopy and understorey species. The second hypothesis could be tested for several age classes. The hypotheses was not rejected for growth and survival of nearest neighbours in several of the seedling size classes. In the case of interactions between adults and conspecific seedlings, 90 per cent of the species tested showed no increase in mortality closer to adults.

Overall Connell *et al.* (1984) demonstrated the existence of some compensation which, if sufficiently strong, could maintain species diversity but had no basis for determining its importance relative to other mechanisms which might also maintain diversity. Neither were they able to identify the detailed mechanisms responsible for compensation.

One mechanism which has been frequently invoked—but rarely proven—in a range of ecosystems, is allelopathy. Webb *et al.* (1967c) have shown that in *Grevillea robusta* an allelochemical interaction between mature trees and seedlings prevents establishment close to established trees. The active principle is apparently released from living roots of mature trees or by micro-organisms associated with roots, and causes seedling mortality. Such a mechanism would explain the non-gregarious distribution of canopy species, but evidence for it occurring in other species is sparse. Webb *et al.* (1967c) regarded the *Grevillea* inhibitor as species-specific, but Bevege (1968) showed that in experimental conditions a root exudate from *Flindersia australis* inhibited growth of *Araucaria cunninghamii* seedlings. Growth of *A. cunninghamii* was also inhibited by an exudate from mature conspecifics.

5.2.6 Epiphytic tree establishment

A feature of complex mesophyll and notophyll vine forests is the presence of strangling figs which, at maturity, form massive emergents. These figs germinate on branches of the host and eventually the anastomosing network of roots encases the host trunk. The seeds of these species are small and are effectively dispersed by birds and fruit bats. Hopkins and Graham (1983) found that strangling fig seeds were frequent in the soil seed-bank in north Queensland rainforest, from which they inferred that there were similarities between the germination requirements of pioneer species and strangling figs. Although occurring in primary forest, strangling figs can be regarded as colonizers.

Epiphytic establishment also occurs in a number of species which characteristically germinate on the trunks of tree ferns. This is commonly seen in *Quintinia sieberi* in *Nothofagus moorei* stands, and is the main method of seedling establishment of *Eucryphia moorei* (Helman 1987; Floyd 1990), and *Atherosperma moschatum*. Seeds germinating on the ground tend to suffer high mortality from grazing, chiefly by wallabies. Olsen and Lamb (1988) reported that in cool subtropical (complex notophyll) rainforest *Caldcluvia paniculosa* and *Geissois benthamii* seedlings are often found on tree fern trunks. Following extensive storm damage, these species established on the organic mounds created by epiphytes fallen from the canopy.

5.2.7 Multi-stemmed trees in rainforest

The ability of many Australian rainforest species to coppice following damage is well documented. Johnston and Lacey (1983) have reported that in some species multistemmed stools can develop by basal sprouting from undamaged stems. This phenomenon is well developed in microphyll and simple notophyll rainforest and occurs in *Eucryphia moorei* (well illustrated in Helman 1987), *E. lucida*, *Nothofagus cunninghamii*, *N. moorei*, *Atherosperma moschatum*, *Doryphora sassafras*, *Ceratopetalum apetalum*, and *Schizomeria ovata* (Johnston and Lacey 1983).

The significance of the development of multiple-stems by uninjured trees is not obvious (Johnston and Lacey 1983), but would enable species to maintain themselves over long periods without seedling regeneration.

Johnston and Lacey (1983) suggest that the development of the multi-stemmed form requires a structure similar to the lignotubers of *Eucalyptus* and other sclerophyll genera. Lignotubers in sclerophyll species are considered to represent adaptations to low nutrient availability, drought, or fire and to have evolved relatively recently (Johnston and Lacey 1983). It is not certain that the structures in temperate rainforest species are anatomically homologous with lignotubers in *Eucalyptus*, but if they are, then Johnston and Lacey (1983) suggested either an ancient origin of lignotubers not primarily connected to either fire or drought, or separate episodes of convergent evolution.

5.2.8 The longevity of rainforest trees

A full appreciation of the significance of the various reproductive strategies shown by rainforest plants requires a knowledge of the longevity of different species. In the case of multi-stemmed trees, several generations of the same genetic individual may be produced over a very long period of time, although there are few data which would enable ages to be estimated.

For maiden trees, there is equally a paucity of information. Strikingly large individuals have excited much speculation and claims of great antiquity, but rarely on the basis of data.

Estimates of age may be obtained by radiocarbon dating, from extrapolation of short-term incremental growth studies, or from studies of tree rings. Ogden (1978, 1981) has investigated the potential for dendrochronological studies of Australian trees, including rainforest species. For rainforest species, ring production may not have a regular relationship with time, thus limiting the value of tree-ring studies. Ogden (1981) indicates that in rainforest gymnosperms, growth rings do seem to be approximately annual. This is the case in temperate rainforests, and also for *Callitris macleayana* (Ash 1983*b*), but for *Agathis robusta* and *Araucaria cunninghamii*, variation in cambial activity within and between trees may complicate detailed examination of tree rings (Ash 1983*a*). Deciduous emergents, such as *Toona australis*, are candidates for annual ring production (Ogden 1978, 1981), although Ogden (1981) reported preliminary studies in which the ring numbers in plantation-grown *Toona* did not agree with the known age. Ogden (1981) also reported unpublished data of J. Ash, suggesting that a large number (roughly 20 per cent of genera) of rainforest trees in tropical Australia may produce annual rings.

The longest-lived rainforest species in Australia, on presently available data, is *Lagarostrobos franklinii* (Huon Pine) in Tasmania, with some individuals being radiocarbon dated to more than 2500 years old (Ogden 1978). Estimates from growth incremental studies of Queensland rainforest suggest that some tropical species may be of comparable age (Ogden 1981), but such extrapolations may be unreliable. In northern New South Wales a mature *Argyrodendron trifoliolatum* has been radiocarbon dated at more than 800 years old, while individuals of *Lophostemon confertus* have been shown to be more than 1000 years old (Turner 1984).

5.3 DISTURBANCES TO RAINFOREST

Regeneration of rainforest is a function of the disturbance regime. Disturbance factors operate at different temporal and spatial scales, and with different intensities, to create the complex mosaic of different phases of the regeneration cycle which constitutes the present-day rainforest vegetation. Various agents of disturbance are documented below (human influence is discussed in Chapter 9).

Over geological history (Chapter 6), rainforest has been subjected to major environmental changes; the survival of rainforest through these changes may well reflect the adaptations of the biota to the continuous background of local disturbance upon which continental and global changes were superimposed. Of the disturbances currently acting on Australian rainforests, most are similar in kind, if not necessarily in frequency or intensity, to those which have prevailed over much of recent geological history.

Johns (1986) has provided a review of disturbances to New Guinea rainforests, while Floyd (1990), Hopkins (1990), and Stocker and Unwin (1989) discussed disturbance in subtropical and tropical rainforests in Australia.

5.3.1 Fire

It is clear (Chapter 4) that fire is a major factor determining the position of rainforest boundaries. A contrast can be made between fire prone, non-rainforest vegetation, in which the structure and composition of communities is controlled by fire, and where fire may be required for regeneration—and rainforest which is neither fire prone nor fire requiring. On most occasions when fire burns into a rainforest stand, the change in microclimate and fuel characteristics are sufficient to prevent spread (Bowman and Wilson 1988), although repeated fires may cause a steady attrition of the boundary, and contraction of the rainforest.

Nevertheless, rainforest is not immune from major fires. After prolonged drought (which in itself would be a disturbance and could cause decline or death of some species), the capacity of rainforest stands to extinguish fire would be much reduced, rendering the vegetation vulnerable to the extensive spread of fire. Johns (1986) records several instances of major fires affecting rainforest in New Guinea during droughts. In Kalimantan, three million hectares of rainforest were burnt during a drought in 1983 (Malingreau *et al.* 1985; Leighton and Wirawan 1986; Whitmore 1988, 1989*a*); the extent of the fires largely reflecting ignition by humans and most affecting recently logged forests. In Australia, small stands of rainforest have been burnt through in drought years and stands in gullies, normally protected from fire, may be vulnerable even in normal rainfall years if the wind direction is atypical. Large areas of rainforest on ridge lines in north-east Queensland were burnt during severe drought in 1946 (Hopkins 1990). Serious fires in rainforests in south-east Queensland were reported by Ridley and Gardner (1961) and in

New South Wales rainforests by Baur (1989) and Floyd (1990). Rapid spread of fire through *Nothofagus moorei* cool temperate rainforest on Mount Boss State Forest under drought conditions in 1957 has been attributed to the flammability of dry epiphytic mosses and lichens (Baur 1989). Epiphytes are generally sparse or absent from rainforest stands with a history of repeated fire (Floyd 1990). A number of extensive fires have affected rainforests in Victoria (Chesterfield *et al.* 1990) and Tasmania (Hill 1982). Given the variability of the Australian climate, it is easy to envisage that in the past very large rainforest fires may have occurred. During the late Tertiary, the decline of rainforest (Chapter 6) reflects both the direct impacts of a changing climate on the flora and, probably more importantly, the indirect effects of climate in promoting fire. The recovery of rainforest in the aftermath of a fire may well have been prevented by the more rapid response of *Eucalyptus* spp. Earlier in geological history, even if fire occurred (Kemp 1981*b* provides evidence of pre-Quarternary fire in Australia), the recovering rainforest may not have faced competition from non-rainforest taxa.

Rainforests may be damaged by much lower-intensity fires than would cause impacts on sclerophyll forest. The shallow rooting habit and the masses of rootlets in the litter layer render many rainforest species vulnerable to damage from low-intensity surface fires (Adam 1987*b*). Even if the damage does not cause immediate death, subsequent pathogen attack may be fatal.

In relatively low-intensity fires, the susceptibility of species to damage is at least partly related to bark characteristics. In New South Wales, *Endiandra sieberi*, *Caldcluvia paniculosa*, and *Trochocarpa laurina* appear to gain a degree of protection from fire by the presence of thick, corky bark. These relatively fire-tolerant species are often prominent in the ecotone, where rainforest appears to be advancing into wet sclerophyll forest (Adam 1987*b*). In simple notophyll vine forest, *Ceratopetalum apetalum* with thin smooth bark is extremely fire sensitive, and in more fire-prone sites is often replaced by the thicker-barked *Doryphora sassafras* (Floyd 1990).

Across the north of Australia, seasonal drought occurs annually and vine forests there are vulnerable to regular burning. Aboriginal burning practices (Chapter 9) may have limited the incursion of fire into the rainforest stands, but are unlikely to have completely prevented such occurrences. Russell-Smith and Dunlop (1987) pointed out that a large proportion of the flora of the Northern Territory rainforests is capable of vegetative reproduction following fire. However, as fire frequency is increased, the ability of rainforest species to recover is much reduced and replacement by non-rainforest species (mainly grasses and species of *Acacia*) occurs.

In north Queensland, a high proportion of trees have the ability to regenerate vegetatively (Stocker 1981) and so might recover after fire, but little is known about their response to repeated fire. It is possible that the ability to recover may decline if fires become a regular occurrence.

The warm temperate rainforests in southern New South Wales and Victoria are often dominated by *Acmena smithii*, which is one of the most fire-resistant rainforest species, and in its recovery after fire shows a number of features more characteristic of eucalypts: particularly the early development of a lignotuber, which provides a reservoir of vegetative buds from which regrowth after fire can occur (Fig. 5.7). These warm temperate rainforests experience rainfalls which are relatively low in comparison to those in subtropical rainforests, and drought may be relatively frequent, and the possibility of fires correspondingly higher, than in many other rainforests. The species paucity of southern warm temperate rainforests may, in part, reflect climatic (temperature) limitations of species' distribution, but may also indicate the influence of fire.

Chesterfield *et al.* (1990) have reported on the recovery of a warm temperate rainforest stand from fire. The 80 ha stand, at Jones Creek in East Gippsland in Victoria was burnt by a hot canopy

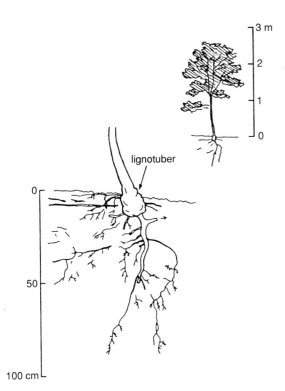

Fig. 5.7 Development of a lignotuber in seedling *Acmena smithii* (from Ashton and Frankenberg 1976).

fire on 8 February 1983. The steep slopes and gullies surrounding the site would normally confer protection from fire damage but the conditions were extreme. There had been prolonged drought and the climatic conditions were conducive to fierce fires across the region (46 °C, relative humidity 17 per cent, winds gusting to 60 knots—Chesterfield *et al.* 1990).

Prior to the fire, the major canopy species were *Acmena smithii* and *Acacia melanoxylon*, with some *Pittosporum undulatum* and a shrub layer of *Eupomatia laurina*. The understorey was almost entirely burnt and much of the canopy was removed, but in a few small areas the canopy was less severely burnt and some recovery was possible.

Despite the extensive loss of canopy, mortality of *A. smithii* was less than 10 per cent, with recovery by coppicing; in addition there was considerable seedling recruitment. Other rainforest canopy species were killed in the fire, and seed regeneration has been sparse. The majority of vascular epiphytes were lost. *Eupomatia* recovered very vigorously by basal coppicing. Vine species (*Cissus hypoglauca*, *Marsdenia rostrata*, *Sarcopetalum harveyanum*, *Smilax australis*, and *Rubus hillii*) also recovered well by vegetative growth from old rootstocks and formed locally dense partial canopies.

In addition to the recovery by rainforest species, there was a significant increase in the number of seedling eucalypts, particularly *Eucalyptus cypellocarpa* and *E. fastigata*. There was also an increased number of *Acacia melanoxylon* seedlings, although growth of a large number of these was suppressed by vines. Chesterfield *et al.* (1990) suggested that, despite the obvious invasion of sclerophyll species, their success was reduced by the abundant growth of herbs immediately after the fire. These grew from seed in the soil, and formed a matted ground cover, inhibiting germination and establishment of light-demanding species. *Calystegia marginata* was particularly important in this respect.

The recovery from the Jones Creek fire indicates that *A. smithii* has considerable resilience to a single severe fire. Protection of vegetative buds in *A. smithii* is provided by the bark. Ashton and Frankenberg (1976) calculated that it would take 30–35 years for maximum bark thickness to develop. Two fires less than 20 years apart would probably cause eradication of *A. smithii*. The loss of epiphytes and many minor rainforest species is a considerable change to the community, which might be restored over many years in the absence of fire. However, fires at more frequent intervals than 40 years would lead to further loss of rainforest species and promotion of eucalypts. Chesterfield *et al.* (1990) argued that the composition and population structure of the Jones Creek rainforest was indicative of a number of fires in European times, and suggested that changes to the regional fire regime have increased the vulnerability of the rainforest to fire damage during drought periods.

However fierce a fire, its effects are never uniform; there will always be small areas in local fire shadows which experience lower fire intensities and species differ in their flammability. After a fire, there will be a complex mosaic of patches displaying different degrees of damage, as has been illustrated for Tasmanian rainforest by Hill (1982), and in Victoria by Chesterfield et al. (1990). Islands of lightly damaged, or undamaged, vegetation will be potential sources of seeds for the re-establishment of fire-sensitive species.

Many rainforest stands are at risk from the burning practices of European Australians. In northern Queensland, fires escaping from the burning off of sugar-cane have entered rainforest (Webb 1958; Unwin 1989). Most sclerophyll forests are subject to control or hazard-reduction burning, in which extensive areas are burnt, with low-intensity fires, in order to reduce fuel loads with the intention of reducing fire hazard during the wildfire season. These hazard-reduction burns, often carried out on a very broad scale with ignition achieved by aerial application of incendiaries, can disrupt long-established fire regimes and hence the relationships of sclerophyll and rainforest communities in the vegetation mosaic. The weather conditions under which control-burns are conducted are very different from those prevailing in the wildfire season, and winds may blow from different directions. Rainforest stands topographically protected from wildfires driven by wind may not be so sheltered from a control-burn under wind from a different direction. Developing fire management strategies for rainforest/sclerophyll forest mosaics is one of the major challenges facing land management agencies responsible for rainforest conservation.

Prior to the arrival of the Aborigines, the most likely source of ignition would have been lightning strikes. The present incidence of lightning strikes in rainforest has not been quantified, but most do not result in fires, although lightning does create localized gaps in the canopy (in the order of 400 m^2—Hopkins 1990). Johns (1986) suggests that in New Guinea a single strike can result in the death of a group of trees, possibly because they are interconnected via root grafts.

5.3.2 Volcanoes

Volcanic action has major impacts on rainforest in New Guinea (Johns 1986), the most severe effects being localized to the vicinity of the eruption, but ash showers and tsunamis may extend the influence far and wide. One of the consequences of volcanic eruptions could be extensive forest fires.

Although no volcanic activity has occurred in Australia since European settlement, many rainforest stands are associated with basaltic soils and, in the past, major volcanic eruptions have occurred within rainforested landscapes. The most recent activity within rainforest was on the Atherton Tableland within the last million years. The large extent of past volcanic activity in eastern Australia suggests that the ability to withstand major disturbance will have been selected for within the rainforest biota.

5.3.3 Cyclones and storms

Tropical cyclones are an important feature of the climate of northern Australia. Cyclones form over warm water (>27°C), which is typical of the Coral and Timor Seas (Gentilli 1986). The main cyclone season is late summer/early autumn (January–April). Cyclones are associated with intense rainfall as well as strong winds. After crossing the coast, cyclones often become rain-bearing depressions which produce heavy rainfalls over inland Australia, well south of the tropics.

Webb (1959) described the effects of cyclones on rainforest in north-east Queensland, with particular reference to the damage caused by Cyclone Agnes in March 1956, and emphasized the importance of topography in determining their local impacts. In relatively sheltered sites, where the major damage was defoliation rather than windthrow, the response to the increased light in the understorey was a dense growth of the lawyer vine *Calamus australis* about 3 m tall. On exposed spurs, the rainforest frequently has *Acacia aulacocarpa* emergents. Many *Acacia* spp. have germination stimulated by fire, but Webb (1958) suggested that there was no evidence that fire had occurred sufficiently frequently to explain the distribution of *A. aulococarpa*, but that it may have taken advantage of canopy gaps created by cyclones.

In sites subject to extreme damage in coastal north-east Queensland, the result is a dense tangle of vines which extends upwards over the uneven canopy, further inland this response is less common. Baur (1965) records similar dense vine growth at some sites subject to heavy logging damage. Webb (1958) pointed out that although vines were a feature of post-fire recovery, the vine species involved after cyclones and fires were different, *Calamus* being fire-sensitive.

If there are periods of hot, dry, windy weather following a cyclone, then the risk of fire is high because of the large amount of potentially flammable litter. In north Queensland, at present, one of the most likely ignition sources is fire escaping from sugar farms. This occurred following Cyclone Agnes in 1956 (Webb 1958) and Cyclone Winifred in 1986. Unwin *et al.* (1988*a*) suggested that the heavy accumulation of litter may also impair seedling regeneration.

Webb (1958) noted the flowering of *Archontophoenix alexandrae* immediately after Cyclone Agnes and suggested that flowering might have been triggered by cyclones. Hopkins and Graham (1987*a*) reported a large number of species flowering simultaneously following Cyclone Winifred. They suggested that defoliation in the cyclone, followed by an unusually hot, dry spell, exposed the understorey to increases in temperature, insolation, and water stress, changes which may have promoted the synchroneity of flowering in a larger number of species than would normally flower simultaneously. It is possible that long-term changes in the composition of forests affected by Cyclone Winifred may result from the coincidence of seed availability and the presence of open regeneration sites.

Cyclones are normally accompanied by very heavy rain (Cyclone Agnes was unusual in being dry—Webb 1958). The saturation of the soil may render trees more liable to windthrow than they would be under dry conditions. Limbs with heavy epiphyte growth may become sodden, and the combination of extra weight and strong wind may result in damage.

Webb (1958) and Unwin *et al.* (1988*a*) suggested that in north Queensland few areas of rainforest would escape some major damage for more than 40 years, so that cyclones are the major widespread and intensive disturbance agent. The impact of Cyclone Tracey on rainforest stands in the Darwin area was described by Stocker (1976). Damage was generally severe, with a high proportion of windthrow and bole damage. Windthrow was particularly severe where soils were saturated, while damage was less at well-drained sites.

Cyclones have their main influence in the tropics, although on occasion their effects may be felt in rainforest further south in southern Queensland and northern New South Wales. However, considerable disturbance can also be caused by local storms. For example, Olsen and Lamb (1988) described the effects of a severe thunderstorm on 8 September 1983 on the vegetation of the Lamington Plateau in southern Queensland. The storm was accompanied by hail and strong winds up to 150 km hr^{-1}, and caused severe damage in a swathe 2 km wide and 15 km long, although the intensity of the damage varied locally with topography and other factors (Fig. 5.8).

5.3.4 Frosts and snow

Many of the regions in Australia in which rainforest occurs are, even if infrequently, subject to brief periods of low temperature. Even in north Queensland, overnight light frosts are not infrequently experienced on the Atherton Tableland and, very rarely, in the lowlands. On a global scale, occurrence of such low temperatures at

Fig. 5.8 Damage to complex notophyll vine forest from the September 1983 storm on the Lamington Plateau, south-east Queensland.

only moderate altitudes in low latitudes is unusual. In closed-canopy forest, frost is unlikely to occur, but in gaps and clearings it may cause severe damage to seedlings (see Fig. 5.3), and can be an important factor influencing regeneration.

Snow may be experienced, if infrequently, in higher-altitude rainforest as far north as south Queensland. Damage to microphyll rainforest is slight, but in notophyll vine forest there may be loss of leaves and small branches of canopy trees.

5.3.5 Floods

In lowland areas in New Guinea, the changing courses of river channels and frequent floods constitute a continuing disturbance to rainforest (Johns 1986). In Australia, rivers flowing through rainforests are subject to rapid fluctuations in level, and the riverine fringe is subject to frequent flooding. The species in these communities must be tolerant of brief periods of waterlogging, while more severe floods can cause substantial damage, either from undermining or because the weight of floodwater and debris break trunks. Floodwaters may be an important agency for the continued introduction of weed propagules into rainforest (Fox 1988).

A number of tree species are particularly associated with river bank habitats. Melick (1990*a,b,c*) has investigated the ecology of *Tristaniopsis laurina*, a characteristic riparian species in Victorian rainforests. *T. laurina* co-occurs with *Acmena smithii*, and Melick has shown that the two species have different physiological tolerances. *T. laurina* has a relatively high light requirement for seedling establishment, and the seedlings respond to increased nutrients, the seedlings require a moist environment, and drought tolerance does not increase as trees mature. *T. laurina* is thus adapted to flood-disturbed environments in the rainforest, where flooding creates openings and deposits silt, and proximity to permanent rivers ensures reliable water supply throughout the year. *A. smithii*, although capable of surviving flooding, has seedlings with high shade tolerance which can regenerate in undisturbed rainforest away from river-banks.

Inland Australia is characterized by low topographical gradients and the major rivers have very broad floodplains. During the Tertiary when rainforest was widespread across Australia (see Chapter 6) the floodplains may have supported extensive periodically flooded rainforest, a habitat now of very limited extent.

5.3.6 Landslips

In both New Guinea (Garwood *et al.* 1979; Johns 1986) and South America (Veblen and Ashton 1978), landslips are a frequent major disturbance in rainforest and may be integral to the regeneration cycle of particular communities.

In Australia, in its present condition of tectonic quiescence, large-scale landslips are rare. However, on steep slopes heavy rain may generate local instability and minor landslides (often initiated along the boundary between two geological strata). Small openings in the rainforest canopy created in this fashion are a feature of upper slopes in the Barrington Tops (Adam 1987*b*) and at Gradys Creek in the McPherson Range, a particularly large gap was produced by a landslip (Floyd 1977, 1990) which, in the early stages, was recolonized very slowly.

5.3.7 Pathogens

Pathogens are the major proximal cause of many small gaps in the canopy, trees having suffered minor damage to limbs are then subsequently attacked by fungi. Over larger areas, damage and disturbance may render forest more susceptible to pathogens. Here, however, we are concerned with the possibility of pathogens being in themselves the prime cause of extensive canopy death.

Examples, such as Dutch elm disease in Europe and chestnut blight in eastern North America, demonstrate that pathogens can (although possibly only with a degree of human intervention) cause the selective eradication of species over very large areas. In south-western Australia, dieback of jarrah (*Eucalyptus marginata*) is caused by *Phytophthora cinnamomi*.

The role of *P. cinnamomi* in rainforest is uncertain. Sampling of a large number of rainforest sites in north and central Queensland showed that the fungus is widespread, although there are doubts about its native status (Department of Forestry, Queensland 1983). Most of the sites with *P. cinnamomi* displayed no signs of dieback, although there was a high frequency of isolation of the fungus from sites with patch death of trees, although demonstration of causality is lacking (Brown 1976). Patch deaths were associated with soils of low nutrient status. The sampling also revealed the presence of nine other *Phytophthora* species which have been shown to be pathogens in horticulture.

P. cinnamomi is a water-borne fungus and tends to spread downhill from the initial focus of infection. In order to retard possible spread of *P. cinnamomi*, the Queensland Department of Forestry has introduced restrictions on access to high hazard areas, and has limited the use of machinery during wet periods where spread of the pathogen is most likely to occur. It has been suggested that feral pigs may be involved in the spread of *P. cinnamomi* in unlogged rainforests (Department of Forestry, Queensland 1983).

Recent surveys in Tasmania (Podger and Brown 1989) have shown that *Phytophthora cinnamomi* is widespread in disturbed implicate rainforest (along roadsides or in fire-damaged areas), and less frequent in the disturbed rainforest types. However, the fungus was not isolated from undisturbed rainforest or from sites above 900 m altitude. *P. cinnamomi* was isolated from specimens of 39 species showing disease symptoms, and about 50 per cent of the rainforest vascular flora was assessed as susceptible to attack (Podger and Brown 1989). The relationship between disturbance and infection is related, not only to the transport of spores, but also to the warming of the soil following opening of the canopy. Infection requires soil temperatures above 15°C. A variety of evidence, including the fact that all isolates belong to a single isotype, suggests that *P. cinnamomi* is a recent introduction to Tasmania. The longer-term impacts are difficult to predict, but Podger and Brown (1989) have suggested that, at least at the local scale, there may be declines in the relative abundance of the most susceptible species.

In simple notophyll forest in New South Wales, crown die-back following logging has sometimes been extensive (Baur 1989). The pathology of crown die-back is complex and still poorly understood but pathogens do not appear to be the primary cause although they may be opportunistic invaders of stressed trees. Although die-back may appear to be very dramatic in the year immediately following logging, recovery does occur (Horne and Mackowski 1987).

Myrtle wilt, affecting *Nothofagus cunninghamii* in Tasmania, is one of the first cases reported of widespread pathogen attack in Australian rainforest. Affected trees show wilting of the canopy, followed by leaf fall. Other symptoms include discoloration of the sapwood, from the roots to the upper branches. Although diseased trees have been observed over many years, they have only been subject to scientific study in the last twenty years. Because of the lack of detailed study, there are insufficient data to determine whether the disease is increasing (Beckmann 1987*a*), although this seems to be a strong possibility.

The disease was first discussed in the literature by Howard (1973*b*, 1981) who observed that trees displaying wilt symptoms and recently dead trees, had numerous underbark tunnels made by a small bark-boring beetle *Platypus subgranosus*. She noted that a fungal mat developed on cut stem surfaces and hypothesized that the beetles carried fungal spores into the tunnels where they germinated into mycelia used to feed developing larvae. The fungal growth blocked the tree's vascular system, resulting in wilt and eventual death. She suggested that young seedling trees withstood beetle-attack, but that coppice regrowth was very susceptible.

Elliott *et al.* (1987) ground surveyed Tasmanian rainforests to assess the incidence of attack by *Platypus* beetle; experience suggested that aerial photographs provide inadequate assessment of disease distribution. Although a very small proportion of trees die without signs of *Platypus* beetle attack, in general attack by beetles and development of myrtle wilt are synonymous, so that beetle damage can be used to assess incidence of disease (Elliott *et al.* 1987).

The proportion of *Nothofagus cunninghamii* trees with signs of damage varied considerably but averaged nearly 25 per cent. Incidence of attack decreased with increased altitude and was higher in callidendrous than in thamnic and implicate forests (see p. 66). In mixed forest (eucalypt–rainforest mixtures), the incidence of attack increased as the density of *Nothofagus* increased (Elliott *et al.* 1987). Elliott *et al.* (1987) confirmed Howard's observation that attack was greatest on old trees and was rare in small stems. Attacked trees tended to be clumped.

The fungus responsible for the wilt symptoms is a hyphomycete *Chalara australis* (Kile and Walker 1987). As far as is known, this fungus is indigenous to Australia. *C. australis* can infect *N. gunnii* but rarely does so in the field. It does not appear to attack other genera.

Is *Platypus* beetle responsible for spread of *Chalara* as postulated by Howard (1973*b*, 1981)? Evidence assembled by Kile (reported in Beckmann 1987*a*) would appear to suggest not. Extensive study of beetles have failed to show that it carries fungal spores. Cut logs rapidly become infected with *Chalara* without any sign of beetle attack. Infection is probably due to fungal spores carried by wind or water. Kile hypothesized that beetles were attracted to trees already infected by fungus (possibly responding to volatile chemicals released by the fungus). Attack by beetles could create further sites for fungal entry and exacerbate the disease. Clump deaths may be related to transmission of the fungus through root grafts.

Myrtle wilt may be a factor in the normal regeneration of *N. cunninghamii* through creation of small canopy gaps. However, the incidence seems to increase following road-building and logging, possibly as a result of more effective dispersal of the fungus by human agency. Large numbers of dead or dying trees could increase the risk of fire, leading to further changes in forest composition; even without fire, the regeneration following creation of large gaps could change the overall structure and composition of the rainforest.

The overall mortality rate recorded by Elliott *et al.* (1987) was 2.4 trees ha^{-1} per year or 1.6 per cent of trees per year, indicating that *Chalara* is a major disturbance factor in Tasmanian rainforests.

In more complex rainforests, pathogens may be a factor promoting species diversity, and attempts to grow commercially valuable species in plantations may be thwarted by disease. Attempts to establish rainforest conifer plantations in north Queensland failed because of damage caused by the root-rotting fungus *Phellinus*.

Augspurger (1984) and Augspurger and Kelly (1984) found that seedlings in canopy gaps in neotropical rainforests suffered less pathogen induced mortality than those in shade, which had lower resistance to attack. The role of pathogens in the seedling mortality of Australian species in the field remains to be explored.

5.3.8 The role of insects

Insects play a variety of roles in rainforest, as consumers, as pollinators, as possible vectors for pathogens, and as a food resource for other animals.

As consumers they may, in some circumstances, be regarded as pathogens. The caterpillars of the cedar tip moth, *Hypsipyla*, prevented the successful establishment of plantations of *Toona australis* (Chapter 9); at low densities of *Toona* a balance between cedar and *Hypsipyla* is reached which permits the establishment of some trees.

Mass defoliation of trees as a result of insect attack has been reported from a number of forests with species-poor canopies, for example eucalypt forest (Elliott *et al.* 1981) and *Avicennia marina* mangrove forests (West and Thorogood 1985). Such events may be unlikely in more diverse forests where different species may vary in their susceptibility to attack by any particular insect species. Matthews and Kitching (1984) suggested that several of the major features of complex rainforest, particularly the dispersed distribution of most canopy trees, the large number of secondary metabolites in the foliage, and infrequent flowering and fruiting have evolved as 'defences' under pressure from herbivores, principally insects.

The role of insects as herbivores in 'undisturbed' rainforest has been the subject of considerable study by Lowman (1982*a,b*), 1984, 1985*a,b*). Lowman demonstrated that assessment of the potential impact of insect herbivory is very strongly dependent on methodology (Lowman 1984, 1985*b*). She suggested that many published accounts underestimate the amount of leaf material consumed; the variations between different methods of study make meaningful comparison of the degree of herbivory in different areas difficult on the basis of the literature alone. Lowman (1985*b*) argued that an accurate measure of herbivory requires more than simple calculation of the area of holes in leaves (Lowman (1987) showed that changes in area of holes and changes in leaf length during growth were highly correlated, permitting extrapolation back to the size of the holes made in young leaves from measurement of holes in old leaves); it also demands information on the longevity of leaves (which can be considerable, Rogers and Barnes (1986) showed that the mean half-life of leaves of the rainforest shrub *Wilkiea macrophylla* was 6.8 years), the amount of damage to different ages of leaves (Lowman and Box 1983 and Selman and Lowman 1983 showed that most damage occurred to young leaves), and the proportions of different leaf cohorts that make up the canopy. Some young leaves are entirely consumed, many estimates do not account for these total losses. It is also necessary to adequately sample the full range of habitats within the various strata of the forests (Fig. 5.9), in a number of species, shade leaves characteristically have higher rates of herbivory than sun leaves (Lowman 1985*b*). Of the different rainforest types studied by Lowman (1985*b*), the highest rate of herbivory (26 per cent leaf area grazed per year) was recorded from *Nothofagus moorei*-dominated forest (Selman and Lowman 1983); in subtropical complex notophyll rainforest the long-term rate of herbivory was 14.6 per cent, while simple notophyll (*Ceratopetalum apetalum*) forest was intermediate (22 per cent).

These figures are substantially higher than many reports from extra-Australian rainforests, but as argued above, any conclusion about fundamental differences between continents would be unwarranted unless comparable data were available. If the difference does turn out to be real, it may be related to differences in the composition of the fauna. Lowman (1982*b*) suggested that the proportions of herbivores, particularly Coleoptera, in the Australian canopy fauna is greater than elsewhere, although comparisons are again fraught with difficulty due to variations in methodology.

Some species characteristically show very high rates of herbivory. In Lowman's studies, *Dendrocnide excelsa* consistently showed very high levels of damage (see Fig. 5.9 and 5.10), and in some instances can be virtually defoliated over large areas (Le Gay Brereton 1957). Although *D. excelsa* is amply provided with stinging hairs which can inflict severe injury on humans, they provide little deterrent to insect grazers. *D. excelsa* is a long-lived pioneer species; the high rate of defoliation is consistent with Swaine and Whitmore's (1988) suggestion that higher rates of insect herbivory may be a general feature of pioneers.

It would appear that, in most circumstances, defoliation by insects is insufficient to constitute a disturbance of sufficient intensity to affect regeneration.

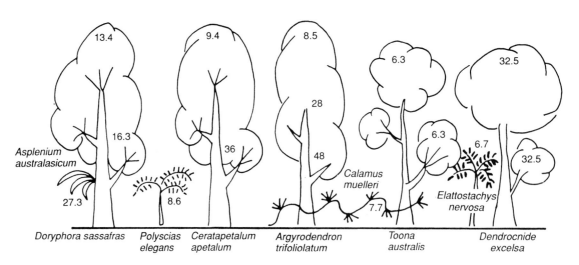

Fig. 5.9 Estimates of the percentage of damage from herbivory in different species and in different strata in complex notophyll vine forest (from Lowman 1985*b*).

Fig. 5.10 Leaves of young *Dendrocnide excelsa*, showing characteristic damage from herbivory, Billilimbra State Forest, New South Wales

Attack on seedlings, with limited reserves to compensate for defoliation may be more significant but have not been investigated. Insects as seed predators may also have a significant impact on regeneration (Connell *et al.* 1984), but again studies are limited.

5.4 THE PROMOTION OF RAINFOREST TAXA IN SCLEROPHYLL COMMUNITIES

In Chapter 9, the role of humans in reducing the extent of rainforest is discussed. However, in some areas, disturbance of sclerophyll communities has promoted the spread of rainforest species. Most sclerophyll communities in Australia are found on nutrient-poor soils (often with very low phosphorus levels), and species richness is inversely correlated with fertility (Adam *et al.* 1989). In the Sydney region, much urban expansion has taken place on sandstone ridges, but steep valleys have been too difficult to develop, and remain as bushland. These valleys have a more mesic microclimate than exposed ridges and more fertile soils (in many cases the valleys have eroded outcrops of more nutrient-rich shales; even in the absence of shale, drainage results in nutrient accumulation). The vegetation was therefore more mesophyllous than that on the ridges, wet sclerophyll forest rather than dry. As a result of urbanization, the valleys have received substantial inputs of nutrients from runoff and deliberate dumping, they are also more consistently damp as a result of increased general runoff and discharge from stormwater drains. These modified environmental conditions, coupled with with fire suppression, have promoted the expansion of rainforest elements (Clements 1983; Lambert and Turner 1987), most particularly *Pittosporum undulatum*, but also *Omalanthus populifolius* and a number of aggressive aliens—*Cinnamomum camphora*, *Ligustrum lucidum*, *L. sinense*, and *Lantana camara*. These species also have fleshy fruit, and dispersal may be significantly aided by the increased suburban bird population. It is likely that these habitats have been permanently changed, and even if the exotic species were controlled there would be a trend towards the development of rainforest. Similar spread of rainforest species, although on a lesser scale, is seen adjacent to many developed areas.

6 ORIGINS AND HISTORY OF AUSTRALIAN RAINFORESTS

> Imagine Australia as a clear pool of water into which two muddy streams enter—a black stream from the north—a white stream from the south. The former represents the Malayan element, the latter the Antarctic element—most evident in Tasmania, and the higher mountain ranges of Australia—both elements due to land connection in the prehistoric past. This Malayan element is especially traceable from Cape York, through central Queensland, and New South Wales, where it finds its southern evidence in regions like Ourimbah and the Illawarra districts of New South Wales
>
> C. Hedley, quoted by Carter (1933)

In previous chapters, the present-day composition and distribution of Australian rainforests were discussed. In very broad terms, the distribution of rainforest can be related to contemporary ecological factors (rainfall, temperature, soil types). For individual species also, distribution can be modelled using climatic parameters as predictors, and such models allow predictions of the response of species to environmental change (Busby 1988). However, they do not provide a complete explanation for current distribution patterns; the possible distribution pattern generated by the model rarely shows absolute coincidence with the known distribution (see examples in Busby 1988). The anomalies may arise in two ways (acting either singularly or in combination); either the models are too simple and require more factors to be considered, or present distributions reflect to some extent historical rather than contemporary environments.

Even if the present distribution of rainforest could be completely explained in terms of control by the present environment, the question of the history and origins of the biome would be of great intrinsic interest.

Over the last two decades, the history of Australia's rainforest has been completely rewritten. The development of the theory of plate tectonics has provided new palaeogeographical insights which have demanded new perspectives in the interpretation both of the fossil record and present distribution patterns.

In this chapter, the evidence of environmental change is examined, before the fossil record is discussed in the light of these changes. A more extensive review of Australian vegetation history was provided by Kershaw (1988), while Truswell (1990) discussed the history of rainforest in detail.

6.1 THE PLATE TECTONIC HISTORY OF AUSTRALIA

The development of plate tectonic theory and the acceptance of the reality of continental drift has revolutionized biogeographical interpretation. The history of Australia's movements over geological time has been reviewed in Veevers (1984), and by Audley-Charles (1987) and Frakes et al. (1987).

Past continental positions are established largely on the basis of geophysical evidence. Determination of the timing of events and matching boundaries requires careful stratigraphical control, which is often provided by palaeontological evidence. However, stratigraphical comparisons are not dependent on environmental reconstructions, so that the plate tectonic history can be regarded as independent data which can be used to assist palaeoecological interpretation of the fossil record, rather than the two interpretations being circularly dependent one on the other. Figure 6.1 shows the plate tectonic movement of Australia from the late Jurassic onwards.

It is convenient for present purposes to take as

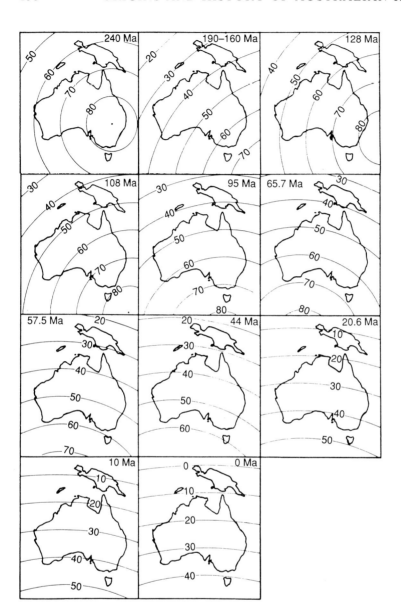

Fig. 6.1 Movement of Australia from the late Jurassic to the present (modified from Embleton 1984).

a starting point the beginning of the break up of Gondwana in the Jurassic.

At the start of the Jurassic, Gondwana compromised an amalgamation of what are now termed Africa, South America, Madagascar, Australia, India and Antarctica. In the late Jurassic (c.160 Ma) West Gondwana (Africa and South America) started to separate from east Gondwana (Antarctic, Australia, India, and Madagascar), although close connection between southern South America and Antarctica was to remain for many millions of years.

In the earliest Cretaceous (about 128 Ma), India split away from West Australia and moved to the north-west.

Around 95 Ma east Gondwana became further fragmented, with two further plates, Lord Howe Rise/New Zealand and Antarctica, coming into existence. Ocean-floor spreading to the east of Australia, involving the separation between Australia and New Zealand, was fairly rapid from about 80 Ma, but the separation of Australia and Antarctica was slow until $c.45$ Ma, when Australia began a fairly rapid movement northwards and it is probable that the developing ocean between them was initially fairly shallow. At about the same time, spreading between India and Australia ceased and the two separate plates coalesced to form the Indo-Australian Plate (Veevers 1984). Australia continued its northwards drift, colliding with the South-east Asian plate from around 15 Ma.

From the above broad-brush overview, it would be possible to gain an impression that Australia was effectively isolated from the rest of the world for about 30 million years, from 45 Ma to 15 Ma, and that 15 Ma marked the first contact between Australia and South-east Asia. This view has wide currency and has certainly coloured the reinterpretation of the historical biogeography of Australian rainforests. However, more detailed studies of the events to the north of Australia render this simple view untenable. Although the detailed history of the region is still subject to various controversies, it is now clear that various fragments of the northern margin of the Australian plate became detached and drifted north to become embedded in what is now South-east Asia. Parts of Burma, western Thailand, Malaya, and Sumatra were all originally part of Gondwana (Audley-Charles 1987). The timing of the splitting-off of these microcontinental fragments remains uncertain. The process may have started in the Permian but Audley-Charles (1987) argued that it was a Jurassic event, perhaps as late as 160 Ma. The various blocks were relatively isolated in the Tethys Ocean between Gondwana and Laurasia for, perhaps, 60 million years until the Cretaceous, about 100 Ma, by when Burma and west Thailand had collided with the Laurasian Indo-China block (east Thailand and Indo-China) (Audley-Charles 1987). Audley-Charles (1987) suggested that during the drift north, most of these continental fragments provided terrestrial habitats and that throughout the Mesozoic there was an achipelago of terrestrial habitats between south east Asia and Australia.

It can be seen that during and subsequent to the break-up of Gondwana, Australia has undergone substantial latitudinal change (Fig. 6.1). During the late Jurassic, there was a movement to the south but from the Cretaceous onwards movement, which continues today, has been steadily northwards.

6.1.1 Topographic change

The present distribution of rainforest in Australia is closely related to topography; the major occurrences along the east coast are associated with 'ragged mountain ranges'. It is of interest therefore to investigate the history of major topographic features and changes they may have undergone.

The most extensive topographic features in Australia today are the Great Dividing Range and, to the east, the Great Escarpment. The watershed of the Great Dividing Range is not a dramatic feature, but the slopes east of the Divide are abruptly terminated by the Great Escarpment with rugged spurs and deep valleys (Ollier 1986). The steep slopes of the Great Escarpment are a major habitat for rainforest. In New South Wales, the Great Divide and the Great Escarpment are, for the most part, close together but in Queensland they are, for much of their length, separated by several hundred kilometres (Fig. 6.2).

There is still debate about the detailed history of the Eastern Highlands (cf. Jones & Veevers 1984; Ollier 1986). There is general agreement, however, that there has been an upland zone in eastern Australia throughout the Tertiary and

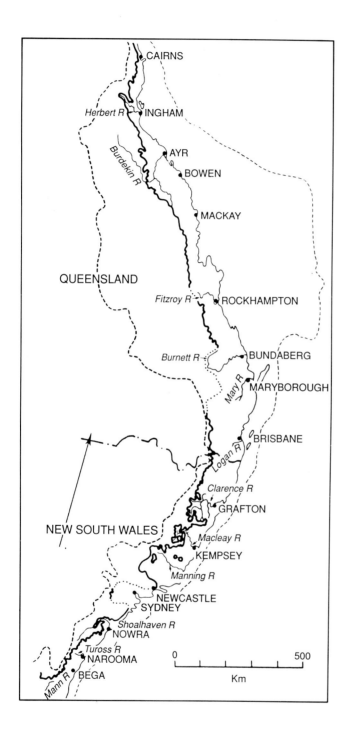

Fig. 6.2 Location of the Great Divide (----) and Great Escarpment (——, · · · · · where ill defined) in eastern Australia (from Ollier 1982).

perhaps longer. In comparison with other continents, the highest parts of the Great Divide (a little over 2000 m in the Snowy Mountains) are not particularly high, and towards its northern extremity the Divide dwindles to a low elevation. Some biogeographers have invoked previously higher mountains in eastern Australia (e.g. Herbert 1950), but while there may have been periods when parts of the Eastern Highlands were somewhat higher than at present, there is little evidence for major mountains considerably higher than those of today. Elsewhere in Australia, there are numerous topographical features referred to as ranges. In terms of ruggedness, defined by local relative relief, Australia is a rugged, even mountainous continent (Costin 1983), but the absolute elevation of the various inland ranges is low, and does not appear to have been appreciably greater at any time in the Tertiary.

The Great Escarpment is one of the most striking topographic features in Australia, although its continuity over virtually the entire eastern margin of the Highlands has been recognized only recently (Ollier 1982, 1986). Its age and origin are controversial. Ollier (1982) regarded it as having formed through the retreat of the scarp of a pre-continental-drift rift valley, so that the initiation of the Escarpment would have been at the onset of the opening of the Tasman Sea. Jones and Veevers (1984), disputed this interpretation, pointing out that while the Escarpment is both parallel to, and close to, the continental margin in New South Wales (thus fitting with Ollier's suggestion) neither condition applies in Queensland (see Fig. 6.2).

Regardless of the debate as to their origin, the Eastern Highlands and the Escarpment have been major landscape features for many millions of years, and will have been an influence on climatic patterns within the continent over this period. As the highest part of the continent, and situated close to the eastern seaboard, it is likely that the Highlands have always experienced relatively high rainfall.

Another factor which will have had important consequences for the biota is the proportion of the continental plate which, at any one time, was dry land. Most of the present continental land surface has been dry land since the Cretaceous. The Cretaceous is believed to have been a time of high global temperatures and sea-levels (Truswell and Wilford 1985), and most of Australia was then covered by shallow seas with the land surface divided into three major 'islands'. The extent of these shallow seas indicates that prior to the Cretaceous, Australia may have been already a continent of very low relief (Ollier 1986). During the Tertiary, there were marine incursions in southern Australia, flooding what is now the Nullarbor Plain and the lower Murray Basin. Changes in sea-level along the continental margins in more recent geological times involved relatively small areas, but may have had important biological consequences.

The volcanic activity in eastern Australia has had considerable consequences on rainforest. There are extensive exposures of volcanic rocks of Cainozoic age, from north Queensland to Tasmania (Fig. 1.5). These have weathered to produce some of the most fertile soils on the continent, which at the time of European settlement supported some of the most extensive stands of rainforest. During the time that the rocks were erupted, there must have been regionally extensive disturbance to both landforms and vegetation. Subsequent erosion has produced some of the most dramatic scenery in Australia (Adam 1987b). The most recent eruptions were in western Victoria–South Australia at $c.1400$ BP, while in north Queensland there were eruptions within the large volcanic field up to 13 000 BP (Veevers 1984). The various volcanic areas can be divided into two major classes: central volcanoes and lava fields (Wellmann and McDougall 1974). The age of central volcanoes is significantly correlated with latitude, with the oldest in the north at $c.30$ Ma, and the youngest in the south. The correlation suggests a southward movement of central type volcanism at a rate comparable with the calculated rate of the drift northwards of Australia and has been explained as the migration of Australia over a fixed hot-spot in the underlying mantle (Wellman and McDougall 1974; Ollier 1986).

6.2 CLIMATIC HISTORY

Australia's climate has changed considerably since the breakup of Gondwana. Climatic reconstructions are based on a range of evidence; lithological, geomorphological, ocean palaeotemperatures derived from oxygen isotope studies (ocean temperatures provide evidence of global temperature trends and also, because of the relationships between ocean temperatures and terrestrial climates, provide input into palaeoclimatic models), and palaeontological. Interpretations of palaeontological evidence suffer the danger, in some circumstances, of degenerating into circularity but in many cases the relationships between climate and biota can be discussed against a background of independently derived climatic reconstructions.

The Jurassic was a time of globally warm temperatures, and at latitudes within 45° of the equator, conditions appear to have been generally arid. At higher latitudes rainfall was probably high, as evidenced by coal deposits which indicate extensive swamps. The Cretaceous was also a time of warm temperatures, oxygen isotope data suggest that the Albian Stage (towards the end of the early Cretaceous) was possibly the warmest interval in the whole Phanerozoic (Truswell and Wilford 1985). For most of the Cretaceous, there was a belt of what would now be regarded as tropical to subtropical climates between 45°N and 60°S. At higher latitudes, conditions were temperate to cool temperate. The prevalence of warm, wet conditions at high southern latitudes was a consequence of the configuration of the continents. The massing of the continents into Gondwana permitted warm oceanic currents to flow to high latitudes, producing mild climatic conditions even in southern polar regions (comparable to the Gulf Stream, a warm ocean current which today ameliorates the climate of north-west Europe). Towards the end of the Cretaceous, there is evidence for a slight cooling of the climate.

In the early Tertiary, Australia was warm and wet, despite its high latitudes, and rain-bearing winds extended well inland (Kemp 1978). South of Australia (between 70° and 80°S), Kemp (1978) postulated a westerly air circulation but, over Australia itself, winds were probably erratic. There was little or no ice in Antarctica.

In the Eocene, the sea-way between Australia and Antarctica had begun to widen, but was still shallow and the South Tasman Rise was a barrier to westerly circulation. Oxygen isotope data indicate a relatively small ocean temperature gradient from equator to high latitudes. The climate over Australia was still warm and wet, but Kemp (1978) suggested that there may have been seasonal effects, possibly with heavier winter rains in the south-east.

In the Oligocene, circumpolar oceanic circulation developed, reducing heat transfer between equatorial and high latitudes, with a consequent increase in the latitudinal temperature gradient. Within Australia, rainfall may have been lower than in the Eocene, but evidence to support the suggestion of aridity in the north and north-west is not available (Kemp 1978).

During the Miocene, major changes in Australia's climate occurred, with a pronounced drying commencing about 15 million years ago, associated with the build-up of the Antarctic ice-cap (Fig. 6.41). As the ice-cap expanded, the patterns of southern hemisphere wind circulation approached those of the present. The aridity, which developed about the Miocene/Pliocene transition, probably spread from the south as a belt of anticyclonic circulation developed at high latitudes and migrated north (Bowler 1982; Truswell and Wilford 1985). The present pattern of seasonal variation in climate probably dates from about 2 Ma (Truswell and Wilford 1985).

Much is still to be learnt about Australia's climatic history, particularly about climatic gradients within the continent and seasonal variation in conditions. A major influence on Australia's climate has been the size of the Antarctic ice cap about which, unfortunately, there is still considerable controversy.

6.3 VEGETATION HISTORY

The fossil record provides much data on the vegetation history of Australia although interpretation is frequently difficult (Lange 1982). Both macro- and microfossils have been studied, and at least for some time periods give rather different pictures of the vegetation (Lange 1982). Fossil epiphyllous fungi on macrofossils have also been examined, and Lange (1982) suggests that they may be sensitive indicators of environmental conditions.

Although the number of sites which has been studied is large, their temporal and spatial coverage is uneven and, overall, sparse. At the present day, the composition and structure of vegetation may vary considerably over short distances. The fossil record is unlikely ever to afford insight into this heterogeneity; for most time periods the existing record is inadequate even to permit clear reconstruction of regional differentiation in vegetation.

A perennial problem with the fossil record is that of determining the affinities of fossil taxa. Interpretations of fossil assemblages may be biased by concentration on those elements for which affinities can be established. Even when affinities can be assessed with a degree of confidence, extrapolation of the indicator value of present taxa back to fossil precursors involves many, normally untestable, assumptions.

The macrofossil record offers advantages, in that data on leaf size, margins, and textures can be utilized in palaeoclimate reconstruction (Wolfe 1979) independently of taxonomic interpretation, although such reconstructions assume that relationships observable today also held in the past. Macrofossils also provide a more localized picture of vegetation than microfossils.

Microfossils provide much of the palaeobotanical evidence for rainforest history, many of the data coming from boreholes. The likelihood of a particular species contributing to the pollen record will be a complex function of the abundance and spatial distribution of the species in the vegetation, its pollen productivity, and the dispersal characteristics of the pollen. For many species in present-day rainforests, the likelihood of representation in the pollen rain is very small. If past rainforests had similar patterns of diversity and a similar frequency of pollination syndromes, then many species are likely to be absent from the record. While occurrence in the record is conclusive evidence that at a particular time certain taxa existed, absences are far more difficult to interpret.

In this section, a broad chronological overview of the fossil record is presented (see Fig. 6.3 for the geological time-scale). In subsequent sections, the contribution of the fossil record to the resolution of particular questions is considered.

The Jurassic flora of Australia appears to have been very similar to that in other parts of the world, although this impression of a cosmopolitan flora and vegetation with little regional differentiation may reflect the lack of resolution in the fossil record rather than reality. Evidence comes from a number of localities, the most notable being the Talbragar Fish Beds in northern New South Wales (White 1986). The picture that emerges from there is of a forested landscape, dominated by *Agathis*, but also with podocarps, and in the understorey cycadophytes (*Pentoxylon australica*), ferns, and tree ferns. Around the lake, there appears to have been a zone of seed ferns (*Pachypteris crassa* and other species). White (1986) suggested that analogues of this vegetation still exist in Australia: 'This reconstruction of a landscape for the Talbragar Fish Beds Lake of 175 million years is particularly interesting because relict *Agathis* forests on the Atherton Tableland have Podocarps growing among the Kauri Pines and a tree-like Cycad, *Lepidozamia hopei*, is present in the understorey. Thus, we see a modern assemblage of plants with the same basic composition (though at a more evolved stage) . . .'

During the Cretaceous, the fossil record

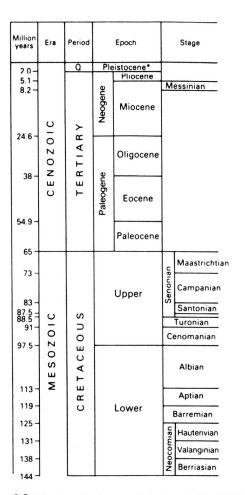

Fig. 6.3 Geological time-scale (adapted from Whitmore 1987).

in Australia and by other species in New Zealand and South America, and *N. fusca*, represented in Tasmania by *N. gunnii* and by other species in New Zealand and South America.)

An outstanding problem in reconstructing the vegetation of southern Gondwana (Dettmann 1989) in the Jurassic and Cretaceous, is finding an explanation for the survival of evergreen mesic forests at latitudes around 70°S. While a variety of evidence indicates that temperature and precipitation were relatively high, winter day-length today at this latitude is very short (Lange 1982). However, neither fossil flora nor fauna show clear evidence of adaptation to long, dark winters. However, as Frakes *et al.* (1987) have argued, the physiology of modern adaptations is not easily used as a restraint on ancient situations; physiology does not leave a fossil record, and it would be unwise to extrapolate present physiological syndromes to the distant past, particularly as the combination of temperature and day-length variation current during the Jurassic and Cretaceous periods is not found at the present time.

In the Palaeocene, the gymnosperms still dominated the pollen record, but the diversity of angiosperms had increased. The Proteaceae were represented by up to 23 pollen types, which were present in some abundance. The interpretation of these high Proteaceae levels is problematical (Martin 1987). At the present day, the Proteaceae are not high pollen producers and are under-represented in pollen spectra. Unless there was a major change in pollen productivity in the family (which, presumably, would also indicate a change in pollination mechanism), it would appear that the Proteaceae were a major, if not the dominant, element in the vegetation. If this were the case, then there would be no comparable vegetation in Australian rainforests today. The evidence of an abundance and diversity of cryptogams (Lange 1982), and the continued importance of gymnosperms, suggest that conditions remained moist.

Eocene pollen assemblages indicated marked changes in the vegetation and provide the first evidence of a greater diversity of angiosperm taxa (Martin 1978). Early Eocene spectra are similar to

suggests continued dominance of gymnosperm forest across the continent, although from the Albian Stage onwards, microfossils of angiosperm origin are also represented. In the late Cretaceous a number of these microfossils (Martin 1978) can be related to extant taxa—Proteaceae, Winteraceae, *Ilex* and *Nothofagus brassii* type (within *Nothofagus* three different pollen taxa are found: *N. brassii* with no living taxa in Australia but still occurring in New Guinea, *N. menziesii*, represented today by *N. moorei* and *N. cunninghamii* those of the Palaeocene, although all three *Nothofagus* pollen types are represented (nowhere does this occur today, again indicating that in many respects the vegetation has no modern analogues). In the mid-Eocene, *Nothofagus* dominated the pollen spectra, and this must represent an increase of *Nothofagus* in the vegetation, although probably not dominance, as *Nothofagus* today is considerably overrepresented in the pollen rain. The diversity of Proteaceae declined over this time. Several taxa with, at present, tropical/subtropical distribution patterns, made their appearance around the late Palaeocene–early Eocene transitions. Most notable among these were *Anacolosa* (Olacaceae, currently represented in Australia by a single species in far northern Queensland), the Cupanieae tribe of the Sapindaceae, and the saltwater palm *Nypa*, found in both western Australia and Bass Strait and interpreted as indicating the occurrence of mangroves (Lange 1982). The main impression from the palynological record is of a widespread cool to warm temperate moist closed forest, but such a reconstruction is difficult to reconcile with the presence of the *Anacolosa–Cupanieae* element which was widespread across southern Australia by the late Eocene. Whatever the implications of this tropical/subtropical element for environmental reconstruction, its presence demonstrates that this floristic element has a long and widespread history in the continent.

Fossil leaf assemblages from Eocene sites in southern Australia fail to show a predominance of the *Nothofagus*–gymnosperm–Proteaceae assemblage suggested by the pollen record. The leaf beds indicate a diverse tree flora, and unlike the uniformity suggested by pollen, considerable variation between sites (Christophel 1981; Lange 1982). On the basis of leaf-size distribution and other features, Christophel (1981) proposed that the vegetation was simple to complex notophyll vine forest, which he regarded as evidence of a tropical/subtropical climate. The comparison of Eocene and modern leaf floras was examined further by Christophel and Greenwood (1988). They showed that the physiognomic signature of fossil material from the Eocene deposit at Anglesea in Victoria was similar to that of leaf litter from simple notophyll vine forest in northeast Queensland, while that from Golden Grove in South Australia was closer to complex mesophyll vine forest. Christophel and Greenwood (1988) suggest that the taxonomic affinities of a number of species identified from southern Australian Eocene sites are with species restricted today to north-east Queensland, and that the closest modern analogue to the vegetation at Anglesey is provided by rainforest similar to that at Noah Creek, north of the Daintree River.

Although the taxonomic evidence suggesting climatic analogies between tropical Queensland today and early Tertiary southern Australia is persuasive, the matching of physiognomic signatures is not necessarily conclusive support. The similarities demonstrated by Christophel and Greenwood (1988) are strong, but notophyll vine forest today is widespread. Would leaf litter from a simple notophyll vine forest on the central coast of New South Wales match the physiognomic signature of the Eocene Anglesea deposit more or less closely than that from north-east Queensland? Clearly more comparative data are required.

The apparent discrepancy between climatic reconstruction based on palynological and macrofossil evidence may, in part, be based on misinterpretation of the significance of the gymnosperm element in the pollen record. The Podocarpaceae have been assumed commonly to indicate cool-temperate conditions based on their presence in Tasmania and New Zealand, but the Eocene macrofossils include taxa with affinities to genera now associated with tropical montane and lowland rainforests (*Dacrycarpus*, *Falcatifolium*, *Podocarpus*, and *Prumnopitys*) (Christophel and Greenwood 1988).

The Oligocene is a period which palaeoclimatic interpretations suggest would have been one of considerable environmental change (Kemp 1981*a*). Palynological evidence indicates continued dominance of *Nothofagus* in south-east Australia but a decline in floristic diversity. *Anacolosa* disappeared from the fossil record in south-east Australia (Martin 1978), but persisted further north. Lange (1982) suggested, on the basis of fossil fruits, that by the Oligocene, if not earlier, there had been considerable diversification among the Leptospermoideae (capsular fruited Myrtaceae).

The pollen record of the early Miocene indicates a widespread closed forest across south eastern Australia with *Nothofagus* and a number of gymnosperms as major elements. Subtropical mesothermal elements were still represented at least locally even as far south as Victoria. This closed forest appears to have persisted in coastal regions throughout the period. Inland, however, the mid-Miocene was a time of major environmental change (Fig. 6.4) with a rapid drying of much of the continent. This caused an equally rapid decline in rainforest, although some rainforest elements may have persisted in gallery stands along creeks.

The most detailed studies of the decline and fragmentation of rainforest are provided by Martin (1978, 1981, 1984, 1987, 1990), who has studied a large number of sites in western New South Wales (sites which today receive less than 250 mm annual rainfall). The record from the

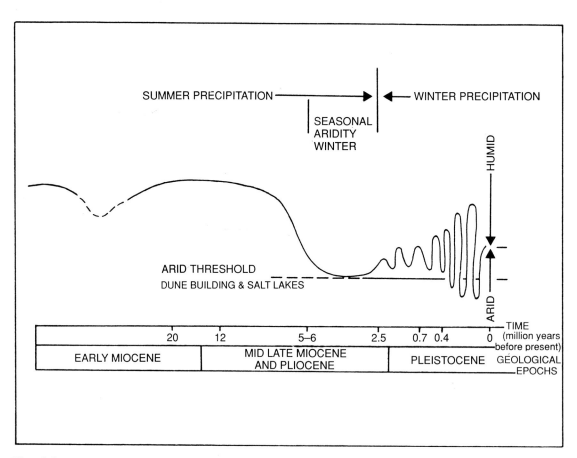

Fig. 6.4 Diagrammatic summary of changes in humidity in southern arid and semi-arid regions of Australia since the early Miocene (from Adam 1987b, after Bowler 1982).

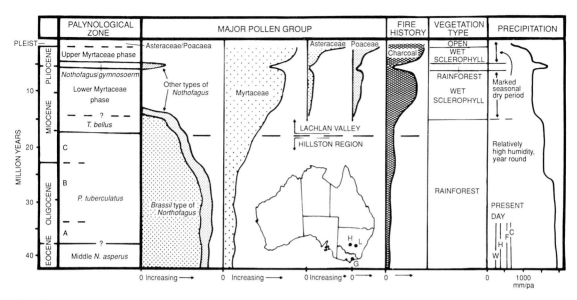

Fig. 6.5 Environmental change in the Lachlan Valley, New South Wales, as shown in the pollen and charcoal record. On the inset map H = Hillston, L = Lachlan Valley, G = Gippsland Basin. Present-day rainfall levels are for Wentworth (W), Hillston (H), Forbes (F), and Cowra (C). The last two are in the Lachlan Valley (from Martin 1990).

Lachlan valley (Martin 1987, 1990, Fig. 6.5) indicates a widespread closed forest, with high representation of *Nothofagus* (*brassii* type), but with a considerable number of other rainforest taxa, all with low representation, from the late Eocene through to the late Oligocene–early Miocene, when there was a rapid decline in *Nothofagus* and its replacement by Myrtaceae. The Myrtaceae are represented by a diverse assemblage of taxa, most being related to rainforest or wet sclerophyll forest taxa, including *Acmena*, *Backhousia*, *Syzygium*, *Lophostemon*, *Decaspermum*, *Austromyrtus*, and *Rhodamnia*, although *Myrtaceidites eucalyptoides* which corresponds with the *Angophora*–bloodwood eucalypt type, currently dry sclerophyll taxa (Martin 1987), was also present. There is a number of other rainforest taxa also recorded. Martin (1987) interpreted the vegetation as still being a closed rainforest. In the early Miocene, silicified woods in the upper Lachlan Valley have been identified as Myrtaceae (with affinities to *Eucalyptus*), *Acacia* and *Nothofagus* (Bishop and Bamber 1985).

By the mid-Miocene, *Nothofagus* had completely disappeared (Fig. 6.5), and in the mid–late Miocene, Myrtaceae were dominant, although a range of rainforest taxa are also represented. A coincident rise in charcoal is interpreted by Martin (1987) as signifying increased fire frequency. Martin (1987) suggested that the vegetation was a wet sclerophyll forest (the actual cover was likely to have been a mosaic with rainforest elements concentrated on creek lines or south-facing slopes); rainfall may have been 1000–1500 mm per annum with a definite dry season.

For a brief period around the Miocene–Pliocene transition there was a return of *Nothofagus* (but only the *menziesii* and *fusca* pollen types) and rainforest gymnosperms, with a corresponding decline in Myrtaceae and charcoal. This phase is interpreted by Martin (1987) as the re-establishment of rainforest, reflecting an increase in rainfall, but rainforest cover was probably patchy. The wet sclerophyll forest was soon re-established. In the late Pliocene, the rainforest element disappeared and the fossil record indicates increasing dryness and transition from wet sclerophyll forest to open woodland.

Martin (1987) suggested that during the late Miocene there was a retreat of rainforest eastwards to refugia in the Highlands. The Miocene/Pliocene rainforest re-establishment phase shows clear evidence of spread from the east, with *Nothofagus* values being higher towards the head of the valley, declining to the west. The influences of climatic change and of fire in the decline of rainforest were interrelated, and the charcoal record strongly parallels that of sclerophyll taxa. A number of rainforest taxa can be cultivated today under much lower rainfall conditions than in their natural habitat, perhaps suggesting that in the absence of fire the decline in rainforest simply as a response to climatic change would have been less than indicated from the fossil record.

It is unlikely that the rainforest retreated to form a continuous belt along the Eastern Highlands. Low stretches of the highlands would have also become dry and may not have supported rainforest, thus segregating the rainforest into regional blocks. One of the largest of these dry gaps would have been in central Queensland, separating the rainforest of the south-east from those of the wet tropics (regions A and B in Webb *et al.*'s 1984 scheme). The notophyll forests of regions A and B have some species in common, but there is a large number of shared genera with differentiation at species level, a segregation which may reflect the effects of spatial separation since the late Miocene.

6.4 INTERPRETATION

After this brief overview of the tectonic, climatic, and vegetation history of Australia, it is possible to reassess what was the conventional view of Australian rainforest until the late 1970s.

The first overview of Australian phytogeography, by Hooker (1860), recognized the distinctiveness of the rainforest flora. Hooker recognized three elements in the Australian flora; an autochthonous Australian element characteristic of sclerophyll forest and woodland, an Indo-Malayan element in the tropical and subtropical rainforests, and an Antarctic element in temperate rainforest and montane areas. The most detailed phytogeographical analysis of the Australian flora was undertaken by Burbidge (1960). Her analysis elaborated on the groupings defined by Hooker stressing the difficulty of assigning taxa to sharply defined groupings, but the basic framework established a century earlier was still clear. She suggested that the autochthonous element recognized by Hooker may have been an early immigrant flora. This initial stock diversified to produce a pan-Australian flora of scleromorphic and xeromorphic species with high generic endemism. Later invasions from the north and south introduced the Indo-Malayan and Antarctic elements. These later arrivals were restricted by competition with the established scleromorphic flora to habitats to which they were pre-adapted and so underwent little diversification. The Indo-Malayan element was thus restricted to tropical mesic and monsoonal habitats, while the Antarctic elements were constrained to cool-temperate and alpine habitats (Barlow and Hyland 1988). Burbidge (1960) identified a number of geological periods when invasion (and movement of taxa out of Australia) may have occurred; the most important was the Cretaceous when, on her interpretation, exchange between South-east Asia and Australia was possible. At various times in the Tertiary, exchange may have been possible via an exposed Sahul Shelf and New Guinea, the final exchanges occurring at times of low sea-level in the Pleistocene. Burbidge (1960) did not consider that arguments dependent on continental drift could be sustained, and argued for invasion via various land-bridges. She argued that the distribution of species indicated that eastern Australia was a long-standing corridor for the migration of species from the Indo-Malayan region.

A problem with such explanations was that there was little geological evidence for the land connections required as migration routes. Even if land-bridges had existed they would have had to provide appropriate habitats for migrating taxa. Herbert, a strong critic of the invasion model, argued that it was extremely unlikely that habitats capable of supporting rainforest would have occurred on the postulated land-bridges (Herbert 1950) and (e.g. Herbert 1932, 1950, 1967) championed a different theory. He argued that the modern floristic elements, and their distribution, were the result of climatic and edaphic sifting of an ancient palaeotropical flora which was established in Australia as early as the late Mesozoic.

The development of plate tectonic theory rendered hypothetical land bridges unnecessary but, at least for rainforest, the earliest tectonic reconstructions were not incompatible with the classic biogeographical model. Invasion of the Indo-Malayan element would have been possible from 15 Ma, while the Antarctic element could have invaded prior to 45 Ma. The postulated early invasion of the autochthonous element (Burbidge 1960) was, however, at variance with early formulations of the continental drift model. The more recent reconstruction (Audley-Charles 1987) which identifies a number of microcontinental fragments between Laurasia and northern Gondwana, permits the suggestion that a stepping-stone migration route has existed between Australia and South-east Asia since the Jurassic.

The palaeobotanical data now available from Australia allow the relative merits of the invasion and ecological sifting models to be assessed.

The invasion model suggests the early establishment of the autochthonous element and later appearance of the Indo-Malayan element. While the Proteaceae and Myrtaceae (both major components of the autochthonous element) have a long history in Australia (Proteaceae recorded from the late Cretaceous, Myrtaceae from the Eocene), the context suggests occurrence in closed forest rather than in xeromorphic vegetation. Some taxa assigned to the Indo-Malayan element have an equally long history. The first angiosperm in the Australian fossil record which can be equated with an extant genus is *Ilex* (Martin 1977, 1978). Currently the only species in Australia is *I. arnhemensis*, restricted to the north and a member of the Indo-Malayan element. A greater diversity of *Ilex* species is found in New Guinea and South-east Asia. On the basis of these distributions, it would be a reasonable hypothesis to suggest that *Ilex* is a recent invasive element in the Australian flora, but the fossil record shows that the genus *Ilex* was widespread in southern Australia in the late Cretaceous and early Tertiary. While it is impossible to refute at a species level the possibility of *I. arnhemensis* (or its ancestor) being a relatively recent reinvader of the north, the early presence of members of the Indo-Malayan element is well established. Dettmann (1989) showed that the distribution of *Ilex* fossils in the Cretaceous favours an Antarctic origin for the genus. Although the present distribution can be described as Indo-Malayan, this does not indicate possible ancestry.

The late Cretaceous fossil record from the southern hemisphere provides strong evidence that a number of lineages invaded Australia and New Zealand from Antarctica (Stevens *et al.* 1988; Dettmann 1989). These include groups which appear to have evolved in Antarctica— *Nothofagus* (Dettmann *et al.* 1990), *Ilex* and several genera in the Proteaceae (Dettmann 1989)—and others which probably evolved in northern Gondwana but reached Australia from South America via Antarctica, including Myrtaceae (Dettmann 1989).

The record of both pollen and macrofossils demonstrates the widespread co-occurrence of taxa now assigned to Antarctic and Indo-Malayan rainforest elements throughout the Tertiary. Although taxa appear at different times, (Martin 1978; Dettmann 1989), it is difficult to determine whether or not these arrivals represent invasions; the concentration of studied sites in the south-east and relative paucity of suitable sediments in the north and west make it unlikely that a migration into Australia will ever be traced in detail. Invasion from the north during the Cretaceous and early and mid Tertiary is likely to have been by island-hopping and to have been infrequent; detection of invasive taxa will be like searching for a needle in a haystack. After 15 Ma, passage into (or from) Australia is likely to have been much easier and any substantial invasion is more likely to be detectable.

The probability of a post-Miocene invasion from the north has been accepted on the basis of taxonomic similarities between northern Australia and Malesia (see Barlow 1981; Barlow and Hyland 1988); Thorne (1986a) suggested that 'no well informed field botanist doubts the heavy impact of these "intruders" from abroad upon the rainforest, monsoon and littoral plant communities of eastern Australia' but the fossil record provides little confirmation of such an invasion. Truswell *et al.* (1987) found great similarities between the pre- and post-contact sequences in post-Miocene records from the Atherton Tableland in north-east Queensland and early sequences from Victoria (Fig. 6.6). The lower values of *Nothofagus* and higher values of *Araucaria* in Queensland are taken to indicate responses to climatic conditions rather than the direct influence of continental drift. While the taxa represented in Fig. 6.6 are obviously only a small proportion of those in the vegetation, and invading species may have left little or no mark in the pollen record, the similarity of spectra pre- and post- the Miocene collision does provide evidence that there was no major alteration in composition or structure of northern rainforests as a result of any invasion which might have occurred.

In assessing the possibility of a major invasion of rainforest taxa during the Miocene, it has to be remembered that this was a period in Australia when the distribution of rainforest was contracting. Immigrant taxa would not only have encountered competition from established inhabitants, but would also have faced an environment increasingly inimical to the survival of rainforest.

6.5 REGIONAL VARIATION IN RAINFOREST VEGETATION

The fossil record demonstrates that while many present-day rainforest taxa have a long history in Australia, they occurred in very different combinations at various times. Unfortunately, the pollen record provides a generalized regional picture of the vegetation rather than permitting the detailed reconstruction of communities, and the macrofossil record, while providing information about the vegetation at a local scale, is too discontinuous in both space and time to do more than provide tantalizing glimpses of community composition.

At the present day it is possible to recognize broad regional differentiation of rainforest vegetation (Webb *et al.* 1984, see Fig. 2.3). For most of the Tertiary the fossil record is inadequate to confirm or deny the existence of comparable regional variation. It seems reasonable to suggest that there will have been local variation, related to such factors as altitude, aspect, drainage, and soil fertility, at a scale undetectable with certainty from the pollen record, but also, given the size of the continental land mass, large-scale patterns of variation which, with adequate spatial and temporal coverage in the fossil record we might hope, one day, to detect. If we cannot yet resolve past patterns of variation, does the fossil record allow the study of the history of those elements which characterize the different regions seen today?

6.5.1 Temperate rainforest

Taxa which now characterize temperate rainforest (region A of Webb *et al.* 1984), such as *Nothofagus* and *Lagarostrobos*, were, for much of the Tertiary, widespread across much of Australia (Martin 1990; Truswell 1990), but apparently within much richer forests which also contained many species now regarded as subtropical elements.

At the present day, the most extensive areas of

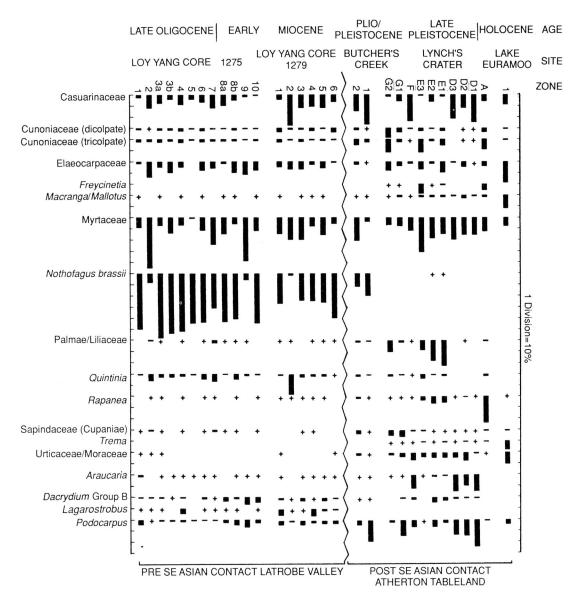

Fig. 6.6 Pollen spectra from sequences before and after the collision between the Australian and Asian plates at 15 Ma. Post-collision specta from northern Queensland, pre-collision spectra from the Latrobe Valley in Victoria. Average values are shown for taxa in each identified zone in the original pollen diagram, expressed as percentages of the forest woody-plant pollen sum. Only woody plants which achieve 5 per cent in any one zone are included (from Truswell et al. 1987).

temperate rainforest are in Tasmania, so that an examination of the fossil record of that state may be particularly revealing of the history of the temperate rainforest. The vegetation history of Tasmania has been studied by Hill and his colleagues (see Hill 1990*a,b*).

The rainforest vegetation of Tasmania today is microphyll fern/moss forest, but the fossil record clearly demonstrates that this is of fairly recent origin (Hill and Gibson 1986). In the early Tertiary, the vegetation had a marked subtropical character, with many broad-leaved angiosperms and estuarine mangrove swamps. In the mid-Tertiary, taxa with temperate affinities were well established, with *Nothofagus* abundant in both the micro- and macro-fossil record, but the subtropical element was still represented, including Araucariaceae and Lauraceae (Hill and Macphail 1983). By the late Tertiary, the rainforest flora was essentially similar to that of western Tasmania today and the subtropical component was absent (Hill and Macphail 1985; Hill 1990*b*).

One of the key genera defining temperate rainforest is *Nothofagus* (Poole 1987). Hill (1983, 1991) studied fossil *Nothofagus* leaves from Tertiary and Quaternary deposits in Tasmania. Leaves intermediate in size and in other features between the extant evergreen species *N. cunninghamii* and *N. moorei* are found in mid-Tertiary floras, but in the late Tertiary the fossil leaves closely resemble those of living *N. cunninghamii*. Hill (1991) suggested that *N. cunninghamii* and *N. moorei* may have shared a common ancestor in the early Tertiary (which has not yet been found), during the cooling that occurred towards the end of the Tertiary reduction of leaf size took place, leading through a series of taxa to the evolution of *N. cunninghamii*. It is suggested that *N. moorei* evolved in a sequence through *N. tasmanica* (Fig. 6.7) and that its present distribution is a result either of migration northwards or of range contraction. There are insufficient data to test these hypotheses but the presence of *N. muelleri* fossils in late Eocene deposits in northern New South Wales (Hill 1991) may indicate that the hypothetical ancestral taxon to both the *N. cunninghamii* and *N. moorei* lineages has a wide distribution. However, while the cool temperate element in Australia has evolved from stocks with a long history on the continent, the form of the present taxa is recent.

There are a number of genera which have species in both Tasmanian rainforest and in mainland cool temperate rainforests. Hill and Read (1987) and Hill (1990*a*) showed that the northern representatives all had larger leaves than the Tasmanian species, a similar pattern to that in *Nothofagus*. (The only exception being in *Eucryphia*; *E. moorei* in New South Wales has much larger leaves than in Tasmanian *E. lucida* and *E. milliganii*, but an undescribed species on Mount Bartle Frere in north Queensland has very small leaves.) This pattern could be explained by similar hypotheses to those advanced in the case of *Nothofagus*, although there is an absence of fossil evidence.

Hill *et al.* (1988), Read and Hill (1989) and Read *et al.* (1990*b*) have shown that the segregation of temperate rainforest into southern (colder) and northern (warmer) components is also reflected in physiological characteristics.

The northern species all have higher-temperature optima for photosynthesis than the southern species, except for *Eucryphia moorei* which is intermediate, reflecting its geographic distribution which 'fills the gap' between the range of *Nothofagus cunninghamii* and *N. moorei*. The ordering of taxa on the basis of their frost resistance is not identical to that based on temperature optima for photosynthesis (Read and Hill 1989), but is consistent with geographic distribution and species' niches.

Read and Hill (1989) suggested that the physiological differences between the northern and southern species are not simply a reflection of recent adaptations to the climatic conditions of their geographic range, but are of older origin. Evidence for this view is provided by the fact that *N. cunninghamii* from Donna Buang in Victoria is more frost tolerant than *N. moorei* at Barrington

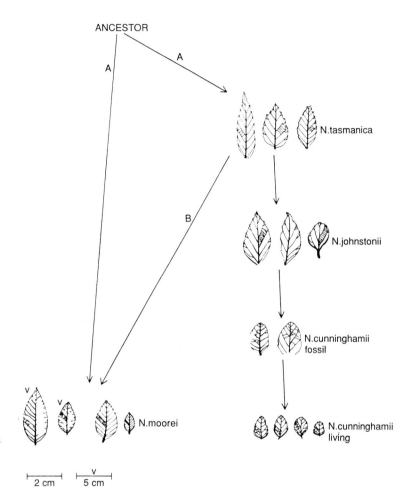

Fig. 6.7 Hypotheses to explain leaf changes during the evlution of *Nothofagus cunninghamii* and *N. moorei* (from Hill 1983).

Tops, despite Barrington Tops experiencing a harsher climate, while *Atherosperma moschatum*, a southern species which has outliers further north, has greater frost resistance than *Doryphora sassafras* although both species occur on Barrington Tops (Read and Hill 1989).

Kirkpatrick and Brown (1984) have suggested that Quaternary glaciation has been a major factor shaping the Tasmanian flora, leading to its distinctive nature when compared with mainland Australia, but Hill argued that the greatest changes to the rainforest vegetation occurred earlier in the mid to late Tertiary (Hill and Gibson 1986; Read and Hill 1989). The early Tertiary fossil record in Tasmania contains many taxa now absent from the island but whose closest relatives are now found further north (Hill 1983; Hill and Macphail 1983; Bigwood and Hill 1985). The development of colder conditions around the Miocene (Kemp 1978), may have caused the migration north, or restriction of previously more extensive ranges of those taxa intolerant of these conditions. Some taxa, now endemic to Tasmania, appear to have changed little during the Tertiary, others have become extinct but persist further north e.g. *Araucaria*, *Agathis*, Lauraceae and

Nothofagus spp. producing *N. brassii* pollen type (Hill 1987; Hill and Bigwood 1987; Hill and Macphail 1983; Bigwood and Hill 1985). In other cases, evolution has occurred, leading to small-leaved, low temperature-adapted species in Tasmania, with taxa more similar to the mid-Tertiary ancestor becoming restricted to lower latitudes. However, while the segregation of northern and southern temperate floras may have taken place in the late Tertiary, the severe climatic changes in the Quaternary may have caused further changes to the rainforest flora. During the coldest periods, alpine vegetation occupied much of Tasmania and any rainforest habitat would have been very limited (Kirkpatrick 1986). Opportunities for migration northwards across the Bass Strait would have been limited in view of the prevailing cold, dry conditions at times of low sea-level. At least one genus (*Quintinia*), now restricted to more northern rainforest, was present in Tasmania up to the start of the Quaternary (Hill and Macphail 1983).

The species listed by Hill *et al.* (1988) as occurring in northern temperate rainforest (Table 6.1) are varied in their distribution. *Nothofagus moorei* is the dominant species in cool temperate (microphyll fern/moss) rainforest from the Barrington Tops to southern Queensland; *Eucryphia moorei* is dominant in southern New South Wales. *Vesselowskya rubifolia* is a small tree virtually restricted to *N. moorei* stands. *Atherosperma moschatum* occurs with both *E. moorei* and in the southern *N. moorei* stands. *Quintinia sieberi* is found in both cool temperate and warm

Table 6.1 Temperate rainforest species in Australia, grouped on the basis of distribution and evolutionary relationships. Group 1 contains species endemic to southern temperate rainforests, without related species in northern temperate rainforests; Group 2 contains genera with at least one species in the northern and southern temperate rainforests, but with no overlap in distribution; Group 3 contains species which occur in both northern and southern rainforests; and Group 4 contains species which are present in northern temperate rainforests but are absent from the southern temperate rainforests (from Hill *et al.* 1988).

Group	Southern temperate rainforest	Northern temperate rainforest
1	*Athrotaxis selaginoides* *Phyllocladus aspleniifolius* *Lagarostrobos franklinii* *Anodopetalum biglandulosum*	
2	*Nothofagus cunninghamii* *Eucryphia lucida* *E. milliganii* *Anopterus glandulosus* *Acradenia frankliniae*	*N. moorei* *E. moorei* *A. macleayanus* *A. euodiiformis*
3	*Atherosperma moschatum* *Parsonsia straminea*	*A. moschatum* *P. straminea*
4		*Doryphora sassafras* *Ceratopetalum apetalum* *Vesselowskya rubifolia* *Quintinia sieberi*

temperate (in the New South Wales sense) rainforest. (Interestingly another species of *Quintinia*, *Q. quatrefagesii*, is found in montane microphyll fern thickets on high mountain summits in northeast Queensland). The vine *Parsonia straminea* is widespread in a range of rainforest types. *Anopterus macleayanus*, a small treelet is found in subtropical (complex notophyll vine) forest and in *N. moorei* stands. *Acradenia euodiiformis* is a small tree, most common towards the edge of rainforest stands in both subtropical and 'warm temperate' rainforest.

Doryphora sassafras and *Ceratopetalum apetalum* are found in *N. moorei* stands, while *D. sassafras* is co-dominant in some *E. moorei* stands. Both species are also found in complex notophyll vine forest, as well as in the simple notophyll vine forest type classified as warm temperate in New South Wales. As discussed on p. 75, most of these simple notophyll forests stands are best regarded as edaphically controlled derivatives of subtropical (complex notophyll vine forest) rainforest rather than a climatically determined form.

The distribution of *A. macleayanus*, *A. euodiiformis*, *D. sassafras*, and *C. apetalum* supports the suggestion that temperate rainforest, despite its apparent distinctiveness, evolved from a more widely distributed 'subtropical' rainforest type. The inclusion of temperate and subtropical rainforests in the same ecofloristic region A, in Webb *et al.*'s (1984) classification is also strong evidence in favour of the hypothesis, and provides strong reasons for regarding subtropical rainforest not as depauperate tropical rainforest but as an important vegetation type in its own right. From the fossil record, subtropical rainforest has had a long history as a vegetation type in Australia, and as environmental conditions have changed, it has diversified and evolved to give rise to temperate rainforest as well as itself surviving at low latitudes.

Not all of the temperate rainforest taxa may share the same history. *Nothofagus gunnii*, a deciduous species, is a Tasmanian endemic. It is found as a tree in higher-altitude rainforests and as a shrub in more exposed alpine and sub-alpine sites. The leaves of *N. gunnii* are very similar to those of the Chilean *N. pumilio* which occupies a similar niche. Hill and Gibson (1986) suggested that *N. gunnii* and *N. pumilio* had a common ancestor which migrated from South America, probably reaching Australia in the early Tertiary. At this time, the environment in coastal Antarctica was probably suitable for the growth of this group of *Nothofagus* (Truswell 1982). Macrofossils indistinguishable from *N. gunnii* have been found at two mid-Tertiary sites in Tasmania (Hill and Gibson 1986). *N. gunnii* produces pollen referred to the *N. fusca* type, and this has been recorded in Australia since the Palaeocene (Hill and Gibson 1986). It is therefore possible that at least in restricted areas, probably at higher altitudes, some form of temperate rainforest has occurred in southern Australia for much of the Tertiary, and what is now regarded as a single element in the flora has not shared a common history.

6.5.2 Monsoonal forests

The deciduous and semi-evergreen vine thickets of ecofloristic region C (Webb *et al.* 1984) are floristically and structurally distinctive. Their distribution is centred on the seasonal tropics but extends southwards to northern New South Wales. Are the species of this group an ancient lineage with a long separate history in Australia, or are they a late Tertiary segregate of the wider rainforest flora which had the ability to evolve in response to the increasingly drier conditions which developed in the Miocene?

Unfortunately, there is a paucity of fossil evidence from northern Australia with which to address these questions, so discussion must be based largely on inferences drawn from present distributions.

The strong similarities (structurally, physiognomically, and floristically) between the deciduous woody vegetation in monsoonal Australia and forests in climatically similar areas of Asia, were commented upon by Hooker (1860) and have since been emphasized by other authors (Webb and Tracey 1981a,b; Barlow and Hyland 1988; Specht 1988). The incidence of endemism at the generic level in ecofloristic region C is markedly lower than in the other two regions (Webb and Tracey 1981a; Webb et al. 1986). Webb and Tracey (1981a) and Webb et al. (1984, 1986) argued that these distinctive features of ecofloristic region C point to its defining floristic element as having a long independent existence, and agreed with Herbert (1960, 1967) that the element is derived from an ancient palaeotropical monsoon forest flora. Whether such a flora existed in the Cretaceous or whether it evolved in the early Tertiary when interchange between Africa, India, South-east Asia, and Australia may still have been possible, as postulated by Barlow and Hyland (1988), remains uncertain. Even if the monsoonal flora dates back only to the early Tertiary, the subsequent very limited diversification at generic level is, nevertheless, remarkable. If the element has existed in Australia throughout the Cainozoic, it implies that, at least in limited areas, a relatively dry and probably strong seasonal climate has prevailed throughout this length of time.

However, not all the species in ecofloristic region C may be part of this palaeotropical monsoon element. Webb and Tracey (1981a) pointed out that there is a significant difference both structurally and floristically within the region, which is correlated with soil type. The deciduous stands occur on the most fertile soils and contain many genera with links to Asia and Africa. The evergreen and semi-evergreen species are found on less-fertile soils associated with permanent water, and are related to simple notophyll and microphyll evergreen vine forests on oligotrophic soils in eastern Australia. These stands may occupy refugia which support elements of the once more widespread mesic flora which has otherwise retreated to the eastern highlands.

In northern Australia, there is a number of species interspersed in open sclerophyll communities—these species are closely related to rainforest taxa (Webb and Tracey 1981a; Gillison 1985, 1987; Table 6.2). Among these species there has been convergent evolution of small leaves, mostly with a narrow linear shape (Webb and Tracey 1981a). It is probable that these species evolved from ancestors in closed forest, in response to the late Tertiary drying. The interspersion of rainforest elements in open woodland is most pronounced in the monsoonal tropics, where there is predictably high rainfall during the wet season. In the south-east, where temperatures are lower and rainfall less predictable, there are few rainforest elements in the open eucalypt woodlands.

One of the striking indicators of fragmentation of the closed forest is provided by the palm genus *Livistona*. As well as occurring in eastern rainforest, there are a number of species across northern Australia: *L. humilis* is widespread in savannah woodland, but other species are restricted to very limited areas (e.g. *L. mariae* near Alice Springs and *L. eastonii* on the Mitchell Plateau) (Figs. 6.8, 6.9). The evolution of these narrow endemics in widely disjunct locations is likely to have occurred in limited favourable habitats after the retreat of the more continuous closed forest.

Accepting that at least part of the dry palaeotropical element has been in Australia for a long period, the question of when and how the present distribution was established has been discussed by Gillison (1985, 1987). He stressed the similarity between the inland monsoonal forests and the closed littoral rainforests on sand dunes. These coastal communities have strong structural similarities throughout the tropics, and the Australian examples have floristic analogues in Malesia and the Pacific. Gillison (1987) suggested that the coastal communities were the habitat of the ancestral species which evolved to give rise to the deciduous inland element, and identified a number of vicariant taxa in littoral and inland floras. He postulated (Fig. 6.10) a number of corridors through which migration could have occurred. Given the stability of the Eastern Highlands, some of these routes would have been available throughout the Tertiary. The existence of a number of widely separated, relatively dry, low regions of the

Table 6.2 Genera with species in both sclerophyll communities and rainforest (modified from Gillison 1987).

Family	Rainforest genus	Species in open woodland
Anacardiaceae	Buchanania	B. obovata
Apocynaceae	Wrightia	W. saligna
Bombacaceae	Bombax	B. ceiba
Boraginaceae	Ehretia	E. saligna
Burseraceae	Canarium	C. australianum
Capparidaceae	Capparis	C. lasiantha
Celastraceae	Siphonodon	S. pendulus
Combretaceae	Terminalia	T. canescens
Euphorbiaceae	Excoecaria	E. parvifolia
Gyrocarpaceae	Gyrocarpus	G. americanus
Lecythidaceae	Barringtonia	B. acutangula
Meliaceae	Owenia	O. acidula
Mimosaceae	Archidendropsis	A. basaltica
Moraceae	Ficus	F. opposita
Myrtaceae	Syzygium	S. eucalyptoides
Oleaceae	Notelaea	N. microcarpa
Palmae	Livistona	L. muelleri
Papilionaceae	Erythrina	E. vespertilio
Pittosporaceae	Pittosporum	P. phillyraeoides
Proteaceae	Grevillea*	G. parallela
Rhamnaceae	Alphitonia	A. excelsa
Rubiaceae	Gardenia	G. vilhelmii
Rutaceae	Flindersia	F. maculosa
Sapindaceae	Atalaya	A. hemiglauca
Sapotaceae	Planchonella	P. pohlmaniana var. vestita
Solanaceae	Duboisia	D. hopwoodii
Sterculiaceae	Brachychiton	B. diversifolius
Verbenaceae	Clerodendrum	C. cunninghamii

*Grevillea spp. are widespread components of southern sclerophyll communities.

Dividing Range has long been recognized, and their biogeographic significance has been the subject of speculation—see, for example, Cambage (1914), who referred to them as geocols. Webb and Tracey (1981a) stressed the importance of these gaps as barriers, separating major blocks of rainforest on the eastern seaboard and permitting the regionalization of the rainforest flora, but this physiographic feature may have at the same time functioned as a link between the coast and inland regions, although the hypothesis is not testable with currently available data.

6.6 PRIMITIVE ANGIOSPERMS IN AUSTRALIAN RAINFORESTS

The origin of angiosperms is a topic which has long excited interest and speculation. The conclusion from the fossil evidence reviewed above is

Fig. 6.8 *Livistona eastonii* on the Mitchell Plateau in the Kimberley region of north-west Australia

Fig. 6.9 An, as yet, undescribed *Livistona* sp., Victoria River, Northern Territory

that the earliest flowering plants in Australia were denizens of rainforest, and that the present rainforest vegetation contains lineal descendents of some of these species. An idea has grown up, and achieved currency in the popular media, that Australian rainforest was the cradle of the flowering plants. Can such a claim be supported?

Comparative morphological studies have led to widespread agreement that a number of angiosperm taxa can be regarded as primitive (Takhtajan 1969). Referring to living taxa as primitive does not necessarily imply that they are ancestral, but rather that they possess a larger number of primitive traits than other taxa. Equally, primitive does not imply inferior. The continuing existence of these primitive taxa indicates the degree to which they are adapted to the environmental conditions they face.

Takhtajan (1969) stressed the concentration of primitive angiosperms in the Western Pacific and argued from this that flowering plants had originated somewhere in this region. More recently, he has suggested that the centre of angiosperm evolution may have been on the continental fragments lying to the north of Australia in the early Cretaceous (Takhtajan 1987).

Of the fourteen families of angiosperms regarded by Takhtajan (1969) as the most primi-

Fig. 6.10 Postulated direction of speciation from littoral to inland semi-evergreen and deciduous vine forest/thicket environments. Arrows on the east coast of Australia indicate dry lowland corridors through the Great Dividing Range. Other arrows indicate possible pathways via older shorelines from early marine transgressions (from Gillison 1987).

tive, eight occur in north-east Australia, including two endemics, Austrobaileyaceae and Idiospermaceae, and one, Eupomatiaceae, which, outside Australia, is found only in New Guinea. At the family level the concentration of primitive angiosperms in Australasia is the highest in the world (Barlow and Hyland 1988). The occurrence of these primitive flowering plants has been an important argument for the conservation of north-east Queensland rainforests (see Chapter 9).

However, the survival of these taxa cannot be taken to indicate that Australia was, or was necessarily close to, their centre of origin. It has been argued by a number of authors (see e.g. Raven and Axelrod 1974) that angiosperms first arose in west Gondwana (west Africa), in the early Cretaceous, and rapidly spread across the supercontinent. If this was the case, the surviving concentrations of the descendants of these early angiosperms thus represent refugia and not the sites of origin.

There is little evidence to suggest that angiosperms originated in Australia. The oldest angiosperm pollen type yet recorded in Australia is the *Clavatipollenites* morphotype, from the early Albian of Queensland (Truswell *et al.* 1987). This occurrence is some ten million years later than the first records in Africa and Europe. Given the imperfections of the fossil record, the absence of angiosperms in Australia, until long after their establishment elsewhere, must remain conjectural, but the evidence is not incompatible with an invasion of early angiosperm taxa, although the route by which migration could have occurred remains uncertain. The early angiosperms may have been streamside or coastal plants, opportunistic colonizers of frequently disturbed habitats, and migration along the coast may have been favoured (see Truswell *et al.* 1987). The high global sea-level in the early Cretaceous resulted in marine transgressions and extensive coastlines, so the opportunities for coastal migration may have been particularly favourable at this time. In the late Cretaceous a major invasion route into Australia appears to have been from South America via Antarctica (Dettmann 1989).

The families regarded as most primitive in present floras are not well represented anywhere in the fossil record, but the available evidence suggests that they were widespread in the late Cretaceous and early Tertiary. Winteraceae pollen has been found in the Cretaceous of Israel and the early Tertiary of southern Africa, and Eupomatiaceae pollen in the Cretaceous of California (this family, now restricted to Australia and New Guinea, has yet to be recorded from Australian sediments) (Truswell *et al.* 1987). Such records, plus the early widespread distribution of families such as the Myrtaceae, is suggestive of a widespread, relatively homogeneous, early angiosperm flora, for which Australia provides a refugium.

The refugial status of primitive taxa is also apparent in their present distribution within Australia. Nix (1982) has demonstrated that the number of primitive genera within a region is correlated with the seasonality of climate; the highest numbers occurring in regions with low seasonality indices. It would appear that these primitive genera with a concentration of conservative traits survive best in those environments which retain climatic conditions that have persisted since at least the early Tertiary. Barlow and Hyland (1988) stressed the importance of the Great Escarpment in providing refugial conditions for primitive taxa since the early Tertiary. The major habitats for primitive taxa are complex and simple notophyll vine forests, rather than either mesophyll or microphyll forests, suggesting that subtropical environments may have been widespread for much of the Tertiary.

6.7 THE QUATERNARY

The Quaternary has been a time of frequent, rapid and massive environmental change globally. During the last 1.8 million years, there have been at least 17 glacial–interglacial cycles, each of about 100 000 years (Williams 1984). The interglacials were relatively short, about 10 per cent of each cycle. The rate of climatic change may have been up to a thousand times faster than changes of comparable magnitude during the Tertiary (Williams 1984). These climatic changes were responsible for the most recent sifting of the rainforest biota.

A broad overview of environmental conditions during the Quaternary is provided by Williams (1984) (see Figs. 6.11, 6.12, 6.13). The last glacial maximum was between 25 and 17 000 years ago, at which time sea-level was about 150 m lower than the present. Tasmania and New Guinea were joined to the mainland and much of the Sahul Shelf was exposed. The lower sea-surface temperatures reduced evaporation and hence rainfall. The incidence of tropical cyclones would have been reduced. In the south-east of the Continent, there were areas of permanent snow and ice on the Alps and in Tasmania. Conditions in at least the preceding six glacial maxima were probably similar.

In Tasmania, summer temperatures on the west coast at the height of glaciation may have been at least 5 °C lower than at present. Alpine and sub-alpine communities were extensive (Kirkpatrick 1986), and rainforest communities would have been very limited in area.

After the glacial maximum, conditions became warmer and wetter. About 9000 years ago (Fig. 6.12), temperatures were probably warmer than the present, but sea-levels had yet to reach their present position and Australia and New Guinea were still joined. The sea-level rise largely ceased about 6000 years ago.

On the mainland, there is little detailed evidence for rainforest from the Quaternary except for north-east Queensland. In the south-east, the long pollen record from Lake George, situated on the southern Tablelands at about 670 m altitude near Canberra (Singh and Geissler 1985) provides tantalizing glimpses of rainforest. The surrounding vegetation is predominantly low open forest and woodland, but with areas of tall open forest (wet sclerophyll forest) about 30 km to the south-east. The nearest rainforest stands are closer to the escarpment edge further to the east. The detailed pollen record covers about 730 000 years, and spans eight glacial–interglacial cycles (Singh and Geissler 1985).

The earlier interglacial, wetter, periods appear to have been dominated by *Casuarina*, accompanied by several rainforest taxa. In the glacial intervals the vegetation is interpreted as an open herbfield.

This interglacial vegetation type is without a modern analogue; *Casuarina* spp. today are only forest dominants in restricted habitats. The fossil record does not permit localization of the rainforest element; was it interspersed through the forest, or concentrated in restricted favourable

ORIGINS AND HISTORY OF AUSTRALIAN RAINFORESTS

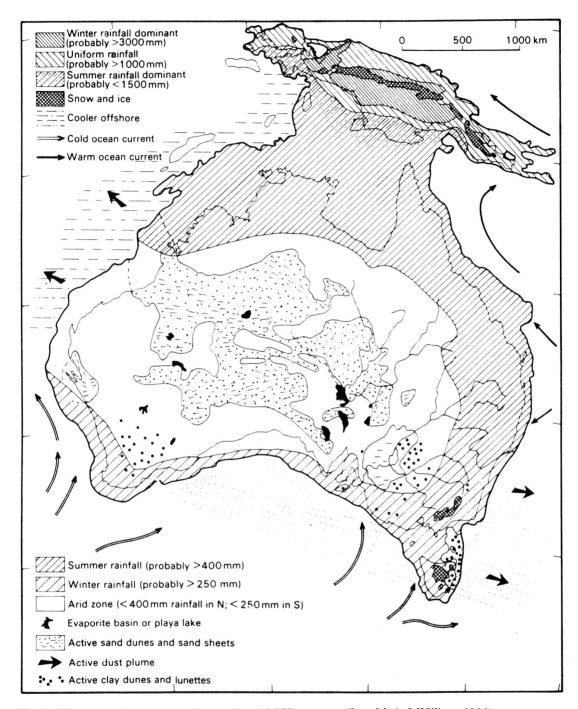

Fig. 6.11 Morphoclimatic map of Australia at 18 000 years ago (from M. A. J. Williams 1984).

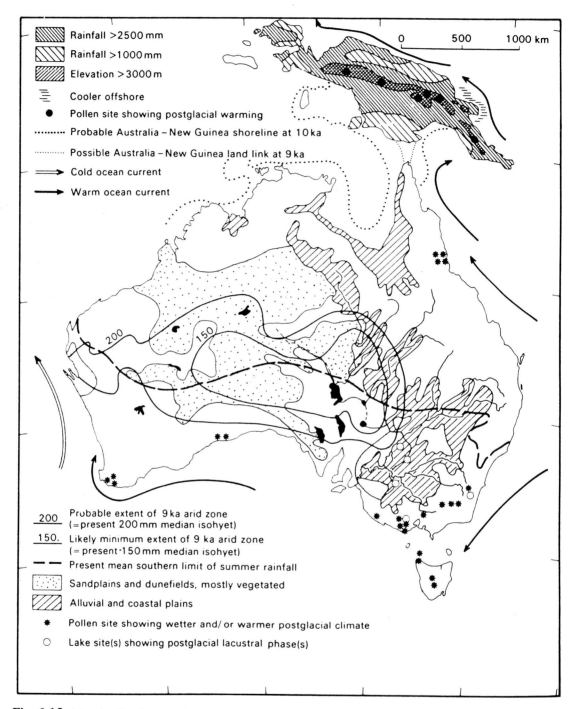

Fig. 6.12 Morphoclimatic map of Australia at 9000 years ago (from M. A. J. Williams 1984).

ORIGINS AND HISTORY OF AUSTRALIAN RAINFORESTS

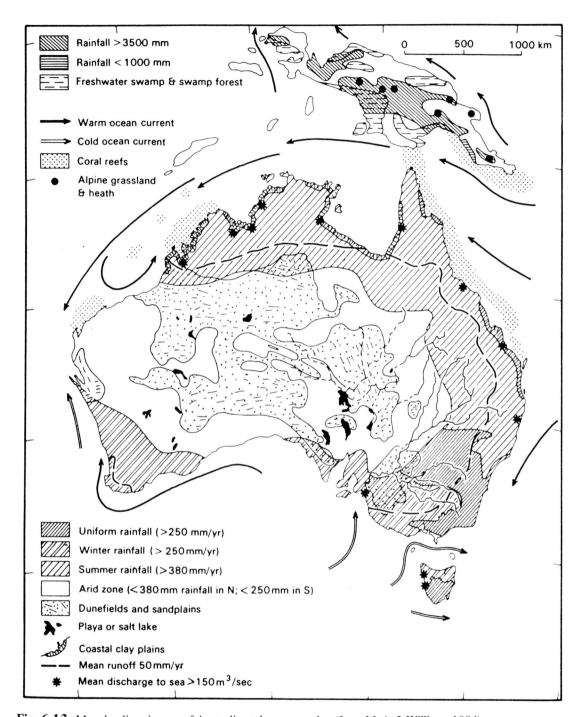

Fig. 6.13 Morphoclimatic map of Australia at the present day (from M. A. J. Williams 1984).

niches? Numerically, the largest component regarded by Singh and Geissler (1985) as being from cool-temperate rainforest taxa in the fossil record is provided by *Lycopodium* and ferns (*Cyathea, Dicksonia, Histiopteris, Pteris,* and *Gleichenia*). These understorey species could equally indicate vegetation similar in structure to present-day wet sclerophyll forest, but with a *Casuarina* rather than *Eucalyptus* overstorey. *Nothofagus*, which is a very high pollen producer, is present in very low amounts. *N. menziesii* and *N. fusca* types disappeared early in the record, but *N. brassii* type (produced by a taxon now absent from Australia) is recorded sporadically until less than 25 600 years ago. It is possible that this *N. brassii* pollen (and that of some other rainforest taxa) is reworked from older sediments, although Singh and Geissler (1985) suggested, on the basis of the condition of the grains, that this is unlikely. If *N. brassii* was present, then it seems to have been in low amount but it is surprising that it apparently survived longer than the other *Nothofagus* groups.

There is a marked change in the pollen record from what is interpreted as the start of the last interglacial (*c.*128 000 years ago) (Zone F of Singh and Geissler 1985), when there was a great expansion of *Eucalyptus* and myrtaceous shrub species, while *Casuarina* was only poorly represented. The interglacial was also characterized by a great increase in fossil charcoal particles. *Casuarina* was regarded by Singh and Geissler (1985) as being more sensitive to fire than *Eucalyptus* and, parallel to the decline in *Casuarina*, other fire-sensitive taxa present in previous interglacials are poorly represented after 128 000 BP. Ladd (1989) has questioned the assumption of fire sensitivity of the Casuarinaceae, and has suggested that Singh and Geissler's (1985) interpretation of the decline in the Casuarinaceae at Lake George may require modification.

Prior to this date, charcoal has been found associated with interglacials, but only in small quantities. In the glacial periods, with open herbfields, there was probably insufficient fuel to support major fires. During the forested interglacials, fires were infrequent and probably caused by lightning. The great increase in charcoal at 128 000 BP at Lake George is presumed to indicate an increase in the frequency (and/or intensity) of fires. As there is no reason to believe that the incidence of lightning would have changed between interglacials, Singh (Singh *et al.* 1981; Singh 1982; Singh and Geissler 1985) suggested that the greater number of fires was initiated by human agency. If this was the case, then the human occupation of Australia must have commenced at least 130 000 years ago.

The Quaternary history of north-east Queensland has been reviewed by Kershaw (1981, 1985, 1986). The long pollen record from Lynchs Crater (Fig. 6.14) covers about 190 000 years, although the dating is less secure than that for Lake George. The record covers two glacial–interglacial cycles.

Lynchs Crater is a volcanic crater at about 650 m altitude, towards the eastern edge of the Atherton Tableland. Annual rainfall is about 2500 mm, with highest rainfalls during summer. At the time of European settlement the Crater was surrounded by complex notophyll–mesophyll vine forest, but much of this was rapidly cleared towards the end of the last century (Chapter 9).

The pollen record cannot provide direct evidence of the structure of vegetation. On the basis of the species composition, inferences as to structure can be made, and these are listed in Fig. 6.14.

The maximum development of rainforest is shown in zones L, I, G, and A. However, the order of importance of taxa varied between zones (Table 6.3). This temporal variation could indicate different environmental conditions in each zone, or reflect the complexity of the mosaic of species composition in rainforest and the small catchment for the pollen rain. These four zones in total account for about 40 000 years out of the last 200 000. For the most of the period, the vegetation is interpreted as araucarian notophyll or microphyll vine forest with *Casuarina* well represented and, on the basis of comparison with modern analogues, is interpreted has having

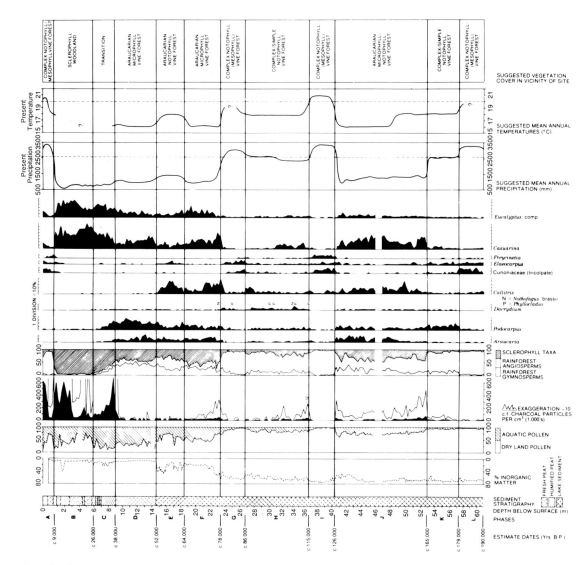

Fig. 6.14 Selected pollen records from Lynchs Crater, together with estimated time-scale, inferred vegetation, precipitation, and temperature (from Kershaw 1985).

occurred under much colder and drier conditions than the present, possibly with a less-seasonal climate. Rainfall is estimated as being between 800 and 1400 mm with mean annual temperatures between 18°C and 16°C. From about 38 000 BP there was a decline in *Araucaria* and a marked increase in *Eucalyptus*, a change correlated with an increase in charcoal. Kershaw (1981, 1986) interpreted the increased charcoal with increased burning and suggested that Aborigines may have first arrived in the area about 38 000 BP, and their burning was responsible for

Table 6.3 Relative importance of the five most commonly represented rainforest pollen taxa within the complex rainforest phases A, G, I, and L of Figure 6.14 (from Kershaw 1985).

Order of importance	Phase A	Phase G	Phase I	Phase L
1	*Rapanea*	*Cordyline*	*Syzygium* sp. B.	Cunoniaceae
2	Cunoniaceae	*Elaeocarpus*	Cunoniaceae	Palmae
3	*Acmena*	*Syzygium* sp. A	Palmae	*Elaeocarpus*
4	*Freycinetia*	Cunoniaceae	*Freycinetia*	unknown tricolpate
5	*Elaeocarpus*	*Rapanea*	*Elaeocarpus*	Sapindaceae

converting the araucarian forest to a sclerophyll forest similar to that prevalent over much of Australia today. However, the influence of the Aborigines does not seem to have prevented the post-glacial expansion of moister rainforest.

The vegetation in zones K and M is not easily interpreted and may not have modern analogues. A feature of zone M is the relatively high representation of *Dacrydium* (*sensu lato*), which is now absent from mainland Australia. There are scattered records older than 78 000 BP of *Phyllocladus* and *Nothofagus brassii*, which are also both absent from the mainland today.

The Lynchs Crater record clearly demonstrates the lack of fixity in rainforest composition (Table 6.3), and that the climatic changes during the Quaternary were responsible for continuing the process of sifting of the flora which had been under way since at least the early Miocene. It seems probable that the first glacial cycle would have been responsible for a major change in the rainforest flora but, unfortunately, no known site provides evidence of these changes. Subsequent glacial periods, including the most recent, nevertheless caused further reductions in the flora.

6.8 THE EVIDENCE FROM PRESENT DISTRIBUTION

Webb *et al.* (1984) showed that the present distribution of species permitted Australian rainforest vegetation to be divided into three ecofloristic regions (see p. 30). Although the regions can be characterized climatically as well as floristically, Webb *et al.* (1984) argued that the difference in their floristic composition reflected ancient differentiation of the Gondwanan stock rather than simply climatic segregation of a single flora. Webb *et al.* (1984) acknowledged that some invasion of taxa from the north into the wet tropics of north Queensland occurred in the late Tertiary (although emphasizing that the taxa concerned cannot be identified), but stressed that most of the wet tropical forest should be regarded as derivatives of an inheritance from the early Tertiary period. Nix's (1982) climatic reconstructions (Fig. 6.15) suggest that areas with markedly different climatic regimes would have existed within Australia throughout the Tertiary; it seems reasonable to postulate a corresponding regional differentiation in the flora and vegetation, although the detection of such geographical variation is not possible from the presently available coarse fossil record. The utility of continuing to recognize Antarctic, Indo-Malayan, and autochthonous elements in the flora is doubtful. While the flora is divisible into groups on the basis of distribution and ecological preferences, the connotations of the names in terms of the invasion hypothesis make it difficult for them to be used as merely neutral labels.

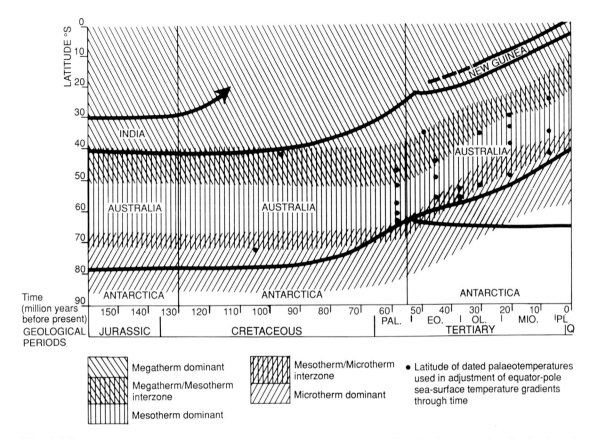

Fig. 6.15 Inferred consequences of continental movements and global climatic change on the distribution of primary thermal response groups in the flora of East Gondwana (from Adam 1987b; after Nix 1982).

6.9 A SYNTHESIS

The conclusions reached on the basis of plate tectonics, the fossil record, and the present distribution of species, result in a synthesis very similar to that held for many years by Herbert. In his 1960 review, he suggested that if the present distribution of rainforest species was considered:

we could come to the conclusion that invading tongues of once alien plants pushed in from different directions, and that in the case of our tropical and sub-tropical rain forests the attenuation southwards and westwards is due to the more recent penetration of a northern element. Yet examination shows that Australian tropical and sub-tropical rain forest is occupying all the sites that are suitable climatically and edaphically and that it has a high degree of endemism...

If we could discard ideas of Australian sea-isolation and regard this continent as a co-inheritor of an ancient and wide-spread flora that had always been differentiated by ecological factors, we might arrive at a more satisfactory conclusion. The flora of Australia is an ancient one. To the south, and more particularly, though not exclusively, in the temperate rain forests and alpine communities it has developed its own expression of the southern hemisphere or sub Antarctic flora to which it has contributed. This element is also sparsely scattered through the tropical and sub-tropical rain forests and on the tops of some of the north Queensland mountains and extends to New Guinea and beyond. Its main relationships are with South America and New Zealand, and to a lesser extent South Africa. The distribution of parasites, such as *Cyttaria* is evidence of its southern nature (Herbert 1932). To the north and north west lies the great and variable palaeotropical centre of Malaysia, divided by van Steenis (1950) into a hierarchy of divisions, provinces and districts. The Malaysian type shows an abrupt change at Torres Strait, south of which it takes on its Australian aspect. That aspect casts a long shadow into arid regions not exploited by its representatives in Malaysia, where those conditions are non-existent. It is beyond Malaysia, in Australia, India and round to Madagascar that the xerophytic potentialities of the palaeotropic flora are best able to express themselves...

... The thread of relationship of these segregations from the palaeotropic reservoir, as shown by our rain forest Myrtaceae, Proteaceae, Rutaceae and other families and even by genera point to the differences being due to long continued sifting by climatic and edaphic factors...

The concept that the sclerophyll element in the Australian flora was predominant in the earliest angiosperm vegetation on the continent, and that there have been a number of discrete waves of invasion of rainforest, must be laid to rest. The sclerophyll element is certainly autochthonous but arose from rainforest taxa some time in the Tertiary and sclerophyll vegetation in its present form has only existed from the late Tertiary or possibly even the Quaternary. The earliest angiosperms in Australia were minor components of a moist gymnosperm forest but rapidly (in a geological context) became dominant.

Although the present-day Australian flora has a very high proportion of endemic species and genera, there are few endemic families (none of which has more than a few species). Although there are families present in rainforest to the north which are absent in Australia (most notably the Dipterocarpaceae), at a family level the composition of the Australian flora is not markedly different from that of other continents.

Does this mean that, at the time of breakup of Gondwana (or even earlier at the breakup of Pangaea) most angiosperm families had already evolved, with subsequent diversification at subfamily level? The late Cretaceous flora and vegetation appears to have been basically similar across fragmenting Gondwana (Dettmann 1989) but the fossil record of many angiosperms families is poor, and does not extend back beyond the mid-Tertiary, long after Australia had become separated from other continents. Either many families were present earlier but did not become incorporated in the fossil record or dispersal into Australia was easier than has been assumed. There are data which indicate that a few families may have invaded Australia during the Tertiary, and that others may have spread north from Australia (Kershaw 1988; Truswell 1990) but no evidence for migration of many families.

Despite the uncertainty still surrounding the history of many elements of the rainforest flora, the outlines of the history of Australian rainforests are now clear. During the Cretaceous, the gymnosperm-dominated rainforests were invaded by angiosperms, from the south via Antarctica and possibly, although less certainly, across the chain of microcontinents between Gondwana and Laurasia. For much of the Tertiary, angiosperm rainforest (although with a significant gymnosperm component) was the predominant vegetation across the continent (Truswell 1990). This cover was not uniform, either in terms of floristics or structure, there would have been variation reflecting climatic gradients and local soil conditions. It is also likely that there were areas of non-rainforest in particularly harsh local environments (steep, rocky, north-facing slopes, extremely nutrient-poor sites, swamps) which did not support rainforest, although the fossil record does not permit identification of such variation with any certainty. As the climate became generally drier from the Miocene onwards, mesic vegetation became increasingly restricted to spatially limited favourable sites, largely on the eastern seaboard, while sclerophyllous derivatives of the rainforest came to occupy much of the continent. Within the rainforest, climatic cooling throughout much of the Tertiary was a major factor promoting the redistribution of taxa, leading to the present distinction between temperate and subtropical rainforests (Hill 1990*b*).

The current distribution both of rainforest and of taxa within the rainforest, although a product of history, is a phenomenon of the present. The regional differentiation of rainforest described by Webb *et al.* (1984) is but the latest outcome of the climatic and edaphic sifting of the flora which has been taking place ever since angiosperm dominated rainforest became established in Australia.

6.10 RAINFOREST REFUGIA

The fossil record is not sufficiently precise as to permit refugia to be located precisely. However, approximate positions can be determined from present distributions of species; regions with concentrations of species with disjunct distributions or restricted distributions are candidates for having provided refugia. Webb and Tracey (1981*a*) provided a detailed discussion of possible refugia (particularly in north-east Queensland), and identified a number of features of these sites. Even during the driest periods in the Quaternary, the continent was not universally dry. As is the case today, there would have been strong gradients in rainfall related to topography and distance from the coast. The upper slopes of mountains along the Dividing Range would, even during dry periods, have been subject to high rainfall, and could have supported rainforest. Other sites may have retained a moist microclimate even within dry regions because of the availability of groundwater (around springs or beside rivers). In addition to the availability of water, an essential feature of refugia would have been the absence of regular fire. In the wettest sites, permanent moisture would have limited the potential for fire, but in drier sites shelter from dry winds (i.e. on south- and east-facing slopes) or barriers to the spread of fire such as rocky broken slopes may have permitted survival of rainforest species. Survival in more fire-prone regions may also have been favoured on more nutrient-rich soils.

Webb and Tracey (1981*a*) have used plant distribution and environmental data to map possible refugial areas in north-east Queensland (Fig. 6.16) (support for these suggestions is also provided by faunal distributions—Kikkawa *et al.* 1981*a*). However, this figure should not be regarded as indicating the minimal extent of rainforest, rather the refugial areas were regions in which pockets of rainforest survived. Hopkins *et al.* (1990*b*) have shown that in a ridge top site in the central portion of the Mount Windsor Tableland (mapped as an extensive refugial area in Fig. 6.16), a series of buried eucalypt charcoal layers, dated between 12 750 BP and 26 860 BP, demonstrate the presence of dry sclerophyll forest during the last major dry interval. The site is today in the centre of a 20 000 ha block of rainforest. If rainforest survived on the Mount Windsor Tableland during the dry period (and the existence of several restricted endemics strongly suggests that it was a refuge) then it must have been in much smaller areas, possibly on cloudy mountain tops or in fire-proof valleys (Hopkins *et al.* 1990*b*).

It is likely that most refugia were, for at least some periods, very restricted in area. Assuming patterns of species diversity comparable with those in present rainforests, the local populations of many species could have been very low. There are insufficient data on variability within rainforest taxa and on their breeding systems to determine critical population sizes, but it is possible that the attrition of species hinted at in the pollen record reflects the deleterious effects of loss of variation caused by reduction in population size.

6.11 THE VAGILITY OF RAINFOREST TAXA

The rival hypotheses of invasion and the sifting of an ancient pan-Gondwana flora involve different assessments of the dispersability of taxa. The balance of the available evidence is strongly in favour of ecological sifting, although recognizing that angiosperms invaded Australia in the Cretaceous. Nevertheless, the question of the vagility of rainforest taxa is an important one, not only for understanding the origins of Australian rainforests, but also the expansion of rainforest from refugia during the wetter interludes of the last two million years. Unfortunately, relatively little is known about the dispersal characteristics of most modern rainforest species, while as Thorne (1986b) pointed out 'we have no way of estimating the dispersal capacities of the Cretaceous prototypes of our modern angiosperms', and suggested that there has been a tendency to underestimate the ability of some angiosperms to disperse long distances even though their propagules are not obviously adapted for dispersal. He pointed to the floras of volcanic oceanic islands as proof of the occurrence of long-distance dispersal, although Whitmore (1973) demonstrated that very few components of the flora of the continental island fringe in the south-west Pacific had reached those oceanic islands. Webb et al. (1986) stressed the low vagility of most rainforest species, and Thorne (1986b) agreed that poor dispersal is a characteristic of rainforest species.

Although Webb et al. (1986) showed that a high proportion of rainforest species are endozoochores, there is an absence of data on the incidence of long-distance dispersal by birds or bats.

Webb et al. (1986) suggested that rainforest vegetation is highly integrated; piecemeal migration of individual species is unlikely to have permitted the development of complex interactions within and between synusia. The extent to

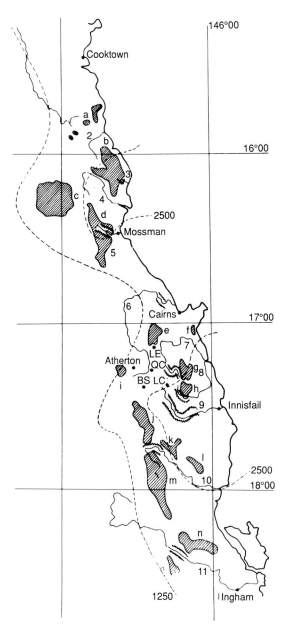

Fig. 6.16 Postulated rainforest refugia in north-east Queensland, (a) Mt. Finnegan, (b) Thornton Peak, (c) Mt. Windsor Tableland, (d) Mt. Carbine Tableland, (e) Lamb Range, (f) Malbon Thompson Range, (g) Bellenden Ker, (h) Bartle Frere, (i) Herberton Range, (j–k–l) Walter Hill Range, (m–n) Cardwell Range, (o) Seaview Range. Permanent rivers: 1. Normanby; 2. Bloomfield; 3. Noah Ck; 4. Daintree; 5. Mossman; 6. Barron; 7. Mulgrave; 8. Russell; 9. Johnstone; 10. Tully; 11. Herbert. Wet coastal gorges occur on rivers 2–11. Gallery forests would have persisted on rivers 1–11 during drier climatic periods in lower rainfall zones. Palynological sites studied by Kershaw (1985) are Lake Euramoo (LE), Quincan Crater (QC), Bromfield Swamp (BS), Lynchs Crater (LC) (from Webb and Tracey 1981a).

which rainforest communities are integrated is, however, debatable and has been little studied (see also Chapter 7). While the fossil record provides evidence for the continuity of taxa, it also shows that assemblages of species in the past have been very different from anything we can observe today (even relatively recently there were combinations of species which do not occur today, for example, 600 000 years ago *Quintinia* was present in Tasmanian rainforest, where it no longer occurs—Macphail 1986). If the rainforest adjusted to such changes, it is difficult to see that the infrequent invasion of one or two new taxa would constitute a major disturbance. What is more probable is that invading taxa would fail to establish because of inappropriate conditions (bearing in mind that the establishment characteristics of species vary (Chapter 5) and that many taxa could not establish in closed stands, while others would not grow to maturity on bare sites).

Despite discounting the importance of long-distance dispersal as an explanation for rainforest species distribution patterns, Webb *et al.* (1986) pointed to the considerable number of species in rainforests throughout much of the Indo-Malesian–Australia region (including *Abrus precatorius*, *Alstonia scholaris*, *Bombax ceiba*, *Ficus racemosa*, *Gyrocarpus americanus*, *Macaranga tanarius*, *Mallotus philippensis*, *Melia azedarach*, *Pongamia pinnata*, and *Toona australis*). The species in this category occur in various stages of forest succession and have various methods of probable dispersal. Are these very ancient species, unchanged since the early Cretaceous, an hypothesis which appears unlikely or are they very well-dispersed ('supernomads'—Webb *et al.* 1986)? If the latter, then why are not more taxa included in the group?

6.11.1 New Zealand and New Guinea

Comparison of Australian vegetation with both that of the presumed long isolated New Zealand, and the much closer New Guinea, part of the same tectonic plate and for extended periods during the last 2 million years with land connections to Australia, may provide some general indication of the likelihood of dispersal of rainforest taxa.

The separation of Australia and New Zealand commenced in the late Cretaceous, at which time the two shared a common flora, much of which had arrived from Antarctica (Dettmann 1989; Dettmann *et al.* 1990; Stevens *et al.* 1988). For much of the Tertiary, the Tasman Sea would have been a major barrier. In addition, New Zealand has experienced much more dramatic changes in landforms than Australia and a much larger proportion of the land surface was covered with snow and ice during the Quaternary.

The immediate impression at the present day is that the flora and vegetation of Australia and New Zealand are very different. This impression is heightened by the absence of eucalypts and of sclerophyll vegetation in New Zealand. The absence of sclerophyll vegetation most probably refects lack of habitat; the geological history ensuring a widespread mantle of, by Australian standards, fertile soils. However, more detailed study shows strong similarities between New Zealand and other parts of Gondwana.

The montane *Nothofagus* forests have similarities, structural, and floristic, with microphyll fern/moss forests in south-east Australia, although overall similarities are probably higher with temperate South America (see Poole 1987). The lowland rainforest is structurally remarkably 'tropical' in appearance (Dawson 1986), and Webb (1987*b*) has demonstrated that in a structural/physiognomic comparison, there is a close relationship between lowland rainforest in New Zealand and lower montane complex/simple notophyll vine forest in New South Wales. Within New Zealand, there is a marked change in leaf-size spectrum from the north, where a few species have macrophyll leaves to the south where microphyll- and nanophyll-sized leaves predominate (Dawson 1986), a trend similar to that in Australian rainforests.

The distribution of taxonomic affinities of lowland rainforest vascular plants have been analysed by Dawson (1986). New Zealand has a rich gymnosperm flora, which is a significant component of the rainforests. All the genera involved are centred in the southern hemisphere today, and they (or their immediate ancestors) were widespread through Gondwana.

What of the angiosperm flora? Takhtajan (1986) places Australia and New Zealand in different floristic kingdoms but this overemphasizes the differences between the two countries. Although New Zealand has a distinctive flora, it is like Australia in that endemism at the family level is slight. In the rainforest, the number of endemic genera is small (Dawson 1986), and it is not clear whether most of these are relics from a once wider distribution, or have evolved in New Zealand. The non-endemic rainforest genera are for the most part shared with Australia and/or Pacific islands (Dawson 1986). At least three hypotheses could be suggested to explain the occurrence of these non-endemic genera in New Zealand:

(1) they represent lineages present before the separation of New Zealand from Australia;

(2) they arrived in New Zealand by long-distance trans-Tasman dispersal;

(3) they reached New Zealand via an 'island hopping' route involving several relatively short steps.

The New Zealand palynological record shows that, in the past, even more groups of plants were shared with Australia. The arrival and subsequent extinction of several groups has been documented (Mildenhall 1980); for example the Casuarinaceae were present in New Zealand from the Palaeocene to the early Pleistocene, *Acacia* from early Pliocene to mid-Pleistocene, and *Eucalyptus*-type from Miocene to early Pleistocene. Based on modern distributions, other taxa may have invaded from the north via New Caledonia, for example, *Beauprea* in the Proteaceae (in New Zealand from mid-Oligocene to early Pleistocene) and *Zygogynum* in the Winteraceae (from Oligocene to Pliocene).

If the palynological record is taken as evidence for Tertiary migrations to New Zealand, then what was the mechanism?

A number of genera are also found on oceanic islands as far away as Tahiti and Hawaii (including *Alectryon*, *Freycinetia*, *Peperomia*, *Planchonella*, *Streblus*, and *Vitex*). Such distributions are presumptive evidence for the feasibility of long-distance dispersal, although the exact mechanism is unknown (Dawson 1986).

Other genera present in New Zealand do not have representatives on oceanic islands (including *Beilschmiedia*, *Corynocarpus*, *Dysoxylum*, *Hedycarya*, *Pennantia*, *Quintinia*, *Tecomanthe*). Dawson (1986) suggested that it was unlikely that these genera could have reached New Zealand by long-distance dispersal across an ocean gap. It is possible that at various times during the Tertiary, lands were exposed along the Lord Howe Rise and the Norfolk Ridge, providing a route for island hopping. Interpretations of the geological history of these now largely submerged features are still controversial, and there is no definite evidence for appropriate islands at a time suitable for migration. Dawson (1986) suggests that the presence of *Araucaria* on Norfolk Island, midway between New Zealand and New Caledonia, may imply the continuous presence of exposed land somewhere along the Norfolk Ridge during the Tertiary as the genus has very limited dispersal ability.

The climatic changes of the Quaternary imposed a dramatic sifting of the flora, and increased the distinctiveness of New Zealand flora and vegetation but affinities to other regions, including Australia, are still apparent.

Whereas New Zealand provides evidence suggesting the vagility of certain rainforest taxa, New Guinea suggests limited dispersal abilities for southern taxa.

Despite its very recent origin, the Torres Strait marks a major biogeographical barrier, at least for vascular plants. The Strait is one of the three 'demarcation knots' (identified by van Steenis (1950)) in the distribution of genera which serve to define Malesia (Fig. 6.17).

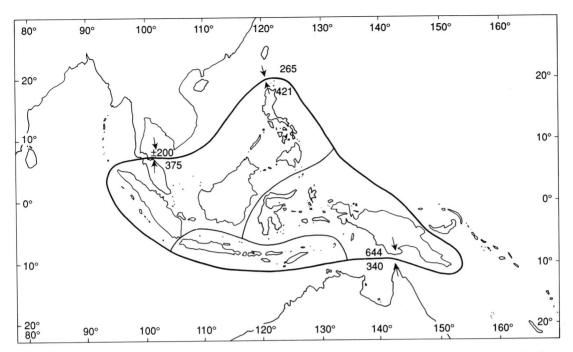

Fig. 6.17 The limits of Malesia. The figures show the number of genera that do not cross the demarcation knots (i.e. 644 genera found to the north of the Torres Strait are absent from Australia (from von Balgooy 1987)).

The distribution of rainforest genera in New Guinea rainforests shows that more genera are confined to Malesia than are shared with Australia (Hartley 1986), and they probably reached New Guinea from the north and west. While some long-distance dispersal may have occurred during the Tertiary, the main opportunities for invasion are likely to have been post-Miocene, after the collision between the Indo-Australian and Asian plates. (The insect fauna of New Guinea is also predominantly Malesian, although the mammals are largely Australian in affinity. The apparently different origin of different elements in the New Guinea biota has long been recognized as a major biogeographical problem—MacArthur and Wilson 1967).

At the time of the collision, New Guinea would have supported 'Australian' plants, but in the wet lowlands these have been largely displaced by the invaders. Why this should have been so is poorly understood, but Hartley (1986) suggested that the original flora may have been adapted to subtropical conditions and was less well-adapted than the invading element to tropical conditions (the evidence from Australia indicates that the core of the Gondwana flora was essentially subtropical/temperate). The southern element in the New Guinea rainforest flora is most prevalent in the montane forests (above 900 m), where they are frequently dominant, although even here Hartley (1986) pointed out a high incidence of genera which are likely to have invaded from the north and west.

Webb and Tracey (1972) cautioned against overemphasizing the differences between Australia and New Guinea floras. They showed that when comparisons between structurally similar vegetation types in similar habitats were made large numbers of genera were shared. However, New Guinea has large areas without analogous

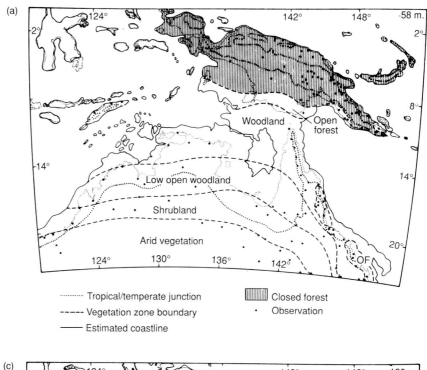

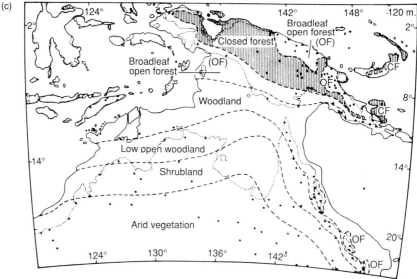

Fig. 6.18 Reconstructions of vegetation distributions in northern Australia and New Guinea over the last 20 000 years.
A. Main structural vegetation types which might have occupied northern Australia, New Guinea and intervening land about 20 000 BP (postulated climatic region—precipitation 0.5 of present, evaporation 0.8 of present, air temperature 3.5 °C below present) sea-level 58 m below present.
B. Main structural vegetation types which might have occupied northern Australia, New Guinea and intervening land about 17 000–14 000 BP (postulated climatic regime—precipitation 0.5 of present, evaporation 1.25 of present, air temperature 1 °C below present) sea-level 120 m below present.

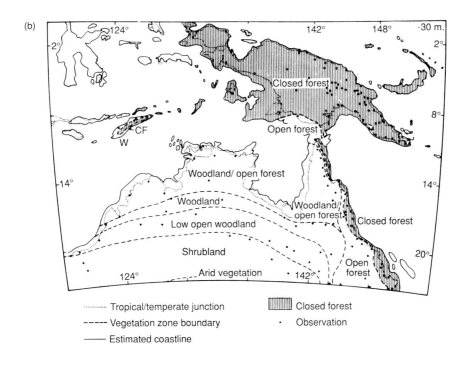

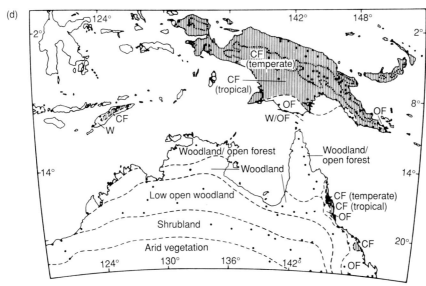

C. Main structural vegetation types which might have occupied northern Australia, New Guinea and intervening land about 8000 BP (postulated climatic regime—precipitation 1.5 of present, evaporation 1.0 of present, air temperature 1 °C above present) sea-level 30 m below present.
D. Main structural vegetation types which occupy northern Australia and New Guinea at present (from Nix and Kalma 1972).

environments in Australia, so it was not unexpected that there is greater richness in New Guinea.

If the invasion of New Guinea occurred from the west, did not at least some of the immigrant taxa spread across into Cape York? Such an invasion has been postulated by a number of authors (see Barlow and Hyland 1988), but is not detectable in the palynological record (Fig. 6.6). It is possible that the Torres Strait land-bridge did not provide conditions amenable to the establishment of many rainforest taxa. Reconstruction of the vegetation of the Torres Strait over the last 20 000 years (Fig. 6.18) suggest that even under optimal climatic conditions rainforest would have been discontinuous on the landbridge. Invasion of Australia from New Guinea by rainforest taxa is likely therefore to have involved 'jumps' between stands of suitable habitat rather than being a continuous 'creeping' process, and relatively few taxa may have been capable of the necessary long-distance dispersal.

There is a very strong similarity between dry, open forest and woodland in southern New Guinea and northern Australia. Some elements of this vegetation may have been present in New Guinea prior to the Miocene, but in view of the evidence for the recent diversification of the sclerophyll flora in Australia, it is likely that much of it spread from Australia.

6.12 THE SCLEROPHYLL FLORA

The fossil record demonstrates that the first angiosperms in Australia were inhabitants of closed, moist forest and that, after flowering plants became dominant, closed mesic forest covered much of the continent between the late Cretaceous and the late Tertiary. Today, most of the forest and woodland vegetation in Australia is referred to as sclerophyll vegetation, with upper canopies dominated by *Eucalyptus*. As has been emphasized, there is a very striking contrast in structure, floristics, and response to disturbance between rainforest and sclerophyll communities.

How, and under what circumstances, did sclerophyll evolve?

Two families very well represented in temperate sclerophyll communities are the Proteaceae and Myrtaceae, both also major components of the rainforest (Johnson and Briggs 1981). The Proteaceae has a very long fossil pollen record in Australia, and in the early Tertiary there was a great diversity of taxa, with a number of lineages subsequently appearing to have become extinct. It is not possible to assign all these pollen taxa to rainforest or sclerophyll types. The Proteaceae is reasonably well represented in the macrofossil record, and in the Eocene of southern Australia a number of leaves has been recorded which can be assigned to the tribe Banksieae (Christophel 1981). These leaves have some sclerophyll features, but the assemblages in which they are found are interpretable as being from rainforest. (At the present day, although *Banksia* is one of the most characteristic genera of sclerophyll vegetation, one species, *B. integrifolia*, is frequently found at the edge of montane rainforest.) The Myrtaceae fossil pollen record is not as long as that of the Proteaceae, but does extend back to the early Tertiary (Martin 1978). Unfortunately, many of the records are not identified below the level of very broad groupings within the family, but more recent, detailed studies of the pollen grains have permitted more identifications to generic level (see, for example, Martin 1987). There are few fossil leaves of Myrtaceae, and in the early Tertiary none which could, with confidence, be assigned to the family (Christophel 1981). However, there are a number of well-preserved fossil fruit from Central Australia which are clearly from the Leptospermoideae, including eucalyptoid forms (Lange 1982). Unfortunately, these fossils are not well dated, but Lange (1982) suggested that major diversification of capsular Myrtaceae had occurred by the Oligocene at the latest. Firmly dated *Eucalyptus* fossils have been described from the mid-Miocene of the Warrambungle Mountains in northern New South Wales (Holmes *et al.* 1982) and fossil wood attributed to *Eucalyptus* has been recorded from Miocene deposits in the upper Lachlan Valley (Bishop and Bamber 1985).

As the rainforest contracted in the Miocene, there was a sharp rise in Myrtaceae pollen, which appears to represent an increase in a number of taxa (Martin 1987). However, reconstruction of the vegetation is still problematical; Martin (1987) suggested a mosaic of wet and dry sclerophyll forest. The long Quaternary records from both the Atherton Tableland and Lake George indicate that sclerophyll forest in its present *Eucalyptus*-dominated form is of very recent origin.

The selection pressures leading to the evolution of sclerophylly have been the subject of discussion for many years. Although many sclerophyll features appear to confer a degree of drought resistance, there is a consensus that paucity of nutrients, particularly nitrogen and phosphorus, was the most important factor. The role of nutrients in the development of scleromorphy was first suggested by Andrews (1916), but the importance of nutrients was clearly shown in the work of Beadle (1954, 1966), and, outside Australia, received confirmation by Loveless (1961).

Much of Australia is a very old landscape, which has undergone no major geological disturbance since the late Mesozoic (Ollier 1986). The soil parent material is also often oligotrophic. For most of the Tertiary, it is likely that there were extensive areas of impoverished soil. Particularly on sites which also imposed water stress (steep, well-drained, north-facing slopes) there may have been strong selection pressure in favour of sclerophylly. The failure to distinguish sclerophyll vegetation in the fossil record may be no more than a reflection of the sparsity of the record and the limited opportunities it affords for the spatial resolution of complex vegetation mosaics.

The sclerophyll element in the present Australian flora is characterized not just by particular leaf features but by its responses to water stress and fire. While low nutrient conditions may have favoured a tendency towards sclerophylly in the closed forest, increased dryness and fire frequency would have imposed new selection pressures. The very different taxonomic spectra of rainforests on infertile soils (for example, *Ceratopetalum apetalum* forests) and sclerophyll forest, suggest that relatively few lineages developed the necessary additional drought and fire tolerance to survive over most of post-Miocene Australia, although those few taxa with those features underwent major radiation.

The present tendency for most rainforest types to be restricted to more-fertile soils may not represent an absolute requirement for high nutrient levels, rather it reflects their greater resilience on more-fertile soils to the disturbance caused by fire (Webb 1968; see Chapters 3, 4, this volume). Nevertheless, even if fire could be excluded it is unlikely that rainforest would spread to cover all the high-rainfall areas, but would continue to be absent from the most nutrient-poor sites. Richards (1967, 1968) demonstrated that under present climatic conditions, *Araucaria cunninghamii* could be grown successfully (at least in seedling and sapling stages) on soils naturally supporting dry sclerophyll forest following the addition of nitrogen. The fossil record indicates that, during the late Tertiary, rainforest occurred over areas now occupied by sclerophyll vegetation. It is possible that the recent increased frequency of fire has been responsible for sufficient loss of nutrients from the poorest soils to exclude the return of rainforest.

The 'tightness' of nutrient cycling in rainforests has been stressed frequently in reviews (see for example Congdon and Lamb 1990), but Lambert and Turner (1989) showed that sclerophyll species were more efficient in retranslocating nutrients (and thus reducing nutrient loss in litterfall) than most subtropical rainforest species (there were some exceptions to this generalization, most notably *Orites excelsa*). This greater ability to recycle nutrients internally may have given sclerophyll species an advantage over rainforests if fertility was declining on poorer soils.

One of the largest and most widespread genera

in Australia is *Acacia* (the wattles), whose pollen record extends only as far as the Miocene (Martin 1978). However, the poor dispersability and low representation of *Acacia* in the contemporary pollen rain (Ladd 1979) suggests that preservation of fossil pollen would have been a rare event, and the absence of *Acacia* from much of the fossil record may not be proof of absence from vegetation. The greatest diversity of *Acacia* is outside rainforest, but a number of wattles are aggressive colonists of disturbed rainforest, for example, *A. elata*, and other species may be characteristic of rainforest margins, although few are frequent members of true rainforest. One of the most important rainforest species is *A. melanoxylon*, which is found from Tasmania northwards. Even in this case, there is dispute as to whether it should be regarded as a true rainforest species. In the absence of a long fossil record, it would be plausible to suggest that the current rainforest taxa are derived from sclerophyll taxa rather than vice versa. *Acacia* was recorded from the late Tertiary Butchers Creek site in north-east Queensland by Kershaw and Sluiter (1982), in what is interpreted as rainforest vegetation. Kershaw and Sluiter (1982) suggested that *Acacia* may have penetrated northern rainforests at an earlier time than in the south-east.

In the tropical open woodlands, there are numerous species with clear affinities to rainforest congeners (Table 6.2). Although these have smaller leaves than the rainforest species (Webb and Tracey 1981*a*), in most cases they are not markedly scleromorphic. It seems reasonable to argue that many of these species diverged from rainforest ancestors during the late Tertiary contraction of rainforest.

6.13 NON-VASCULAR FLORA

The non-vascular flora makes a major contribution to the diversity of rainforest, but is virtually absent from the fossil record. Present distribution patterns are poorly known (Scott 1988), but there are indications that the non-vascular flora of Australian rainforests is, like the vascular plants, a segregate of a once pan-Gondwana flora.

Fungi play an important role in the forest ecosystem—as decomposers which promote nutrient recycling and, in the form of mycorrhizae, in the uptake of nutrients by vascular plants. Few data are available on the incidence of mycorrhizae in Australian rainforest. With the current state of taxonomic and biogeographical information, it is difficult to be certain of the composition of the Australian mycoflora. On the basis of a few species, which may or may not be representative, Horak (1983) has suggested that current distributions reflect on evolutionary origin pre-dating the breakup of Gondwana. A particularly interesting case is provided by *Cyttaria*, a fungus obligately parasitic on *Nothofagus* (Korf 1983; Poole 1987). *Cyttaria* is absent from *Nothofagus* species of the *brassii* group (the oldest group in the fossil record) but is found on both the *menziesii* and *fusca* groups throughout their range. Korf (1983) suggested that *Cyttaria* evolved after the *menziesii* group diverged from *brassii*, possibly in the late Cretaceous.

Lichen floras of cool-temperate (microphyll/ nanophyll) rainforest in Tasmania, New Zealand, and South America are markedly similar (Jørgensen 1983; Galloway 1988). It can be suggested that this similarity has its origins in a pan-Gondwanan flora, although Galloway (1988) has argued that because of the tectonic complexity of the Pacific rim it is important to examine the flora and history of each of the component terrains. Much less is known about the taxonomy and biogeography of lichens in rainforest at lower latitudes.

Bryophytes are frequently a prominent feature of rainforest. The dense covering of bryophytes found on the canopy branches of trees may play an important role in catchment hydrology, absorbing much rainfall, reducing the impacts of torrential rain, and releasing moisture during drier periods. The ability of bryophytes to absorb nutrients from throughfall and to release them as leachates may be an important part of the rainforest nutrient cycle (Pócs 1982; Ramsay *et al*.

1987), but there are no data on these topics from Australia.

Although Scott (1988) emphasized the perils of speculation from imperfect distributional data, he nevertheless concluded that 'it is hard not to believe in a Gondwanaland bryophyte flora which was carried apart on the drifting continents ending up in South America and Australia and to a lesser extent in Southern Africa, for there are so many bryophyte taxa which have broadly similar distributions.' He emphasized in particular the very great similarity between the bryophytes in south-east Australia and New Zealand—'with relatively few exceptions, the bryophyte flora of the cool temperate rain forest of Victoria and Tasmania occurs in comparable habitats in New Zealand—if not identical species then the species so close that they are hard to separate . . .', and suggested that this illustrates a stability of taxa at the species level for perhaps 80 million years, as he regarded colonization between the fragments of the former Gondwana as extremely unlikely.

Ramsay *et al.* (1987) discussed the distribution of bryophytes in different habitats within Australian rainforests, and made the suggestion that hepatics predominate in lowland forests, while mosses achieve greatest abundance at higher altitudes. It is not immediately obvious why this should be so.

7 THE RAINFOREST FAUNA

The vertebrate fauna provides a number of the symbols by which Australia is known internationally—kangaroos, koalas, emus, and budgerigars; these are all creatures of either the arid zone or sclerophyll communities. What of the fauna of rainforests, how distinctively Australian is it, and how is it related to the fauna outside the rainforest?

The extent of knowledge of the fauna varies considerably between taxonomic groups, with vertebrates being better studied than most invertebrates. For a number of major groups, it is known that highest species richness occurs in regions where rainforest is well represented (particularly north-east Queensland and the Border Ranges region). This pattern is shown by, for example, marsupials, birds, frogs (Pianka and Schall 1981), mygalomorph spiders, land snails, and butterflies (Main 1981; Bishop 1981; Kitching 1981). But, by contrast, lizards, for example, show greatest diversity in the arid zone (Pianka and Schall 1981).

7.1 MAMMALS

The mammal faunas of Australia and New Guinea are unique in the dominance and diversity of marsupials and the presence of monotremes. The only native eutherians are bats and rodents, although many others have become naturalized following introduction by humans.

The composition of the mammal fauna of rainforest (excluding bats) north from the Illawarra region has been documented by Winter (1988), who suggested that the species can be assigned into a number of categories determined by the degree of dependence on rainforest habitats:

(1) rainforest specialist species, strictly confined to the rainforest over the whole of their geographical range;

(2) rainforest margin species, confined to the interface between rainforest and more open habitats (referred to as rainforest ecotone species by Winter 1988);

(3) ubiquitous species, occurring in a wide range of habitats including rainforest;

(4) independent species, restricted to a particular habitat which may occur in rainforest stands, but not dependent on rainforest (for example, the platypus *Ornithorhynchus anatinus*).

Winter (1988) recorded a total of 40 species of rainforest mammals (Table 7.1). Categorization of some was difficult, partly because of lack of information on local distribution patterns and ecology. For others, the degree of restriction to rainforest varies with latitude, for example, *Antechinus stuartii* occurs in a wide range of habitats in New South Wales, but in the wet tropics it is restricted to upland rainforest (Winter

Table 7.1 Non-volant rainforest mammals of eastern Australia in relation to their ecological category and biogeographic region (modified from Winter 1988).

Species		Biogeographic region			
		NG	CY	WT	SE
Rainforest specialists					
Strigocuscus minicus	†‡	*	*	—	—
(Grey cuscus)					
Antechinus leo¶			*	—	—
(Cinnamon antechinus)					
Uromys caudimaculatus¶	‡	*	*	*	—
(White-tailed rat)					
Rattus leucopus¶		*	*	*	—
(Cape York rat)					
Pogonomys cf. *mollipilosus*¶		*	*	*	—
(Prehensile-tailed rat)					
Cercartetus caudatus	†	*	—	*	—
(Long-tailed pygmy possum)					
Antechinus godmani¶			—	*	—
(Atherton antechinus)					
Pseudocheirus herbertensis	†‡		—	*	—
(Herbert River ringtail possum)					
Pseudochirops archeri	†‡		—	*	—
(Green ringtail possum)					
Hemibelideus lemuroides	†‡		—	*	—
(Lemuroid ringtail possum)					
Hypsiprymnodon moschatus	‡		—	*	—
(Musky rat-kangaroo)					
Dendrolagus lumholtzi	†‡		—	*	—
(Lumholtz's tree-kangaroo)					
Dendrolagus bennettianus	†‡		—	*	—
(Bennett's tree-kangaroo)					
Melomys hadrourus¶			—	*	—
(Thornton Peak melomys)					
Forest generalists					
Echymipera rufescens		*	*	—	—
(Rufous spiny bandicoot)					
Spilocuscus maculatus	†‡	*	*	—	—
(Spotted cuscus)					
Melomys capensis¶			*	—	—
(Cape York melomys)					
Dactylopsila trivirgata	†	*	*	*	—
(Striped possum)					
Petaurus breviceps	†	*	*	*	*
(Sugar glider)					

Table 7.1 (*cont.*)

Species	Biogeographic region			
	NG	CY	WT	SE
Thylogale stigmatica (Red-legged pademelon)	‡	*	*	*	*
Perameles nasuta (Long-nosed bandicoot)		*	*	*
Trichosurus vulpecula (Common brushtail possum)	†‡	—	*	—
Antechinus flavipes¶ (Yellow-footed antechinus)		—	*	*
Pseudocheirus peregrinus (Common ringtail possum)	†‡	—	*	*
Melomys cervinipes¶ (Fawn-footed melomys)		—	*	*
Dasyurus maculatus (Spotted-tailed quoll)		—	*	*
Antechinus stuartii¶ (Brown antechinus)		—	*	*
Sminthopsis leucopus¶ (White-footed dunnart)		—	*	*
Rattus fuscipes¶ (Bush rat)		—	*	*
Antechinus swainsonii¶ (Dusky antechinus)		—	—	*
Trichosurus caninus (Mountain brushtail possum)	†‡	—	—	*
Cercartetus nanus (Eastern pygmy possum)	†	—	—	*
Rainforest margin				
Potorous tridactylus (Long-nosed potoroo)		—	—	*
Thylogale thetis (Red-necked pademelon)		—	—	*
Macropus parma (Parma wallaby)		—	—	*
Macropus dorsalis (Black-striped wallaby)		—	*	*
Generalist				
Tachyglossus aculeatus§ (Echidna)	*	*	*	*
Canis familiaris dingo¶ (Dingo)		*	*	*

Species	Biogeographic region				
	NG	CY	WT	SE	
Independent					
Ornithorhynchus anatinus§ (Platypus)		—	*	*	
Hydromys chrysogaster¶ (Water rat)	*	*	*	*	
Total	12	15	29	21	=40

NG = New Guinea (many more species occur in rainforests in New Guinea—this table only indicates the faunal overlap with Australia).
CY = Cape York Peninsula
WT = Wet tropics
SE = South-eastern Australia (between the Illawarra region of New South Wales and the wet tropics)
* = present
† = arboreal species (other species may be partially arboreal)
‡ = primarily folivores and/or frugivores
§ = monotreme
¶ = eutherian
all other species are marsupials.

1988). In Table 7.1, it is classified as a forest generalist.

All the rainforest specialist species occur either in Cape York or the wet tropics of north-east Queensland. Forest generalist species occur in all biogeographic regions, while rainforest margin species are largely restricted to the south-east.

The extent to which these distribution patterns can be explained as a response to present-day ecological factors has been considered by Winter (1988). For many taxa, species richness is inversely correlated with latitude. However, while there are more species in total in the wet tropics than further south, the number in Cape York is lower than in either of the other regions. Rainforest specialist species are absent south of the wet tropics. Neither total richness nor that of rainforest specialists is simply related to latitude.

Species richness could be a function of the area of rainforest in each region—but this is not the case. Both Cape York and the wet tropics currently have larger contiguous areas of rainforest than the south-east where distribution is more fragmented, and it might be postulated that this fragmentation is responsible for the absence of specialist species. However, in the south-east, there were, until recently, some large contiguous stands (notably the Big Scrub, see Chapter 9); there is no evidence that, at the time of European settlement, these areas supported any rainforest specialist species, despite their area being much greater than, for example, the total distribution range of *Melomys hadrourus*, a specialist species restricted to c. 24 000 ha in the Thornton Peak uplands of the wet tropics.

Species richness may reflect the diversity of available habitats. A high proportion of the specialist species are folivores and/or frugivores (Table 7.1), so it may be reasonable to suggest that tree species richness is a measure of habitat diversity for specialist species. However, while the number of specialists is highest in the wet tropics which have the richest tree flora, there are no specialists in the south-east even though the tree flora is larger than that in Cape York.

The distribution of rainforest mammals in Australia at the present day is thus not easily explained by contemporary ecological factors.

The restriction of rainforest margin species to the south-east, however, is most simply explained by the greater extent of rainforest/wet sclerophyll forest transitions at higher latitudes (see Chapter 4).

Winter's (1988) analysis did not include data from rainforests south of the Illawarra. However, the general trends reported from further north also apply to the more southern rainforest. There is an absence of specialist species, and the major components of the mammal fauna are forest generalists and rainforest margin species.

A detailed discussion of the ecology and distribution of rainforest species in northern Queensland is provided by Winter *et al.* (1984).

7.1.1 History of rainforest mammals

If present ecology is inadequate to explain the distribution patterns of mammals in rainforest, what insights does history offer?

The palaeobotanical record (Chapter 6) clearly establishes that Australia was largely clothed with rainforest throughout the Tertiary, and that the development of the extensive arid zone was comparatively recent. What is known of the mammal fauna of the Tertiary rainforest, and how is it related to the present-day fauna?

The early record of mammals in Australia is still very sparse. The oldest fossil is of a monotreme, *Steropodon galmani*, from the Cretaceous (about 100 Ma) at Lightning Ridge, New South Wales (Archer *et al.* 1985). Currently the oldest records of both marsupials and eutherians are from the Paleocene/Eocene (Hand *et al.* 1991). The remarkable limestone deposits at Riversleigh in northern Queensland provide a very rich faunal record covering various ages from the late Oligocene/Miocene (25 Ma) to the recent. The Riversleigh fauna is still being intensively investigated, but Archer *et al.* (1989*a*) provided a review of findings to date.

Unfortunately, there is no palaeobotanical evidence which allows reconstruction of the vegetation at Riversleigh. However, Archer *et al.* (1989*a*) suggested a number of reasons why the older fossil faunas there should be interpreted as being from rainforest. These are:

(1) the high species diversity;

(2) the occurrence of what is interpreted as complex feeding guilds of small sympatric mammals;

(3) the high number of sympatric obligate leaf-eaters, which is taken to indicate that many different species of trees occurred in small areas (at present high diversity of arboreal folivores is a particular feature of the wet tropics and is not found in all rainforests—Table 7.1);

(4) the presence of many taxa thought to be restricted to rainforest;

(5) the absence of grazers, although browsers are common.

The species richness of mammals in the Oligocene/Miocene rainforest at Riversleigh was much greater than that in the richest rainforests in present-day Australia (the difference is likely to have been even more marked as the fossil record provides an incomplete sample of the palaeofauna). Comparison of the fossil Riversleigh and the present-day Atherton Tableland faunas (Fig. 7.1) shows that there has been a marked decline in the diversity of marsupials, partially compensated for by a increase in murids and bats. Archer *et al.* (1989*a*) also compared diversity with that in a Pliocene deposit at Hamilton in Victoria and the present mid-montane fauna in Papua New Guinea.

The Hamilton site provides the only known Pliocene record of a rainforest fauna. Between the early Miocene and the Pliocene, there had been a considerable decline in the area of rainforest in Australia (Chapter 6). However, it is not possible simply to account for the differences between Riversleigh and Hamilton, as the effects of the reduction and fragmentation of rainforest *per se*.

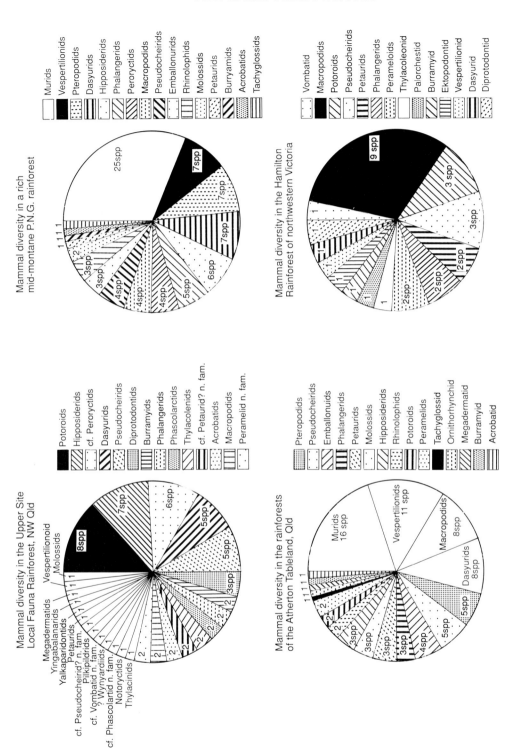

Fig. 7.1 Taxonomic diversity in the mammal fauna of the Atherton Tableland, mid-montane Papua New Guinea, the Hamilton deposit in Victoria (Pliocene), and the Riversleigh limestone (Oligocene/Miocene). (Redrawn from Archer *et al.* 1989*b*). For details of the family-level taxonomy see Aplin and Archer (1987), Archer *et al.* (1988) and Archer *et al.* (1990).

The Hamilton site was a microphyll *Nothofagus* rainforest, different structurally and floristically from that likely to have occurred at Riversleigh. Nevertheless, the Hamilton fauna was more diverse than that of present-day *Nothofagus* forests. The present Atherton fauna is, at the species level, much less rich than that in Papua New Guinea. Eutherians (murids and bats) are proportionately better represented in Papua New Guinea than in Australia (Fig. 7.2).

The various mammal groups present in the Riversleigh fauna have suffered one of three fates (Archer *et al.* 1989*a*):

(1) extinction before the Pliocene;

(2) persistence in rainforest refuges;

(3) adaptation to non-rainforest environments.

Many of the groups which became extinct were relatively small animals (yalkaparidontids, yingabalanarids, pilkipildrids); all the medium-sized (sheep-sized) browsers assigned to the Wynyardidae were also lost.

Groups represented at Riversleigh and still found in Australian rainforests are *Hypsiprymnodon*, peroryctid bandicoots, the cuscus *Strigocuscus*, and the ringtail possum *Pseudochirops*. *Hypsiprymnodon moschatus*, the musky rat kangaroo, is restricted to the wet tropics of northeast Queensland, and is regarded as the most primitive living kangaroo. It is an omnivore and has a grasping big toe and grooved pads on the feet. These are features of possums and are regarded as adaptations to arboreal life, their occurrence in *Hypsiprymnodon* indicating aboreal ancestry. (These features are absent in the tree kangaroos, *Dendrolagus*, which are more recently evolved from terrestrial kangaroos rather than being ancestral to ground-living kangaroos.) *Hypsiprymnodon* is represented in the Hamilton fauna, so until relatively recently was far more widespread and ecologically versatile than it is today.

Among the groups which adapted to non-rainforest habitats, a division can be made into those which also persisted in rainforest and those now totally absent from rainforest (Archer *et al.* 1989*a*). Among the former are extant groups, such as dasyurids, macropodids, and acrobatids but also groups which subsequently become extinct across a range of habitats, such as the

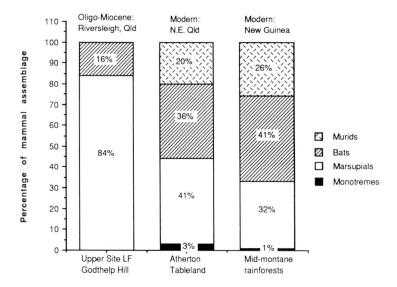

Fig. 7.2 Representation of major mammalian groups at Riversleigh and from present-day Atherton Tableland and Papua New Guinea (redrawn from Archer *et al.* 1989*b*).

browsing diprotodontids and palorchestids and the carnivorous thylacoleonids and thylacinids.

It is difficult to explain why some groups are no longer found in rainforest. The phascaloarctids (koalas) were represented in Riversleigh by at least two species. Archer and Hand (1987) have speculated that early phascaloarctids were associated with ancestral eucalypts on relatively nutrient-poor sites and so were 'pre-adapted' to exploit the nutrient-deficient sclerophyll vegetation which became dominant in the late Tertiary. The marsupial mole, *Notoryctes typhlops*, is a taxonomically isolated species restricted to the arid zone (K. A. Johnson 1983). It has been regarded as a classical demonstration of parallel evolution between marsupials and eutherians and being superbly adapted to its desert habitat (even though, in geological terms, this habitat is of recent origin). One of the most surprising finds at Riversleigh was an abundance of notoryctids in the Oligocene/Miocene in what was almost certainly rainforest. Archer *et al*. (1989*a*) proposed that these were burrowing insectivores/carnivores in the mossy floor of the rainforest. However, at present, bryophyte carpets are rare in most Australian rainforest types except microphyll mossy forest. It is unlikely that the forest at Riversleigh was of this type, so that Archer *et al*.'s (1989*a*) suggestion does not seem likely.

One of the most striking changes between Riversleigh and the present-day mammal fauna is the reduction in the number of large and medium-sized species. This is in contrast to the fauna of Malesian rainforests, which still includes a number of large mammals (although many of these are now regarded as endangered), particularly ground-dwelling mammals (see Fig. 7.3).

Most of the large mammals at Riversleigh were quadrapedal terrestrial browsers (Archer *et al.* 1989*a*), which presumably fed on understorey plants. A number of these groups underwent a major radiation in the arid and semi-arid zone in the late Tertiary, but this arid zone megafauna declined rapidly in the Pleistocene. Whether the loss of the large rainforest animals is to be explained by factors specific to rainforest, or whether it was part of a general megafaunal decline remains to be determined. The timing of the loss of larger herbivores from rainforest is unknown, but one species of diprotodontid (*Hulitherium*) persisted in upland rainforest in New Guinea as recently as 20 000 years ago (Flannery and Plane 1986; Flannery 1987).

In comparison with earlier faunas, and with the current fauna in South-east Asia, there is an absence of large carnivores in Australian rainforests. Until earlier this century, the thylacine included Tasmanian rainforests among the habitats it occupied. On the mainland, thylacines probably occupied rainforest until the late Pleistocene, until displaced by the dingo, currently the largest mammalian carnivore in Australia.

Despite the sparsity of the pre-Miocene fossil record, the textbook story that the mammal fauna of Australia, at the time that the continent became isolated, contained only monotremes and marsupials, and, with the absence of eutherians, marsupials subsequently radiated to fill all available niches, appears to be justified. The major taxonomic groups within the marsupials evolved within rainforest, with some groups subsequently radiating into the developing sclerophyll and arid zone communities.

Two groups of eutherian mammals, the murid rodents and bats, occurred in Australia prior to the arrival of humans, and both groups are well represented in current faunas. The rodents are represented by a single family, the Muridae, with representatives in two subfamilies, the Hydromyinae and the Murinae (Lee *et al*. 1981). The fossil record of rodents in Australia extends back to the Pliocene, but Archer *et al*. (1989*a*) speculated that they may have been present from the early Miocene onwards. Godthelp (in Archer *et al*. 1989*a*) has suggested that the first rodents in Australia were inhabitants of drier non-rainforest communities and invaded the continent via southern Indonesia rather than New Guinea. When the occupation of rainforest occurred is unknown. It is the case that endemic genera like *Zyzomys*, *Pseudomys*, and *Notomys* are absent

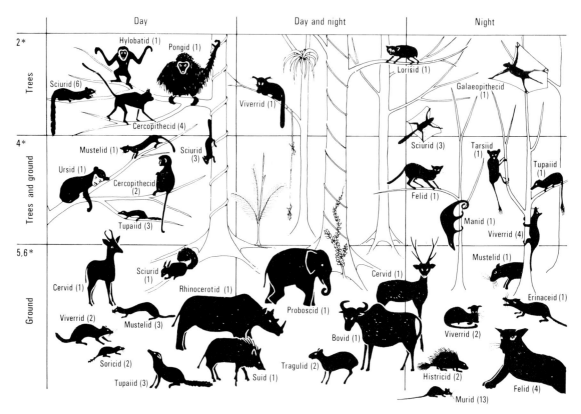

Fig. 7.3 Stratification of non-volant mammals in the lowland rainforest of Sabah (data of MacKinnon, from Whitmore 1984). The diversity of this fauna is in marked contrast to the depauperate mammal assemblages in Australian rainforests.

from rainforest. The genera in rainforest (*Melomys*, *Uromys*, *Rattus*) have affinities with New Guinea and the fossil record suggests they did not reach Australia before the Pleistocene (Archer *et al*. 1989*a*). For these late arrivals, a land connection between Australia and New Guinea would have been available, but if the earlier arrivals did not traverse New Guinea, their mode of arrival in Australia remains unknown. Whenever rodents reached Australian rainforest, there is little evidence that they displaced any marsupials (Archer *et al*. 1989*a*).

Bats have a much longer fossil record than rodents in Australia, and are well represented in the Riversleigh fanua. Nevertheless, there has been an increase in diversity in the rainforest bat assemblage since the Miocene (Archer *et al*. 1989*a*, Fig. 7.1).

Hand (1989) has suggested that bats dispersed to Australia from Asia during the early to mid-Tertiary along the archipelago of continental fragments between the northern edge of the Australian Plate and Asia (Audley-Charles 1987). Possessing the ability to fly, bats may have been able to cross relatively short sea passages comparatively easily. (Rodents must also have utilized a similar route but the mechanism of water crossing is unknown—however, if rodents made the journey, have there been other eutherians which reached Australia but failed to colonize?)

It is unlikely that invasion by bats had much

direct impact on marsupials (which did not evolve groups able to fly). Fruit bats (Pteropodidae), the group most familiar to the general public, are absent from the Australian fossil record until the Quaternary. Eight species occur in Australia (six shared with New Guinea, which has many additional species). Hand (1989) has speculated that fruit bats may have colonized New Guinea before the establishment there of arboreal possums which share similar food resources, or, alternatively, that the failure of many fruit bats in New Guinea to invade Australia reflects the abundance of arboreal possums pre-empting food resources in Australia. The establishment of fruit bats in northern Australia in the Quaternary may have been a consequence of the late Tertiary decline of many small arboreal marsupials. Evidence to test these hypotheses is absent.

If the roots of the Australian mammal fauna lie in rainforest, why is the rainforest specialist fauna so small and restricted to the north?

Winter (1988) argued that the depauperate nature of the fauna is the result of contraction of rainforest during the dry periods of the Pleistocene, when it was repeatedly reduced to small disjunct refugia. If the area of a refugium fell below some critical size, then it would have been unable to sustain viable populations of rainforest specialist species which would therefore become locally extinct. Winter (1988) postulated that major refugia in northern Queensland remained large enough to allow survival of some rainforest specialists, but elsewhere rainforest refugia became too small to support specialist species (some of which had survived in south-east Australia until the Pliocene—Archer *et al.* 1989*a*). Even in northern refugia, only a few specialist species survive; it is probable that the areas of refugia were too small to sustain more species. The total loss of larger rainforest mammals may have occurred before the Pleistocene, but, if not, the contraction of stand area in this period is likely to have inflicted the *coup de grâce*.

Even if rainforest specialists were lost from the south-east, why have new ones not evolved in the last 20 000 years? Could the northern specialists have speciated in this time? These questions are discussed by Winter (1988).

'New' rainforest specialists are likely to evolve from forest generalist species, which include rainforest within their habit range, or from sclerophyll forest species. Whether evolution of specialist species occurs will depend on the time available for evolution and on the selection pressure. In the case of south-east Australia, neither of these factors have been sufficient (as yet) for the evolution of specialists.

If the northern specialist species have evolved within the last 20 000 years, then it would be expected that there would be closely related species in non-rainforest habitats, yet this is not the case. Two of the specialists are in monotypic genera (*Hemibelideus* and *Hypsiprymnodon*), while *Dendrolagus*, *Strigocuscus*, *Spilocuscus*, *Pogomys*, and *Uromys* are all genera restricted to rainforest. For those genera with rainforest specialists and species in other habitats (*Melomys*, *Antechinus*, *Pseudocheirus*) there is no evidence to suggest a very recent origin for the rainforest taxa (Winter 1988).

There is, however, evidence for incipient speciation in the extensive rainforest block of the Queensland wet tropics. This is split into a number of subregions with a narrow link, the Black Mountain Corridor, between the extensive southern block of the Atherton Tableland, and the northern Thornton Peak—Mt. Spurgeon region (Winter 1978, 1988; Winter *et al.* 1984, fig. 9.9). This Black Mountain corridor is susceptible to disruption and rainforest may have been absent from it 25 000–15 000 years ago. The Herbert River ringtail possum *Pseudocheirus herbertensis* is split into two races, differing in colour and chromosome number (Winter 1988); *P. herbertensis cinereus* to the north of the corridor and *P. herbertensis herbertensis* to the south. In the lemuroid possum, *Hemibelideus lemuroides*, the populations at either end of the corridor differ in the proportion of white individuals, suggesting some genetic divergence (Winter 1984). The distributions of the two

Australian tree kangaroos show a similar segregation, *Dendrolagus lumholtzi* to the south and *D. bennettianus* to the north.

Given that the specialist rainforest mammals are restricted to the north of Australia, it would be possible to hypothesize that the refugial model is incorrect, at the driest times all the rainforest specialists in Australia became extinct, with subsequent reinvasion from refugia in New Guinea. Four of the specialist species in northern Australia—*Strigocuscus minicus*, *Uromys caudimaculatus*, *Rattus leucopus*, and *Pogonomys* cf. *molipillosus*—are also found in lowland rainforest in New Guinea. For these species, a recent invasion (or reinvasion) cannot be ruled out. Other rainforest specialists in Australia have congeneric species in New Guinea, so that a history of interchange between the two regions is again possible. However, in the case of endemic genera, any interchange during recent geological time is unlikely. *Hypsiprymnodon*, now restricted to the wet tropics, but once more widespread in Australia, is in a family, the Potoroidae, which is absent from New Guinea (Archer *et al.* 1989*a*). Thus, while some rainforest specialists may be recent invaders, there are certainly others which are relicts from a time of more extensive rainforest.

The rainforest mammals present a number of important management issues. The northern rainforests (the wet tropics and Cape York) have high importance, supporting a specialist fauna, several elements of which are endemic. Given that the absence of a specialist fauna elsewhere appears to be the result of climatically induced habitat fragmentation, it is clearly important that management does not result in fragmentation of the northern forests, or loss of species may occur. While the management for conservation of the wet tropics now seems assured (Chapter 9), the future integrity of the Cape York rainforests is less certain.

Of wider significance is the evidence that Australia provides of the loss of faunal diversity in rainforest with reduction in area of habitat. If the fauna of the world's rainforests is to be conserved, then adequate areas of habitat must be protected as a matter of urgency.

Determination of what is an adequate area is difficult, but detailed studies of distribution patterns in northern Queensland by Winter *et al.* (1984) may provide some indications. In this region, there are numerous isolated rainforest stands, some resulting from climatic changes and probably isolated for some thousands of years, others produced by clearing of more extensive stands, and thus, in their present form, only a hundred years or less old.

In mid and western Cape York, populations of mammals such as *Spilocuscus maculatus* and *Thylogale stigmatica* appear to be distributed at random among patches of rainforest less than 500 ha in size. Winter *et al.* (1984) suggested that stands in this size range were too small to support sustainable populations of a number of rainforest mammals, and that these species are either gradually in decline or there are periodic episodes of local extinction and recolonization. If the latter suggestion is correct, then it may be possible to conserve the mammal fauna in small rainforest stands, provided that there is a network of such stands and that there is a possibility of passage between them. Winter *et al.* (1984) suggested that the minimum size for a single stand to retain the full mammal assemblage was about 2000 ha.

Studies by Pahl (1979) (quoted in Winter *et al.* 1984) on rainforest isolates on the Atherton Tableland, created by clearance early this century, showed that the full suite of arboreal mammals found in large stands was absent in areas of less than 75 ha. Given the short time since clearance, it would be unwise to assume that 75 ha is the critical minimum area for long-term conservation, without longer-term studies on population dynamics.

While there is no evidence that since European settlement any rainforest mammal has become extinct (in contrast to the major loss of small mammals in arid and semi-arid zones), there can be little doubt that continued fragmentation of habitat and competition from introduced species (including cane toad, feral dogs, cats, and pigs) are potential threats to a number of species. Lack of detailed studies on most rainforest mammals makes prediction of population trends difficult.

7.2 AVIFAUNA

To the casual visitor to Australian rainforest, birds are often the most immediately obvious elements in the fauna.

Australian rainforest is utilized by about 150 species of bird, over sixty of which are virtually restricted to this habitat (Schodde and Tideman 1986).

Biogeographic analysis of present distribution patterns (Schodde and Calaby 1972; Schodde 1982, 1986), shows that most of the rainforest species fall into two elements, the Irian and the Tumbunan.

The Irian element is the distinctive element in the rainforest of Cape York, and is shared with lowland New Guinea. Species in the group include the spectacular palm cockatoo *Probosciger aterrimus*, the red-cheeked and eclectus parrots *Geoffroyus geoffroyi* and *Eclectus rotatus*, the glossy swiftlet, *Collocalia esculenta*, red-bellied pitta *Pitta erythrogaster*, tropical scrubwrens *Sericornis beccarii*, green-backed and tawny-breasted honey eaters *Glycichaera fallax* and *Xanthotis flaviventer* and trumpet manucode *Manucodia keraudrenii* (Schodde 1986).

Some members of this element are found in rainforests further south in Queensland (including *inter alia* the cassowary *Casuarius casuarius*, the helmeted friarbird *Philemon buceroides*, yellow-spotted honeyeater *Meliphaga notata*, graceful honeyeater *M. gracilis*, metallic starling *Aplonis metallica*, and yellow-bellied sunbird *Nectarinia jugularis*. This extension of the Irian element southwards obscures the distinctiveness of the major component of the rainforest avifauna, the Tumbunan element, whose distribution is centred on complex notophyll rainforest from southern New South Wales to the wet tropics of north-east Queensland and montane New Guinea (Schodde and Calaby 1972; Schodde 1986).

Tumbunan species include the brush turkey *Alectura lathami*, topknot pigeon *Lopholaimus antarcticus*, large-billed scrubwren *Sericornis magnirostris*, and brown gerygone *Gerygone mouki*. The distribution of this element is not continuous along the east coast, but has several major disjunctions separating southern areas from north-east Queensland. In several species, regional populations can be separated, with races in north Queensland being taxonomically distinct from those in south-east Queensland/New South Wales. This is shown, for example, in White's thrush *Zoothera dauma*, the pale yellow robin *Tregellasia capito* and yellow-throated scrubwren *Sericornis citreogularis*. Sometimes the difference between northern and southern isolates is at species level, although the southern species may also be found in New Guinea. The logrunner *Orthonyx temminckii* is a species which seldom flies but spends most of the time on the forest floor raking through leaf litter in search of invertebrate prey. It is found in New South Wales and southern Queensland, with other races in New Guinea. In north-east Queensland, the same niche is occupied by the chowchilla *O. spaldingii*. A similar distribution pattern is shown by the sooty owls, with *Tyto tenebricosa* in New South Wales and southern Queensland and a different race in New Guinea, and the lesser sooty owl *T. multipunctata* in north-east Queensland.

A striking feature is the relatively large number of endemic species in north-east Queensland rainforests. In addition to *O. spaldingii* and *T. multipunctata*, these include the bridled honeyeater *Lichenostomus frenatus*, Atherton scrubwren *Sericornis keri*, fernwren *Crateroscelis gutturalis*, mountain thornbill *Acanthiza katherina*, and Bower's shrikethrush *Collurincla boweri*. Other endemic species with restricted distributions are found in other rainforests, for example, the Eungella honeyeater *Lichenostomus hindwindii* (one of the most recently discovered Australian birds, not described until 1984), and in northern New South Wales/southern Queensland, Albert's lyrebird *Menura alberti*.

The avifauna of microphyllous fern/moss forests is much less diverse, but a few species can be regarded as specialists in this habitat. In northern New South Wales, these include the rufous scrub-bird *Atrichornis rufescens* and the olive whistler *Platycephala olivacea* (further south this species, although never common, is found in a wider range of habitats, including heathland). The rufous scrub-bird *Atrichornis rufescens* is extremely rare, with a total population of probably only a thousand birds. It is strongly territorial, and Schodde and Tideman (1986) suggested that the limited extent of *Nothofagus* forest in northern New South Wales prevents any increase in population as surplus young are regularly produced. The scrub-birds (Atrichornithidae) are an Australian endemic family of only two species. The other, the noisy scrub-bird *A. clamosus*, is restricted to a very small area of south-western Australia with a total population of only a few hundred individuals. Neither species is a strong flier, and the current disjunct distribution of the family suggests that it once may have been more widespread across the continent before forest cover was fragmented.

Species with a distribution centred on the Tasmanian rainforests are the pink robin *Petroica rodinogaster*, scrubtit *Sericornis magnus*, and Tasmanian thornbill *Acanthiza ewingii*.

Among the rainforest avifauna, three feeding groups are particularly prominent, frugivores, insectivores gleaning prey among foliage, and ground-feeding species which turn over leaf litter in search of invertebrates. Diversity within these groups increases northwards, and even in the richest sites in Australia is far less than in New Guinea (Table 7.2).

Fleshy fruits are relatively uncommon in

Table 7.2 The composition of bird assemblages in four Australasian rainforests. Rare species have been excluded. Mixed feeders take invertebrates and nectar and/or fruit. Thomas recorded all species; other studies were more selective (from Ford 1989).

Location	Tasmania	North Queensland		New Guinea
		Upland	Lowland	
Latitude	42°	19°	17°	9°
Reference	D. G. Thomas (1980)	Frith (1984)	Crome (1978)	Bell (1984)
Herbivores				
Frugivores	0	2	6	27
Nectarivores	1	1	0	6
Granivores	1	0	0	0
Mixed feeders	2	5	6	15
Carnivores				
Invertebrates—ground	3	6	2	11
Invertebrates—air	1	1	0	7
Invertebrates—bark	5	1	1	4
Invertebrates—foliage	7	6	12	11
Invertebrates—generalized	0	0	0	2
Scavengers	1	0	0	0
Total number of species	21	22	27	83

Australia outside mesophyll and notophyll vine forests, which accordingly provide the major habitat for frugivorous birds. Prominent among the frugivores are pigeons, for which Australian rainforests are a centre of diversity (Frith 1982). At any one locality, there are unlikely to be adequate food resources throughout the year, so that frugivorous species are nomadic. In some regions, food may be available all year round in a small area because of different fruiting times along steep altitudinal gradients. In other regions, more extensive dispersal is required. Very small patches of rainforest, even single trees, may be visited briefly by flocks of pigeons. Protection of small rainforest isolates in some regions may be essential for the conservation of frugivorous birds. Unfortunately, little is yet known about utilization of rainforest fragments—whether some are more important than others, whether some food species are more exploited than others, whether there is a critical separation distance which prevents isolates being used, are all questions requiring answers. In some regions, introduced plant species may provide substitute food resources. For example, in northern New South Wales, camphor laurel *Cinnamonum camphora* occurs widely in the area formerly occupied by the Big Scrub rainforest. The camphor laurel fruits are avidly eaten by pigeons and these trees may provide staging posts for birds moving between widely disjunct rainforest remnants. If this is the case, then it presents a problem for land managers. There are pressures to remove camphor laurels as they may repress regeneration of other plant species and have the potential to invade rainforest. However, removal may adversely affect pigeons (which may also be a major dispersal agent promoting further spread of camphor laurel).

Frugivorous birds clearly have the potential to act as dispersal agents, and may play an important role in rainforest regeneration, although this remains to be more fully investigated. The extent to which there is competition for resources between birds and fruit bats is also unknown.

As well as birds like pigeons, which largely consume fruit within the canopy, there are other species which include fallen fruit in their diet and which may also act as dispersal agents (see Chapter 5). These include the megapodes *Megapodius reinwardt* and *Alectura lathami*, and the cassowary *Casuarius casuarius*. The megapodes construct very large mounds in which eggs are incubated. Ground-dwelling insectivores also rake over and disturb the litter layer (most prominently the lyrebirds *Menura* spp.). While the role of birds in litter ecology has rarely been given consideration, it is possible that it is a major one; Adamson *et al.* (1983) have reported that in the Blue Mountains near Sydney the turnover by lyrebirds of litter and soil during feeding may exceed $6000 \text{ g m}^{-2} \text{ yr}^{-1}$.

7.2.1 Origins of the avifauna

The Australian avifauna is rich, diverse, and distinctive. Families prominent in the avifauna of Asia and Africa are absent or only poorly represented in Australia (Schodde 1986). Why is this so?

A priori birds may be expected to be well dispersed and relatively little affected by barriers which would limit the spread of mammals. At the present day, many species (some very small) are known to migrate over very long distances (including extensive water crossings). Nevertheless, there are many other species which have geographically restricted distributions and show little tendency to expand their range despite the availability of apparently suitable habitats. The sea- and shore-birds of Australia are, for the most part, very widely distributed, and are part of a cosmopolitan assemblage. Among the freshwater birds, many are either conspecific or closely related to those in other continents. It is the land-birds which provide the distinctively Australian element in the fauna.

The traditional view of the origin of the birds in Australia suggested that the source was Asian, and that successive waves of immigrants arrived in northern Australia via the Malay Archipelago. Once in Australia, many groups underwent considerable radiation. The most diversified groups, for example honeyeaters and parrots, were thought to have been part of the earliest invasions (Schodde 1982, 1986). If this were so, the distinctive nature of the present avifauna, and particularly the absence of prominent Asian families, would seem unusual, but what other hypotheses can be presented?

Australia has yielded some of the oldest fossil evidence of birds. Fossil feathers have been recovered from Cretaceous sediments in Victoria (c. 125–110 Ma), but, unfortunately, nothing is known about the birds from which they came (van Tets 1986). Various identifiable bird fossils are known from other sites of Tertiary age, but, as yet, little evidence is available to document the history of passerine birds, although the oldest passerine fossils are at least early Miocene in age. The Riversleigh deposits have yielded a great diversity of bird fossils which are currently being studied (Boles 1991). A particularly interesting find is a species of *Orthonyx*, an almost flightless genus now represented by two species in rainforests in Australia and New Guinea (Boles 1991). A feature of the fossil record is giant flightless ratites, the Dromornithidae, now extinct. While a number of species in the late Tertiary were inhabitants of open communities (Rich 1985), others may have occurred in rainforest (Boles 1991).

The occurrence of Cretaceous bird fossils suggests that at least part of the avifauna may have an ancestry stretching back to before the break-up of Gondwana. Identification of the Gondwanan element is still speculative, but probably includes the flightless emus and cassowaries (although the southern origin of ratites has been challenged by Houde 1986), and also the megapodes, the plains-wanderers, the parrots, and possibly pigeons, cuckoos, and rails (Schodde 1982, 1986).

In the case of the majority of the avifauna, it would still seem probable that their origin was outside Australia. The question is: when did the ancestral stocks arrive in Australia? If water crossings were barriers to many land-birds, the conventional story would have been that the major invasion occurred in the late Tertiary, when relatively short hops across Indonesia and through New Guinea would have been the probable route. The fossil record of passerines in Australia earlier in the Tertiary casts doubt on the conventional hypothesis, but the fossil record in itself is inadequate to resolve the problem.

An alternative approach to elucidating the history of birds in Australia is to examine the relatedness both of species within Australia and between Australia and overseas, using electrophoresis to assess degrees of similarity at the molecular level. This allows the construction of a phylogeny which can be dated using similarities in DNA between species (measured by DNA hybridization techniques) and assuming a relatively constant mutation rate.

The results of such studies indicate a radical departure from previous assumptions about origins and taxonomy of Australian birds (Sibley and Ahlquist 1985; Schodde 1982, 1986; Christidis 1987; Beckmann 1987*b*). Relationships among the Australian passerines, as a whole, are closer than those between particular groups and their previously supposed northern hemisphere counterparts. Despite considerable morphological and anatomical similarities, the molecular evidence suggests that, for example, those birds referred to in Australia as flycatchers, tree-creepers, and robins, are very different from those northern hemisphere groups which lent their names to southern species. The immediately obvious similarities must therefore reflect convergent evolution rather than phylogenetic relatedness. Sibley and Ahlquist (1985) argued that Australian oscine passerine birds fall into two major groups, the Corvi which originated in Australia and subsequently dispersed to other parts of the world, and the Muscipapidae which colonized Australia from Asia.

Estimation of the time of various evolutionary branches, using DNA data (Sibley and Ahlquist 1985; Beckmann 1987*b*), indicates that the first major radiation of passerines in Australia occurred at 60–50 MA, suggesting the arrival of the ancestral stock at, or before, this time. This would support the concept of microcontinental stepping-stones between Australia and Asia which provided a migration route for organisms with relatively well-developed dispersal characteristics. While dates derived from the use of DNA sequences are subject to debate (Houde 1986), it is clear that most lineages of Australian birds were established at a time when rainforest was the dominant vegetation. The occupation of the sclerophyll and semi-arid communities occurred through the evolution from rainforest stocks. Schodde (1982, 1986) has argued that the major source of many of the non-rainforest birds was from what now survives in rainforest as the Tumbunan element.

7.3 HERPETOFAUNA

Australia has a very rich herpetofauna, but the proportion of species which is restricted to rainforest is small. Broadbent and Clark (1976) estimated that 4 per cent of reptiles and 18 per cent of frogs were largely confined to rainforest or wet sclerophyll forest. The number of rainforest specialist species declines with increasing latitude (Fig. 7.4); this, in part, reflects winter temperature limiting an essential tropical faunal element (Cogger 1977).

Although the number of rainforest specialist reptiles is small, a much larger number of generalist species include rainforest within their habitat range. These include some of the more frequently seen species which penetrate rainforests along creeks or tracks, and are often encountered basking in clearings.

While reptiles have a long fossil record in Australia and the record from Riversleigh discussed above contains representatives of several families which are found in rainforest today (Archer *et al.* 1989*a*), the current interpretation of the history of the reptile fauna is, that while there are a number of old Gondwanan groups, much of the fauna appears to be derived from comparatively recent invasions from Asia which

have undergone adaptive radiation within Australia (Cogger 1977; Cogger and Heatwole 1981; Tyler 1979). The reptile fauna of Australian rainforests is depauperate in comparison with that of other continents, in contrast to the arid zone, which contains a much-richer reptile fauna than that overseas (Heatwole 1981). Heatwole (1987) has suggested that the rainforest reptiles suffered a particularly high rate of extinction during the dry periods of the Pleistocene. Unfortunately, there are insufficient data on the herpetofauna in rainforest prior to its contraction for this hypothesis to be assessed.

The amphibia, despite their ability to swim, are not candidates for dispersal across marine barriers. Their skin is permeable and immersion in salt-water would cause rapid death from osmotic shock. The two major families of frog in Australia (Hylidae and Myobatrachidae) are also found in South America and are part of the inheritance from Gondwana, although there is

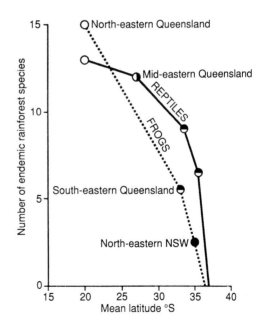

Fig. 7.4 Numbers of rainforest specialist frogs and reptiles in relation to latitude (from Adam 1987*b*, after Cogger 1977).

controversy over the timing of evolution at the species level within the continent (Heatwole 1987).

The Microhylidae are restricted to northern Australia, mainly to rainforests in north-east Queensland, and appear to be recent invaders from Asia via New Guinea (Tyler 1979; White 1984).

The Oligo-Miocene deposits at Riversleigh have yielded at least fifteen frog species (Tyler *et al*. 1990), but are unusual in that a single species, the myobatrachid *Lechriodus intergerivus*, is prevalent, overwhelmingly so in the oldest sites. The decline in abundance of the species in the younger sites is interpreted by Tyler *et al*. (1990) as being a correlate of the decline in rainforest induced by climatic changes.

Rainforest provides habitats for a number of species regarded as being primitive (for example, the pouched frog *Assa darlingtonii*, in which the male broods the tadpoles in pouches along its flanks, and *Taudactylus* spp.).

The introduced cane toad (*Bufo marinus*) is widespread in rainforest and other habitats in Queensland, and is still extending its distribution southward into New South Wales and westward into the Northern Territory. The toads have scapular glands which secrete a toxin lethal to many carnivores, and their impact on native fauna is believed to be considerable.

7.4 INVERTEBRATES

Invertebrates in Australian rainforests far exceed vertebrates in species diversity. Invertebrates play a major role in rainforest ecosystems—as herbivores (see Chapter 5), as agents in the breakdown of litter and wood, as pollinators, and as components of food webs.

Most rainforest invertebrates are restricted to the habitat, although there is some spread into tall open forest with a mesic understorey (Nadolny 1984). Many species have highly specialized habitat requirements (in the case of phytophagous species often being restricted to particular host species, for example the spectacular Richmond birdwing butterfly, *Ornithoptera richmondia*, whose larvae are restricted to *Aristolochia* spp., and in some cases having different requirements at different stages in their life history).

Although there has been considerable interest in rainforest invertebrates in recent decades (see *inter alia* Broadbent and Clark 1976; Keto and Scott 1986; Kikkawa *et al*. 1981*a*; Nadolny 1984; Sands and House 1990), there is no site for which a complete inventory is available, and even in the better-studied groups there is still much taxonomic work to be done.

Most terrestrial invertebrates are poorly represented in the fossil record, although the Riversleigh deposits contain a diverse, although still poorly studied, invertebrate assemblage (Archer *et al*. 1989*a*). In the absence of fossils, the history of the rainforest invertebrates must be reconstructed on the basis of taxonomy and current distribution.

Some invertebrate groups may constitute a Pangaean element, with an ancestory dating prior to the mid-Jurassic breakup of the supercontinent Pangaea. Despite the fact that the current distribution is restricted to the southern continents, Heatwole (1987) argued that the Onycophora is a member of this element. Species from this phylum are widely distributed in litter and decaying logs in Australian rainforests, but are also found in other habitats (Tait *et al*. 1990).

A much larger element is probably of Gondwanan origin. Examples in rainforest include many terrestrial molluscs and much of the litter fauna. There are also many groups which appear to have invaded, and subsequently undergone adaptive radiation, from the north at various times during the Tertiary (Heatwole 1987). Within some higher taxa, there is evidence for more than one entry into Australia. For example, in the north-east Queensland rainforest, there are some butterflies which appear to be relatively ancient (and whose larvae feed on species from primitive plant families) and others which are recent invaders from New Guinea (see Keto and Scott 1986).

Different types of rainforest differ in the species of invertebrates present and also in their diversity. For example, in New South Wales, cool-temperate rainforest appears to have a much smaller fauna than other rainforest types (Nadolny 1984). Low diversity may reflect the small vascular fauna as well as the less-favourable climatic conditions. The fauna of cool-temperate rainforests is, nevertheless, a distinctive one, with a number of Gondwanan elements—for example, the Peloridae found in the litter have a world distribution matching that of *Nothofagus* (Evans 1981).

Rainforest conservation reserves are chosen and managed largely in ignorance of the requirements of the invertebrate fauna. It seems probable that sites selected on other criteria will sample a large proportion of the invertebrate fauna but, in the future, greater attention should be given to ensuring that the high diversity and richness of invertebrate fauna are maintained. There has been some attention given to the conservation of the larger, more spectacular butterflies. For example, a major reason for the dedication of Hayters Hill Nature Reserve in northern New South Wales was to protect the habitat of the Richmond birdwing *Ornithoptera richmondia* and the regent skipper *Euschemon rafflesia*.

At least in general terms, the diversity of the Australian invertebrate fauna is comparable with that in rainforest elsewhere. Crome and Irvine (1986) have suggested that insect communities in rainforest of north-east Queensland may be less diverse than those elsewhere in the tropics, on the grounds that Australian bat and bird communities are known to be of lower diversity. Much more work is required before this assumption can be properly tested. Attention can be drawn to one curious difference. Termites are an important element in the invertebrate fauna of sclerophyll and arid communities, where they are frequently responsible for a major proportion of mineral nutrient cycling and, in the arid zone, may be the major primary consumers. Termites are a major element in the litter fauna of rainforests worldwide (Collins 1988), but Australian rainforests are depauperate in comparison with rainforests overseas and with other Australian vegetation types. Gay (1970) reported only four species as regularly occurring in rainforest and suggested that this low number could relate to the fact that rainforest had never been an extensive habitat in Australia. This explanation can no longer be supported. It would be of interest to know whether there were more termites in the extensive Tertiary rainforests, but the fossil record does not yet shed light on this question.

Whatever the explanation for the limited number of termites in Australian rainforests, the low natural representation of this group may mean that introduced species are a potential threat (Nadolny 1984). Timber and other products are imported into Australia from southeast Asia, and, despite quarantine inspections, there is a strong possibility that various organisms will accidently accompany these imports. Exotic termites have already been recorded from north Queensland but not, as yet, from rainforests.

7.5 AQUATIC FAUNA

Present-day Australia is a dry land, and over much of the continent permanent streams are rare. The rainforests of eastern Australia contain the headwaters of many permanent streams which provide habitats uncommon elsewhere—cool, clear, well-oxygenated waters, often more shaded than streams in sclerophyll forests, and with inputs of leaf litter of higher nutrient content than that from eucalypts. While not part of the rainforest, these streams are very much part of the rainforested land systems and protection of the rainforest serves to maintain the stream environment.

The fauna of these streams is of considerable scientific and biogeographical interest (Adam 1987*b*). The native fish fauna is particularly noteworthy for the very high proportion of migratory (particularly catadromous) species, a proportion far higher than in other continents (Harris 1984). The aquatic invertebrate communities are largely dependent on input of plant litter from the rainforest—the composition of these communities differs markedly from that in the less permanent streams in non-rainforest catchments. A striking element in the fauna is the freshwater crayfish (Parastacidae—a family of Gondwanan distribution) of which a number of species occur, most restricted to particular catchments, for example, in New South Wales *Euastacus polyetosus* and *E. reductus* are restricted to the Barrington Tops, *E. spinosus* to the upper Hastings River catchment, and *E. aquilus* to streams near Point Lookout in New England National Park. Several of these species are semi-terrestrial and may be encountered foraging through rainforest litter many metres from the nearest stream.

7.6 COMMUNITY ORGANIZATION

It is difficult to discuss communities defined by the total fauna, due to the lack of comprehensive studies at any localities. Studies of particular groups of organisms do, however, permit some observations on community structure.

Harrison (1962*a,b*) compared the distribution of species of birds and mammals in feeding zones within rainforest in north-east Queensland and Malaya. He proposed that the species' assemblage could be divided into six 'communities' defined by the level (between ground to above canopy) at which they occur and their range of foodstuffs. Despite the fundamental difference in taxonomic composition and the great difference in numbers of species between the two regions, the proportion of the total assemblage represented in each community was very similar. Further south in Australia where the number of rainforest mammals is much less than in Queensland, the communities may also be fewer (and individually simpler), but this has not been investigated in detail.

The variation in faunal assemblages between rainforest types is partly expressed in broad-scale geographic distribution patterns but, in any one region, is also seen in altitudinal zonations. Winter *et al.* (1984) demonstrated that among mammals in the wet tropics of north-east Queensland there were distinct lowland and upland assemblages. However, among the birds and bats these patterns are partially obscured by movements of many species, both altitudinally and latitudinally. In some species these movements are sufficiently predictable as to be termed 'migration', but in many cases, driven by unpredictability of food resources, are best referred to as 'nomadic'. These movements have important implications for the potential dispersal of plant species and maintenance of vegetation assemblages; conversely the survival of this nomadic and migratory fauna requires the conservation of a range of rainforest sites over altitudinal and latitudinal gradients.

For the vertebrates, species' assemblages in rainforest and tall open (wet sclerophyll) forest have many species in common, and at higher latitudes are not significantly different (see e.g. Kikkawa 1968). There are insufficient data to test whether this also holds for invertebrates but, given that for phytophagous species tall open forest provides niches for specialist species of both rainforest and eucalypts, it is probable that total assemblages are richer than those of pure rainforest. For mammals, the rainforest margin/tall open forest complex is probably the richest of all habitats (Calaby 1966: Adam 1987*b*).

Rainforests include the most species-rich terrestrial communities. Although Australian rainforests, at least for some taxonomic groups, are not as species rich as examples elsewhere in the tropics (Kikkawa 1990), they are still the habitat for a great richness of organisms. It has been suggested in a number of texts that a consequence of species richness would be the presence of a larger number of specialized links between particular species of plants and animals than in less species-rich communities. It has also been argued that such links would reflect the outcome of coevolution. However, the historical evidence (Chapter 6) highlights the lack of constancy in floristic assemblages in rainforest over time, suggesting that linkages throughout the ecosystem will have been subject to reorganization, and that one-to-one species interactions, if they occur, may be of no great antiquity. In terms of seed dispersal, seed predation, and herbivory, the evidence currently available suggests a greater prevalence of diffuse (generalist) relationships than tight obligatory linkages (Jones and Crome 1990) (although there are some examples of specific interactions). It is important that we develop a greater knowledge of the nature of linkages within rainforest systems if we are to understand how best to manage the habitat, particularly where it has been fragmented by geological history and human intervention into isolated remnants.

8 MANGROVES

Mangroves are a major feature on many parts of the Australian coast. Although mangrove is one of the few technical terms of vegetation description which is widely recognized by the general public, it is difficult to formulate a simple universally applicable definition. Mangrove has, in fact, two separate meanings; as a description of a plant community, and as a category of species which occur in mangrove communities—thus *Avicennia marina* is a mangrove and the communities that it dominates are also referred to as mangroves. Several authors have proposed to differentiate between these two meanings by adopting the term 'mangal' to refer to the community (e.g. Macnae 1968; Chapman 1976), but this usage has not achieved any popularity and mangrove continues to be applied in both senses; in most cases the context serves to indicate which meaning is appropriate, and confusion is rare.

Mangroves (in the community sense) can be defined by three criteria; habitat, structure, and floristics. In most cases the criteria co-vary, but there are many instances of stands which are conventionally regarded as mangroves, but which fail to satisfy the definition on at least one criterion.

In general, the habitat of mangroves is the intertidal zone on sheltered soft-sediment shores; structurally the community is a closed forest (ranging in height from tall (30 m+) to low (5 m)). In terms of species composition, a small number of woody species are conventionally regarded as mangroves; in Australia the mangrove flora consists of about 40 species.

Exceptions to the habitat criterion are frequent but are insignificant in terms of area. Scattered individuals of mangrove species (most frequently *Avicennia marina*) are common on intertidal rock platforms, and, at a few localities, small stands of stunted mangrove trees are found on rocky substrata. In north-west Australia, a stand of *Avicennia marina* occurs some 40 km inland (Beard 1967). For the major mangrove habitat, the estuarine intertidal zone, problems arise in defining upper limits, both along the main axis of the river and normal to it. In the upper reaches of an estuary, the waters flooding the intertidal zone may be virtually fresh. Are mangroves restricted to seaward of the limit of saline incursion, or are the fringing forests in the freshwater tidal zone also mangroves? Brackish water tidal forests occupy extensive areas in some parts of the world, and have been recognized as a distinct vegetation type with a characteristic flora as, for example, by Whitmore (1984). The boundaries of the freshwater tidal zone are not fixed, under low river-flow conditions the saline incursions may extend much further upstream, while under flood regimes the freshwater may limit tidal inputs of sea-water. In the lower estuary, the maximum landward extent of fringing mangroves is defined by the high-tide mark. However, its position is not constant, but varies between years on a long-term cycle, and can also be influenced by weather. If mangroves are defined strictly as occurring in intertidal habitat, then there may be an extensive fringe landward of 'true' mangrove, which, in some years, would be mangrove and in others, not. In practice, application of the habitat criterion to defining the limits of mangroves must be tempered by use of the other two criteria, particularly floristics, although the limits of whatever definition is employed must be arbitrary.

Although the height of mangrove stands varies considerably, the presence of woody plants taller than 5 m normally distinguishes mangroves from other intertidal communities. However, at the southern limit of mangroves in Victoria, *Avicennia marina* is a low shrub, in some stands not exceeding 2 m. On adjacent salt-marshes, the shrubby chenopod *Sclerostegia arbuscula* may be equally tall. These communities cannot be differentiated on structural grounds, but differ physiognomically and floristically; *S. arbuscula*, by convention, is not a mangrove, while *A. marina* is.

If the principal mangrove community type is recognized to be a forest occurring within the intertidal zone, then it is a comparatively simple task to document the tree species within such communities. The total number of mangrove species in the world is less than 100, with the number of core species being about 50–60 species. There are additional species which are characteristic of the uppermost fringe of mangrove stands and many more which are casual inhabitants of this zone. Mepham and Mepham (1985) have regarded all these species as mangroves, listing some 400 arborescent species from the Indo-west Pacific region. Such an approach obscures the basic similarity of the core mangrove flora over very large areas.

The question arises as to whether or not mangroves should be regarded as being rainforests.

In structural terms, Australian mangroves vary from closed forests (tall to low) to open shrublands. Under the Specht scheme (Specht 1970, 1981a, see p. 16 this volume), in which closed forest and rainforest are treated as synonyms, mangroves should be regarded as rainforest. However, mangroves have, by tradition, been regarded as a distinct vegetation type, and have not been included in accounts of Australian rainforests. Whitmore (1984) regarded mangroves in South-east Asia as being a particular form of tropical rainforest restricted to the intertidal habitat. It is appropriate to take the same view as Whitmore (1984) and to treat mangroves as one of many rainforest formations, notwithstanding the occurrence of some stands on climatically dry coasts and the shrubland structure shown in some areas.

8.1 AUSTRALIAN MANGROVES IN A WORLD CONTEXT

Mangroves are essentially tropical and sub-tropical in distribution. Mangroves in southern Australia are at the highest latitudes of any mangrove stands in the world.

Biogeographically, mangrove floras fall into two distinct groups, one in the Atlantic and east Pacific and the other in the Indo-west Pacific (Tomlinson 1986). The Atlantic–east Pacific group is small, about eight species, while the much larger Indo-west Pacific assemblage has more than 40 species. Australian mangroves are part of the Indo-west Pacific assemblage, and the total species richness is—by world standards—high, with over 35 species currently recorded.

Not only are Australian mangrove stands species-rich, but they are also extensive. The total area of mangroves is estimated as nearly 12 000 km^2 (Saenger *et al.* 1983); 22 per cent of the coast is mangrove fringed.

8.2 GENERAL FEATURES OF MANGROVES

The canopies of mangrove forests characteristically have low species richness; in many communities there is only a single species. The majority of species have simple notophyll leaves with entire margins. One of the most striking features of mangroves is the highly modified root systems shown by many species—pneumatophores in *Avicennia*, knee roots in *Brugiera*, stilt roots in *Rhizophora* (Fig. 8.1).

The ground layer is normally absent. In some limited cases, salt-marsh herbs are found under mangroves (Adam *et al.* 1988; Bridgewater 1985), dense growths of algae may occur on aerial roots and lower trunks (King *et al.* 1990, Adam and King 1990), and more rarely free-living algae carpet the ground. Towards the landward limit of mangroves, particularly where conditions are brackish, the fern *Acrostichum speciosum* may be common. In many stands, vines are absent or rare but a small number of vines may be common locally. Mistletoes are locally abundant in northern Australian mangroves, although absent from southern Australia (Hutchings and Saenger 1987).

Fig. 8.1 Stilt roots in *Rhizophora stylosa*: Hinchinbrooke Island

Mangroves provide a series of contrasting habitats. The lower trunks and aerial roots are an intertidal marine habitat occupied by epiphytic marine fauna and flora. The upper canopy is a terrestrial habitat with little direct marine influence. (Most of the Australian coast experiences micro- and mesotidal regimes, so that once beyond the seedling stage the foliage of mangrove plants does not suffer immersion. In the northwest, macrotidal regimes prevail, and small trees several metres tall may be regularly submerged.) The epiphytes of the canopy level do not experience the same rigorous environmental conditions as their hosts, a contrast to the inland situation where epiphytes endure greater stresses (particularly nutrient and water) than their host. Mangrove trees support a rich lichen flora (Stevens 1979), although bryophytes are not particularly prominent. In north-east Queensland, frequent epiphytes are *Hydnophytum formicarium* and *Myrmecodia* species which are myrmecophytes with swollen tuberous stems honeycombed with galleries occupied by ants of the genus *Iridomyrmex* (Huxley 1978, 1982). A number of orchid species are also common as epiphytes in north Queensland mangroves.

8.3 DISTRIBUTION OF MANGROVES IN AUSTRALIA

There is a very extensive literature on Australian mangroves (see e.g. Lear and Turner 1977; MacNae 1966, 1968; Saenger *et al.* 1977; Semeniuk *et al.* 1978; Clough 1982; Tomlinson 1986; Hutchings and Saenger 1987).

Mangroves are found in almost all estuaries, and on some sheltered open shores, from Carnarvon in Western Australia clockwise to the New South Wales/Victoria border. To the south of Carnarvon in the west, isolated stands occur on the Abrohlos Islands and, south of Perth, at Bunbury. There are extensive mangrove stands at localities in South Australia between Adelaide and Ceduna. In Victoria there are a number of mangrove stands, the most southerly on Wilsons Promontory, but no mangroves occur on the Gippsland coast between Wilsons Promontory and the New South Wales border.

As far as is known, all mangrove species in Australia are native but, with the possible exception of the recently described *Avicennia integra* (Duke 1988), none are endemic. The mangrove core species are, with a few isolated exceptions, restricted to intertidal habitats and do not have closely related species in inland habitats.

Mangroves, at a global scale, are largely restricted to the tropics, and species richness and stand structural complexity decline with increasing latitude. In Australia, mangroves at high latitudes are species-poor. In Western Australia from Carnarvon southwards, in South Australia, and in Victoria, there is only a single mangrove species: *Avicennia marina*. For much of the New South Wales coastline, only two species occur: *A. marina* and *Aegiceras corniculatum*.

The exclusion of mangroves from temperate latitudes is generally attributed to intolerance of decreasing temperatures, both in summer and, more particularly, in winter. Mangroves are believed to be extremely frost sensitive (Chapman 1976) and the absolute latitudinal limit of mangroves is apparently determined by the occurrence of frost. However, very little is known about the temperature tolerances of mangrove species, and neither are there data on the climatic conditions within mangrove stands: owing to the ameliorating effect of proximity to the sea, the climate within mangroves is likely to differ considerably from that recorded at standard meterological stations.

Across northern Australia, there is variation in the distribution of species between estuaries. Highest species richness is found in north-east Queensland and, in general, mangroves in north-west Australia are less rich than those in the north-east.

Wells (1982, 1983, 1985) described the distribution of mangrove species in northern Australia. He suggested (Wells 1985) that estuaries with very strong freshwater influence were less species rich than those with seasonal fluctuations in salinity conditions. Bunt *et al.* (1982*b*), on the other hand, reported that, in north-east Queensland, estuaries with greater freshwater runoff were more species rich. In north-west Australia, Semeniuk (1983, 1985) showed that mangrove stands with groundwater influence in the upper intertidal zone had more species than those without groundwater inputs. Saenger and Moverley (1985), after investigating the phenology of a number of mangrove species, suggested that the distribution of species on the north and east coasts was controlled by temperature but that in the north-west the influence of temperature was overridden by the effects of variation in rainfall.

These various hypotheses were examined by T. J. Smith and Duke (1987), who carried out an analysis of the distribution of mangroves in 92 estuaries in northern Australia. The estuaries were divided into two groups; those east of Cape York (the eastern group) and those to the west (the western group). The eastern group estuaries experience higher rainfall. The presence or absence of 40 mangrove species were used to define the botanical characteristics of each estuary. In both eastern and western estuaries, temperature was an important predictor of species richness, and variation in minimum temperature produced larger changes in species number than changes in maximum temperatures. Tidal amplitude was also important, with species richness declining with increasing amplitude. T. J. Smith and Duke (1987) pointed out that increased tidal amplitude will result in stronger tidal currents which may prevent the establishment of some species. For a given change in amplitude, the change in species number was greater in the west.

Within the eastern group, increased cyclone frequency and variability in rainfall were associated with a decline in species richness, but these factors are not significantly related to species richness in the west.

Estuary length and catchment area were predictors of species richness in the east but not the west. T. J. Smith and Duke (1987) suggested that longer estuaries provide a wider range of habitats which permits greater species richness. The absence of such a relationship in the west may reflect the generally more unfavourable climate for mangrove growth.

In neither eastern nor western estuaries did rainfall itself appear to have any significant effect on species richness, contrary to the hypothesis advanced by Saenger and Moverley (1985). In the east, runoff had no effect on species richness, but, in the west, increased runoff was associated with decreased species richness.

8.3.1 Distribution within estuaries

In addition to broad geographical patterns in the distribution of mangrove species, there is considerable local variation in the occurrence of species within single estuaries.

Towards the head of an estuary, mangroves are likely to experience greater brackish water influence than those further seaward. Bunt *et al.* (1982*a*) demonstrated that, in north-east Queensland, the distribution of species could be related to the position along the estuary.

Discussion of mangroves has often stressed the zonation of species within stands and suggests that communities are arranged in bands parallel to the shore line (e.g. Macnae 1966). Bunt and Williams (1980) analysed the distribution of species in north-east Queensland estuaries, and showed that patterns were complex and that zonation could not be characterized simplistically in terms of a single invariant sequence of species as implied by Macnae (1966) and other accounts. However, Buckley (1982) emphasized that Bunt and Williams' data indicate that particular mangrove species are associated with different sections of the intertidal zone. There is variation between estuaries, however, in which of the assemblage of 'possible' species occur in each zone. In regions with fewer mangrove species than in north-east Queensland, zonations may be more repeatable.

At many localities, mangroves extend up to the tidal limit, but at other sites mangroves form a seaward fringe with, on their landward side, either extensive mud- or sand-flats (Thom *et al.* 1975; see Fig. 8.2) or salt-marsh (Fig. 8.3). In the upper part of the intertidal zone, flooding tides are infrequent and, between tides, evapotranspiration may result in the development of hypersalinity unless ameliorated by rainfall. While many mangrove species are tolerant of salinities up to, or slightly above, that of sea-water, their ability to withstand prolonged hypersalinity is limited. Where hypersalinity regularly develops, mangroves are absent, being replaced by either clay pans or salt-marsh.

8.3.2 Dynamics of mangroves

Zonation of mangroves reflects differential responses of species to patterns of inundation and soil salinity, but it has also been traditional to interpret the zonation of mangroves as a spatial reflection of succession. In southern Australia the zonation of mangrove and salt-marsh communities has been taken to indicate the replacement of mangroves by herbaceous species (Pidgeon 1940). However, there are very few long-term studies of mangrove stands, and such assumptions are poorly supported by data. Studies of mangroves in the Sydney region have provided no evidence to sustain the successional model (Mitchell and Adam 1989*a*, *b*).

It is widely claimed that mangroves are land builders and that colonization by mangroves promotes sedimentation. The ability of mangroves to colonize exposed shores is limited, and the alternative model advanced by, for example, Thom (1982), which views mangroves as opportunistic invaders of available, relatively sheltered, substrate, appears to better accord with field evidence. Spread of mangroves is thus a response to environmental changes, rather than being an inevitable consequence of mangrove establishment. Once established, mangroves may provide a degree of protection of the coastline against subsequent erosion.

8.4 COPING WITH THE ENVIRONMENT

The mangrove environment differs from the terrestrial in two major respects: in being saline and in having frequently waterlogged soil. Both salinity and the oxygen status of the soil will fluctuate considerably. Mangrove species characteristically display a number of features which are interpreted as adaptations to these environmental conditions. These features have been reviewed by Tomlinson (1986), Hutchings and Saenger (1987), Ball (1988), Adam *et al.* (1990), and in Field and Dartnall (1987).

Mangroves are not obligate halophytes, as can

Fig. 8.2 Extensive intertidal flats with mangroves restricted to narrow strips fringing low-tide channels in a tropical macrotidal estuary. The Ord Estuary, near Wyndham. The most recent high tide-mark shows up as a boundary between wet (dark) mud and dry salt-encrusted (white) flats to the extreme right of the photograph.

Fig. 8.3 *Avicennia marina* forming the mangrove stand with salt-marsh at higher levels on the shore. Towra Point, Botany Bay.

be demonstrated in cultivation; their survival in saline conditions is related to tolerance rather than requirement for salt. The metabolism of mangroves is sensitive to salt concentration and central to tolerance is the segregation of salt into vacuoles in order to maintain low salt concentrations within the cytoplasm. Osmotic balance between vacuoles and cytoplasm is achieved by the accumulation of organic solutes in the cytoplasm (Adam 1990). Regulation of leaf salt content is achieved in a number of ways; restriction of uptake, dilution through growth and succulence and excretion (in some species) through salt glands.

Survival of mangroves in anaerobic mud appears to depend on maintaining aerobic conditions within the roots. Mangrove roots are well provided with aerenchyma, while the modified aerial parts of the root systems allow the subterranean portions to be in connection with the atmosphere for most of the time. Under normal conditions, the aerial roots are submerged only for very brief periods. Modifications to mangrove stands which restrict tidal ventilation and lead to water standing over aerial roots for prolonged periods, lead to death (Fig. 8.4).

8.5 NATURAL DISTURBANCE IN MANGROVES

Mangroves are subject to a range of natural disturbances whose importance has been little studied.

Smith and Duke (1987) demonstrated that in north-east Queensland, the species richness and composition of mangrove stands could be related to the frequency of cyclones. The effects of cyclones on species composition may be related to the ability of different species to recover from damage. Bardsley (1985) reported the effects of Cyclone Kathy on mangroves in the Northern Territory. Overall, the incidence of windthrow was small but there was extensive defoliation and, in some species, considerable canopy damage. Most species, but particularly *Avicennia marina* and *Excoecaria agallocha*, recovered quickly from epicormic shoots. However, insect defoliation of regrowth on *A. marina* was observed, suggesting that long-term re-establishment was not assured. Members of the Rhizophoraceae (*Brugiera*, *Ceriops*, and *Rhizophora*) had no, or very limited, ability for vegetative recovery and suffered substantial mortality. In addition, the

Fig. 8.4 Dead *Avicennia marina* killed following restriction of tidal exchange and permanent submergence of pneumatophores. Homebush Bay, Sydney

seaward fringes of *Rhizophora* showed a high percentage of broken trunks as well as damage from wave surge. Bardsley (1985) suggested that recovery of Rhizophoraceae-dominated stands would be very slow, if it occurred at all. High frequency of cyclones might cause local loss of these species. She pointed out that cyclone damage also results in a substantial change to the mangrove environment; loss of the foliage permits greater light to reach the forest floor and greater evaporation may alter the soil salinity regime. The input of extra litter will increase the availability of nutrients. These changes may favour some species more than others, possibly resulting in a shift in the balance between surviving species. The effects of cyclone damage are shown in Fig. 8.5.

As well as direct wind damage, cyclones may be

Fig. 8.5 Cyclone-damaged mangrove stand near Darwin

responsible for wave damage and erosion, although in the case of Cyclone Kathy, there was little erosion damage (Bardsley 1985).

Both Johns (1986) and Frodin (1987) indicated that lightning strikes may cause considerable local disturbance in mangroves in New Guinea. The extent of similar damage in Australian mangroves remains to be recorded.

The species paucity of mangrove stands suggests that they may be liable to severe defoliation by insects. However, there are relatively few reported examples of large-scale defoliation in mangroves anywhere in the world. Whitten *et al.* (1984) report two cases both involving caterpillars, one affecting *Avicennia* in Thailand and one of *Excoecaria agallocha* in Sumatra; in both cases, the trees recovered. In 1985, large areas of *Avicennia marina* in the Hunter Estuary, New South Wales, were defoliated by caterpillars (West and Thorogood, 1985), but again recovery was swift.

Studies in north-east Queensland (Robertson and Duke 1987) indicate that, in general, the incidence of insect herbivory is low, much lower than that reported in terrestrial rainforests (see p. 134), although there were differences in the degree of losses to herbivores between mangrove species.

The same argument about the potential susceptibility of mangroves to insect attack would also apply to pathogens. Pegg and Foresberg (1981) found that the fungus *Phytophthora* was associated with dead and dying mangroves at Gladstone in Queensland, but was unlikely to be the sole cause of mortality. Invasion by *Phytophthora* was associated with trees which were 'stressed' by a range of other factors in an industrialized estuary. Patchy death of *Avicennia marina* has been widely reported in New South Wales in recent years, and in some instances *Phytophthora* spp. have been isolated, but die-back appears to be the expression of a syndrome of environmental changes (West *et al.* 1983), with opportunistic pathogens attacking weakened trees rather than initiating their decline.

8.6 HUMAN IMPACTS

Nowadays direct exploitation of mangrove forest in Australia is very limited (Allaway 1987). Aboriginals made greater use of mangroves but their total impact was small. Early European settlers exploited mangroves for a number of purposes and, at least locally, may have had considerable impacts. Around Sydney in the early nineteenth century, mangroves were felled and burnt to yield alkaline ash used in soap manufacture (Bird 1981). Later, various species were used in small quantities to provide timber for minor uses and tannin was extracted from several species. There has been no attempt to develop large-scale forestry operations, either for timber or for wood chips, such as practised in Indonesia and Malaysia.

Indirectly, mangroves are exploited through harvesting of fisheries resources (p. 211).

Substantial areas of mangrove habitat have been lost from infilling to provide land for ports, industry and housing. While the greatest extent of losses from these causes has been in south-east Australia, similar infilling has occurred in the north-west (Semeniuk 1987). Some developments, such as constructing causeways for roads, may directly involve limited death of mangroves but, by interfering with tidal exchange, may adversely effect much larger areas.

In recent years, pressures have been for non-industrial uses, marinas, canal estates, and tourism developments. These developments take place away from existing towns and have resulted in considerable impacts on species-rich mangroves in Queensland which, until recently, had been virtually untouched.

In north-west Australia, a potentially major impact on mangrove areas may arise from continued expansion of salt manufacture, when large areas of coastal land are converted to shallow evaporation ponds (Semeniuk 1987). The development of mariculture may also result in the loss of mangrove habitat through its conversion to culture ponds as has occurred extensively in Indonesia and the Philippines.

Mangrove environments are subject to numerous pollutants. Close to urban areas, stormwater drains entering mangroves introduce many pollutants, as well as causing rapid local salinity changes; the background pollutants in many estuaries (heavy metals from antifouling paints and industrial discharges, organic chemicals from industry, domestic discharges) may have considerable impacts on the fauna, although these are poorly documented. Specific effects of pollutants on the flora are not well documented, although patch mortality in mangrove stands in industrialized estuaries is frequently attributed to generalized pollution.

A major threat to many mangroves is posed by oil pollution. Fortunately, to date, Australia has not experienced a major oil spill, although numerous minor ones have been recorded, a number of which have affected mangroves. Allaway (1987) has reviewed the effects of oil pollution on *Avicennia marina* in Botany Bay, near Sydney. Contamination with crude oil causes rapid mortality of seedlings and young plants. Some considerable time after the pollution has occurred there is defoliation and death of mature trees. The mechanism responsible for this delayed response is yet to be elucidated.

One form of disturbance which has had little impact on mangroves is the introduction of alien species. Even in urban areas, mangrove stands, except along the landward fringe, have not experienced invasion by exotic plants. There is considerable concern about the potential threat posed by marine species introduced in ships' ballast water (Hutchings *et al.* 1987) but, although a number of species have been introduced into Australian waters, as yet, they have had little effect on mangroves.

The aggregate effects of human impact on mangroves have been greatest in south-east Australia where substantial areas have been lost. As a percentage of the total area these losses are small, although, as southern mangroves are approaching the geographical limits of this formation, they may be significant from a biogeographic standpoint. Although there have been impacts on mangroves in some of the more remote parts of the continent (Fig. 8.6), Australia has probably the most extensive mangroves not subject to immediate pressures by human impacts. This is in contrast to the Indo-Malesian region where mangroves are exploited for fire-wood, charcoal, poles, or wood-chips and are subject to reclamation for mariculture, agriculture, housing, and industry. There are great opportunities in Australia both to conserve extensive areas of undisturbed mangrove, and to devise management strategies to retain mangroves in the proximity of expanding urban areas which might provide models for management elsewhere.

8.7 FAUNA OF MANGROVES

Mangroves provide a habitat for a diverse marine and terrestrial fauna. Many animals are permanent residents but others are transients, although the availability of mangrove may be crucial for the completion of life cycles.

The fauna of Australian mangroves has been reviewed by Hutchings and Recher (1982) and Hutchings and Saenger (1987). For the marine invertebrate fauna, the distribution of species richness is similar to that of the vascular flora, being low in the south and highest in the north-east. The high species richness in the north-east fauna reflects the increased structural diversity and area of stands. A number of observations support the generality of these claims, but Hanley (1985) has argued that the number of polychaete worms in Northern Territory mangroves is far lower than might be expected. He suggested that the marked seasonality of the climate, and the large tidal range, resulted in fluctuations in sediment salinity beyond the tolerance of most polychaetes. There have been too few studies across a range of taxa to determine whether Hanley's findings are repeated in other groups.

Among the non-marine animals, birds are conspicuous, although population densities of residents are relatively low compared with those in terrestrial forests. The composition and origins

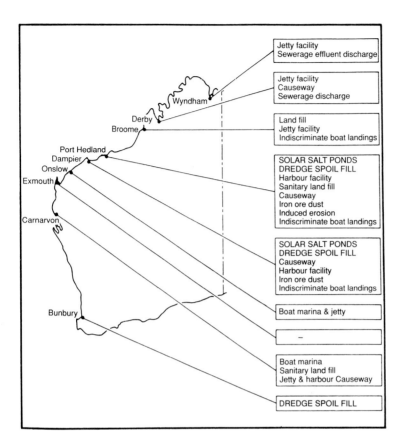

Fig. 8.6 Impacts on mangrove stands along the coast of West Australia. Major impacts are given in upper case (from Semeniuk 1987).

of the mangrove avifauna in Australia have been discussed by Ford (1982) and Schodde *et al.* (1982). Although more than 200 bird species have been recorded at some time or other from Australian mangroves (Saenger *et al.* 1977), only 14 are virtually confined to the habitat, another 12 are restricted to it in part of their range, while around 60 more regularly frequent mangroves throughout the year or at certain seasons (Schodde *et al.* 1982). This mangrove avifauna is large by world standards.

The taxonomic affinities of the Australian mangrove avifauna, the paucity of endemic genera, and the few species from the two largest families (honeyeaters—Meliphagidae and parrots—Psittacidae) provide evidence from which Schodde *et al.* (1982) argued a relatively recent origin from inland rainforest sources. Ford (1982) also regarded the mangrove avifauna as derivative from dryland rainforests.

The number of mangrove specialist birds is significantly greater in the north-west than the north-east despite the greater species richness and structural complexity of the north-east flora. Schodde *et al.* (1982) suggested that the reasonably close proximity of dryland rainforest to mangrove coasts has facilitated interchange between populations in the two habitats, the lack of isolation limiting the opportunity for divergence of mangrove taxa. In the north-west, dryland rainforest has been fragmented for a long period and geomorphological barriers are likely to have resulted in a patchy distribution of mangroves: both are factors which may have promoted speciation and development of a varied mangrove avifauna (Schodde *et al.* 1982).

The other element of the vertebrate fauna of mangroves which excites comment from the general public is far less conspicuous than the birds. Both Australian species of crocodile, the saltwater crocodile *Crocodylus porosus*, and the freshwater crocodile *C. johnstoni*, frequent mangroves in northern Australia; the larger (and to humans much more dangerous) *C. porosus* is the more abundant in tidal waters.

The interactions between the fauna and flora of mangroves are poorly known (Robertson 1987). Wood-boring species may be abundant (Hutchings and Saenger 1987), but whether they have serious impacts on living trees is unknown. Crabs are often a major component of the marine fauna in mangroves. Through their burrowing activities, they may generate environmental heterogeneity, influencing the distribution of other fauna and flora. They may also be major consumers of fallen leaves (Robertson 1987), fruits and seedlings (Smith 1987; Osborne and Smith 1990).

8.8 THE ECOLOGICAL ROLE OF MANGROVES

The destruction of mangroves in south-east Australia did not, until recently, attract much adverse criticism. Mangroves, and wetlands in general, were regarded as unhealthy, insect-ridden places, the reclamation of which was a community benefit. In the last few decades there has been a dramatic shift in public opinion and there is now concern to prevent further loss of wetland habitat.

Many arguments can be advanced in favour of protection of mangroves (see e.g. Adam *et al.* 1985) but those which have swayed the general public relate to the possible role of mangroves in the maintenance of fisheries. Commercial and recreational fisheries in Australia are of great economic value, as well as directly involving a large proportion of the population (recreational fishing is, nationally, by far the largest participation sport).

The importance of mangroves to fisheries arises in two ways; through the provision of habitat and as a source of detritus to aquatic ecosystems.

The value of mangroves as fish habitat is reasonably well established, even though the number of detailed studies is small. In Australia, mangroves are not breeding habitats for many fish, but are important nursery areas for species which, as adults, inhabit adjacent estuaries and inshore coastal waters (Bell *et al.* 1984). Mangrove creeks are mainly shallow, which excludes large predatory fish, and so provide a relatively sheltered habitat for juveniles. They also provide an open, soft substrate, occupied by numerous small invertebrates and algal and bacterial films, which are the food for many of the young fish. Loss of mangrove would be expected, in the long term, to be reflected in declines in fish stocks, although the relationship between mangrove area and fish population size will be complex, and loss of a certain proportion of habitat is unlikely to result in the same proportional decline in the fish population. Nevertheless, the utilization of mangroves by juvenile fish is an important argument in favour of their conservation.

The linking of mangroves to marine detrital food chains is more speculative. Popular accounts and elementary textbooks frequently present elaborate food webs in which mangrove detritus is the major input. As Williams (1986) has pointed out 'Plausible though such diagrams undoubtedly are, they are entirely hypothetical; they are based on no direct evidence whatsoever'. Elucidation of the links in the food web, let alone quantification of the energy flow in any particular link, will be a major task. Nevertheless, it is an important one, as the theoretically derived food chains are indeed plausible and, if correct, would provide a very powerful argument for mangrove conservation.

The state of the art in unravelling energy flows in mangroves is reviewed by Robertson (1987) and other authors in Field and Dartnall (1987) and by Twilley (1988). There are few studies on total primary productivity in mangroves, but a number of estimates of litter-fall have been made (Allaway 1987). Although litter-fall represents only a fraction of annual productivity (not accounting for material incorporated into trunks or roots), it does provide a measure of material potentially available for conversion to detritus. A number of authors have, by implication, suggested that the litter-fall is all potentially available for export from mangroves to adjacent waters. The concept of salt-marshes and mangroves as outwelling systems, exporting nutrients and detritus to offshore waters has wide currency, but is largely unvalidated (Nixon 1980; Robertson 1987; Adam 1990). Although litter-falls may be high, Robertson (1987) has demonstrated that in northern Australia a very large proportion of the fall is rapidly utilized by sesarmid crabs. At higher latitudes, crabs may be lesser *in situ* consumers, but this remains to be determined. It is possible that most mangrove detritus is reprocessed within the stand and the major export is as live biomass (as in juvenile fish leaving nurseries). Given the pressures globally (if not in Australia) for mangrove destruction, it is essential that trophic pathways in mangroves and neighbouring estuarine waters be more fully explored. Australia, with extensive areas of mangroves accessible from major centres of research, could make an important contribution to such studies.

8.9 HISTORY

The mangrove flora of Australia is a part of the mangrove flora of the Indo-Pacific region and displays few features which are uniquely Australian.

A number of mangrove genera have long fossil records (Tomlinson 1986) but there is a paucity of evidence about the history of mangroves in Australia. Churchill (1973) claimed that mangroves were established in what is now south-west Australia during the mid-late Eocene, but Martin (pers. comm.) indicates that these identifications require checking and there are a number of Tertiary records of mangroves on the shores of Bass Strait.

The propagules of many mangrove species are buoyant and appear to be well adapted for long-distance dispersal. However, proof of long-distance dispersal is difficult, and Mepham (1983) has suggested that the spread of mangroves over geological time has been by short hops along contiguous or near contiguous coastlines, rather than by transoceanic drift. The successful introduction of mangroves to Hawaii, and the vigorous growth of planted *Avicennia marina* (from Australia), prior to eradication, in California (Sauer 1988), demonstrate that not all coastlines with climatic conditions capable of supporting mangroves have been colonized by natural means, and suggest that Mepham's (1983) doubts about long-distance dispersal were not unreasonable.

If the distant history of mangroves remains speculative, the more recent past is better understood. Mangroves occupy a limited portion of the intertidal zone. The coastline has been in its present position for only about 6000 years. Over the last few million years there have been a number of major fluctuations in sea-level. At the times of lowest sea-level, corresponding with glacial periods, mainland Australia, Tasmania, and New Guinea were united. At these times, any mangroves would have been many kilometres seaward of their current positions. Different configurations of the coastline would have provided different extents of mangrove habitat. If coastal fish populations are functionally linked to mangroves, then the changes in mangroves over the Quaternary could have been reflected in much wider ecological changes in coastal waters. The Alligator Rivers region of the Northern Territory today contains extensive freshwater wetlands with important but spatially limited mangroves along the tidal river channels and northern coast. This distribution pattern is geologically recent (Woodroffe *et al.* 1985): towards the end of the major post-glacial sea-level rise, the area was occupied by an extensive mangrove swamp. The transition between this mangrove and freshwater wetlands on the Magela floodplain has been studied through palynological investigation by Clark and Guppy (1988). The mangrove forest which developed as the sea-level rose was dominated by *Rhizophora*, until *c.* 3000 BP. *Avicennia* and other mangroves then became more abundant, followed by a decline in all mangroves and replacement by freshwater wetland genera, which was completed about 1300 BP. This study graphically illustrates the dynamic and (geologically) ephemeral nature of mangrove stands.

The history of temperate mangroves during the Quaternary sea-level fluctuations is poorly known. In South Australia, any available habitat would have been at higher latitudes at a time of generally colder climate. If mangroves, a few thousand years ago, survived at higher latitudes, why do they not today? If, on the other hand, mangroves become locally, and temporarily, extinct at high latitudes, how, and how rapidly, did recolonization occur? If recolonization from the north took place, it is difficult to explain the patchy distribution of mangroves at high latitudes and the absence from many apparently suitable sites. If survival in refuges was possible, what special features did these sites possess?

9 THE HUMAN INFLUENCE

Australian rainforests and their biota reflect the consequences of millions of years of selection by, and adaptation to, environmental change. In Chapter 6, it was shown that the trend towards aridity in the centre of the continent which commenced in the Miocene was responsible for a contraction of rainforest towards the east coast and its fragmentation into numerous small areas with especially favourable climatic and/or edaphic conditions. From some time in the late Pleistocene, human influence was added to the list of significant impacts on rainforest.

The interactions between humans and rainforest can be seen in a number of successive phases of activity. The first phase, extending over thousands of years, involved the Aborigines. The later phases, since European occupation, have been compressed into a period of two hundred years which has seen massive environmental change across the continent and major disturbance to many rainforests. The first phase of European involvement with the rainforest was the period of the 'timber-getters', a time of exploration and selective exploitation. In their wake came the farmers interested in rainforest, not as a timber resource, but as an indication of what was thought to be potentially valuable land once the existing vegetation had been destroyed. The rate of destruction of rainforest was slowed as much of the remaining area came under the control of foresters, although more intensive forest management did not develop for many years. In the past two decades, with an upsurge of public concern over environmental matters, large areas of rainforest have been included in national parks, and continuing rainforest logging has aroused bitter controversy.

9.1 ABORIGINAL HISTORY

Many aspects of the peopling of Australia are still poorly understood; exactly where invasion first occurred, the size of the invading group, the path of spread across the continent, and whether more than one invasion occurred, are matters which, at present, can only be speculated upon (Flood 1983). A particularly important question for any assessment of the possible impacts of Aborigines on the environment is—how long have humans lived in Australia?

There are a number of archaeological sites in southern Australia which have been radiocarbon dated to between 30 000 and 40 000 years ago (Flood 1983). In northern Australia, thermoluminescence dating of an occupation site in Arnhem Land has suggested an age of 50 000–60 000 for the lowest stratum with recognizable human artefacts (Roberts *et al.* 1990).

The suggestion has been made that human occupation of Australia long predates 40 000 years ago, and that, even in south-east Australia, Aborigines have been present for more than 120 000 years. This claim is based on environmental reconstruction from the fossil pollen record at Lake George on the Southern Tablelands of New South Wales, some 40 km north east of Canberra (Singh *et al.* 1981; Singh 1982; Singh and Geissler 1985, see Chapter 6), and the interpretation of an increased charcoal frequency after 128 000 BP as reflecting the use of fire by humans.

The great discrepancy between the oldest-dated archaeological remains (*c.* 40 000 BP) and *c.* 130 000 years ago has meant that the claim for the early arrival of humans has been questioned. A number of assumptions made in the interpretation of the Lake George evidence could be questioned: does the increase in charcoal reflect an increase in fire frequency; should an increase in fire frequency be attributed to human influence; is the dating of the event reliable?

It is difficult to derive a relationship between charcoal in deposits and fire frequency or intensity; nevertheless, there is general agreement that an increase in charcoal of the magnitude shown at

Lake George does represent a change in fire regime. A change in fire regime could have many causes; the arrival of humans is one. The change in charcoal frequency and correlated rise in *Eucalyptus* (zone F in Singh and Geissler 1985) occurred beyond the range of radiocarbon dating; the dating of the event must therefore be by indirect methods. Wright (1986) has suggested that Singh's estimate of the date of the beginning of Zone F is in error, and that a more realistic figure would be 60 000 BP. If this were the case, then Zone F does not represent the last interglacial, but rather an interstadial, a warmer intermission within a glacial period. Singh (1986) disputed this interpretation pointing out that, if Wright's extrapolation is continued, it would date the last reversal of magnetic polarity at 233 000 years, a serious discrepancy from the well-documented accepted estimate of 730 000 years.

In the absence of additional evidence, the dating and interpretation of Zone F at Lake George will remain controversial. It seems clear that the zone represents a major change in vegetation, and suggests that much of the eucalypt forest and woodland in eastern Australia is of comparatively recent origin. If Singh's interpretation is correct, much of the development of these eucalypt communities has occurred in 130 000 years, which to ecologists might seem a very short period. If Wright is correct, and human influence was involved in the vegetation change, then the date of c. 60 000 BP is not in conflict with available archaeological evidence but the time frame for development of extensive eucalypt-dominated vegetation is further reduced.

Singh's suggestion of an Aboriginal presence for at least 130 000 years cannot be conclusively denied. In considering the impacts of Aborigines on rainforest, the possibility that these extend back to the last interglacial should be borne in mind.

9.1.1 Aborigines and rainforest

Aboriginal exploitation of rainforest is reviewed by Byrne (1987), Boutland (1989), and Ritchie (1989). Very few archaeological sites have been discovered within rainforests and much of the evidence on Aboriginal use is ethno-historical, being derived from Aboriginal informants or the observations of explorers and early settlers.

Estimates of the size of the population of Australia at the time of the European invasion are based on slender evidence (Flood 1983), but a frequently quoted figure is about 300 000. Whatever the population, it was certainly not evenly distributed; the density was greatest in the coastal zone, reflecting the reliability of resource availability. It seems likely that, regardless of the environmental fluctuations during the Quaternary, the coast would always have supported the highest populations. Much of the land which would have formed the coastal zone during the last glacial period was submerged by the post-glacial sea-level rise (the coastline achieved its present position about 6000 BP), so that evidence of human activity during the glacial period will have been lost with the rising sea-level. Many of the rainforest stands which are now close to the coast may have survived the glacial period, albeit in a reduced, fragmented form. However, they would have been further inland, and so may not have been within the region of highest population density.

Throughout the period of human occupation of Australia, be that period 60 000 or 130 000 years, rainforest has been a minor component of the vegetation cover of the continent; only in northern Queensland was the area of continuous rainforest cover sufficient to support a permanent population. Elsewhere rainforest was just one of the mosaic of habitats exploited by local populations, who were rainforest 'users' rather than rainforest 'dwellers'. Bowdler (1983) has drawn attention to the contrast between north Queensland, where Aborigines colonized the rainforest, and Tasmania, where, she argued that in post-glacial times the expanding colonising rainforest displaced people from previously occupied land.

The rainforest-dwelling Aborigines of north Queensland were distinctive physically (being short in stature), linguistically, and culturally, but genetic studies do not indicate racial distinction from other Aborigines (Flood 1983).

An account of the Aborigines in north-east Queensland at the time of European settlement was provided by Lumholtz (1889). The rainforest Aborigines built semi-permanent waterproof huts (*mia mias*), covered by leaves or grass; these were much more elaborate than the temporary shelters (*gunyahs*) of other groups (Birtles 1982; Keto and Scott 1986).

The dense vegetation of the rainforest prevented the use of boomerangs and spears, common weapons in other areas. Unique to the rainforest Aborigines was a flat stone axe, usually made of basalt and bound to a lawyer vine haft (Birtles 1982). These axes were used to remove bark, for digging, and for chopping open the nests of bees; they were not suitable for felling trees (Birtles 1982). For capturing prey, a variety of nets, pit traps, and traps made from lawyer vine, coiled grass, bark fibre, or plaited human hair was used (Birtles 1982). Flying foxes and birds were caught by smearing the roosting branches with a sticky gum from fig trees (Birtles 1982).

The diet included a large number of plant products. A feature of many of the plant foods is that in the raw state they are extremely toxic, for example the fruits of black bean *Castanospermum australe* (Leguminosae), (McBarron 1976), Queensland walnut *Endiandra palmerstonii*, and yellow walnut *Beilschmiedia bancroftii* in the Lauraceae (Horsfall and Hall 1990), and cycads. To render these, and other foods, edible, sophisticated leaching and cooking processes were used. The frequent use of a range of toxic foods is unique to rainforest Aborigines; in other areas toxic plants were eaten but less regularly (Keto and Scott 1986).

The rainforest Aborigines of north-east Queensland may possess a very long accurate oral history. Legends tell of a time when the coastline was where the Barrier Reef is today, and possibly of volcanic activity on the Atherton Tableland (Keto and Scott 1986). To record such events, the history must date from between 10 000 and 15 000 years ago. The effects of European impact on these Aborigines are described by Birtles (1982), but the few surviving guardians of the culture still live in north-east Queensland, retaining their traditional attachment to the land and knowledge of how to use the resources of the rainforest (Keto and Scott 1986).

Aboriginal use of rainforests in New South Wales has been extensively reviewed by Byrne (1987); the picture of rainforest users who went into rainforests but at other times exploited the coast, or swamps, is probably equally applicable to the rainforests of southern and central Queensland. From an Aboriginal viewpoint, there may have been three types of rainforest with different patterns of utilization: littoral rainforest and gallery stands along rivers, the large rainforest stands of the lowlands, and the highland forests. The gallery forests along the major rivers were easily accessible by canoe and on foot, littoral rainforests were also readily accessible to groups moving along the coast. The large lowland stands were easily exploited around the edges but were, perhaps, only rarely penetrated. The highland forests were in areas with only low population densities (Belshaw 1978) and were probably visited only occasionally by small bands.

From the observations of the early explorers and settlers in New South Wales, information on the use of rainforests can be obtained (Sullivan 1978). A wide range of plants yielded food but, as the resources were dispersed, gathering large amounts would have been time-consuming. As in north Queensland, a number of these plants were toxic, and elaborate processing was required to render them edible (Sullivan 1978).

Although Aboriginal tribal groups had defined territories there were occasions when they travelled long distances. The periodic gathering of large numbers on the Bunya Mountains in southern Queensland to feast on the seeds of the Bunya pine *Araucaria bidwillii* is well attested (Fig. 9.1). Bundock (1898), discussing the Aborigines of the Richmond River region of northern New South Wales wrote:

About 200 miles north of us lies the peculiar strip of country where the Bunya pines grow, and every third year or so the pines bear a profusion of cones, are as large as a man's hand and the seeds are about 1½ inches long and as thick as a woman's fingers; they are sweet and very nourishing and when roasted are like Spanish chestnuts. In the years when these pines bore, word was passed from tribe to tribe, and there was a sort of 'Truce of God' understood for the blacks went there thro' each others territories unharmed and all met together in peace, and feasted on the plentiful provision. They valued the trees very highly, for they would have killed anyone who had cut one down to get the cones, tho' the trees were very prickly and unpleasant to climb. Our tribes used to come back looking very fat and well after an expedition to the Bunyas.

There are a number of accounts which indicate that organized hunting parties took place in New South Wales rainforests (Byrne 1987), particularly aimed at pademelons (*Thylogale stigmatica* and *T. thetis*). These hunts probably concentrated on the rainforest/sclerophyll forest boundary rather than taking place entirely within rainforest.

As well as being a source of food and other resources, many rainforest sites were also of great spiritual significance to Aborigines, and are associated with various myths and legends.

In Tasmania the rainforests would have provided few, if any, resources (few fleshy-fruited or otherwise edible plants) for an Aboriginal population, were frequently inhospitably cold and wet, and were largely avoided.

9.1.2 The impact of Aborigines on rainforest

Given the low population density of Aborigines, even in favoured habitats, it is difficult to imagine that the kinds of resource use discussed above had major impacts on rainforests. In the semi-arid zone, Aborigines may have played a part in causing the extinction of the megafauna (see Chapter 7), although this remains controversial (Flood 1983). By the Quaternary, the megafauna had long since gone from the rainforest, and there is no evidence to indicate that the hunting pressure was sufficient to cause loss of other animals.

However, it is certain that Aborigines had a profound influence on all Australian vegetation types through the use of fire. There is evidence of fire in Australia in the Tertiary (Kemp 1981*b*), long before the advent of humans, but fire regimes changed dramatically with human intervention.

Fig. 9.1 *Araucaria bidwillii* (Bunya pines) at Bunya Mountain, Queensland (photo: B. S. Wannan).

At Lake George, the increase in charcoal, reputedly at 128 000 BP, is used to suggest the arrival of Aboriginals (Singh et al. 1981). More immediately relevant to a discussion of rainforest is the evidence from Lynchs Crater on the Atherton Tableland (Kershaw 1985, 1986, described in Chapter 6 (p. 164)). The almost complete replacement of rainforest by sclerophyll taxa toward the end of the last dry period (from about 38 000 BP) is associated with increased charcoal, and is interpreted as probably indicating the impact of burning by Aborigines.

The early white visitors to Australia saw many large fires, presumably lit by Aborigines. The first settlers also recorded numerous examples of Aboriginal burning practices. Aborigines used fire for many purposes: to drive game, to promote new growth, to clear undergrowth, and in warfare. This burning clearly had a profound impact on sclerophyll vegetation, did it also affect rainforest?

It was argued in Chapter 4 that the transition between rainforest and sclerophyll forest is strongly influenced by fire. This zone is also a habitat particularly rich in utilizable resources; whether by accident or design, it is probable that Aboriginal burning practices served to maintain wide transition zones and to retard the advance of the rainforest.

Despite an understanding of fire behaviour, it is probable that fires did not always proceed according to plan. Once they lit fires, Aborigines did not put them out (Flood 1983), and burning of the edges of rainforest stands, or through small gully rainforests, would have occurred. Such events could have affected rainforest distribution on a local scale.

In Tasmania, at the height of the last glacial period, rainforest was very limited in distribution. Repeated firing of the west coast sedge and shrublands by Aborigines kept the advancing rainforest at bay during the warmer post-glacial period.

The rapid destruction of Aboriginal culture by Europeans on the east coast of the mainland, and in Tasmania, also destroyed the opportunity for detailed recording of burning practices. Although there is a generalized picture of frequent use of fire, the details of the pattern and purpose of burning were not studied. It is only inland and in the north that Aboriginal burning practices can still be observed, although these have undoubtedly been modified by contact with Europeans and take place in environments very different from those of the east coast rainforests; nevertheless such study is instructive and indicates very skilful manipulation of fire, to create environmental mosaics, rather than mindless pyromania.

Haynes (1985) has documented the burning practices utilized in north central Arnhemland by the Gunei people. Control of fires was achieved by the timing and position of ignition, and importance was placed on preventing fire entering rainforest stands, partly because of their spiritual significance but also to protect the important food resources that they provide. Protection is achieved by burning fire-breaks around rainforest stands early in they dry season at one- or two-year intervals. Such regular burning, while reducing the possibility of fierce, late dry-season fires damaging the rainforest, would also prevent any expansion of the stands.

Haynes' study documents the Aborigines' sophisticated understanding of fire behaviour and its relationship to climate. If similar understanding and skills were possessed by other groups, then it is possible that some east coast rainforests may have been protected from fire. However, the aims of fire management, as still practised in Arnhemland, are clearly different from those in Tasmania, where rainforest was repressed rather than protected.

It is important to emphasize that Aborigines did not clear rainforest (or other tree-dominated communities). Shifting cultivation and other agricultural practices, which are a major influence on rainforests in other continents, are unknown in Australia. As Flood (1983) pointed out, Australia remained a continent of hunter–gatherers, and other traits which developed elsewhere, such as the use of pottery and domestication of animals (except for dingoes) were not adopted.

An explanation for the persistence of the hunter–gatherer economy in Australia has long been sought by anthropologists. Prior to the European invasion, the Aborigines had not existed in total isolation from the rest of the world. Before 6500 years ago, there was a land-bridge between New Guinea and Australia (see Fig. 6.18), after this time the islands in the Torres Strait would have formed a series of stepping-stones. In New Guinea, cultivation was established well before the breaking of the land bridge. Domestic pigs, which play an important role in New Guinea culture, had been introduced to the island by 10 000 years ago but were unrecorded in Australia (Flood 1983) prior to the European invasion. The Aborigines in northern Australia also had contact with Macassan fishermen who visited Australian waters every year to harvest trepang.

It is clear that these contacts influenced Aboriginal culture. Flood (1983) documented a number of examples where both ideas and technology (for example, the double outrigger canoes of the Cape York Aborigines) were acquired by contact with outside groups. This willingness to accept aspects of other cultures heightens the mystery of the absence of agriculture from Australia.

Flood (1983) suggested that the main reason for the apparent failure to adopt agriculture was the affluence of the Aboriginal life style. In the northern and coastal regions, Aborigines were able to enjoy a varied diet with a high daily energy intake for a relatively small expenditure of time and effort. Not only was there no pressure to develop agriculture but the diversity and flexibility of the hunter–gatherer economy allowed greater accommodation of the inherently variable environment than would dependence on one or two agricultural crops. Even in the arid interior, food resources were available provided that the people were nomadic, something that would have been difficult in an agriculture system.

9.2 EUROPEAN HISTORY

The European colonization of Australia began in January 1788, with the arrival in Sydney Cove of the First Fleet. This was not the first European contact, which extended back more than a hundred and fifty years earlier, but it marked the beginning of a period of rapid and extensive environmental change.

In the European history of Australia, rainforest has played an important role, although one neglected by many historians. The pattern of the early exploration and settlement of the east coast can be viewed, without undue exaggeration, as the consequence of the search for rainforest and its valuable reserves of *Toona australis*. While exploitation predominated, the rainforests attracted the attention of early colonial scientists and excited particular interest from artists (Ritchie 1989).

The first European colonists found themselves (many reluctantly) in a land very different from any that they had previously experienced. The flora and fauna were strange, indeed bizarre. Despite the predictions, encouraged by Sir Joseph Banks, of a prosperous agricultural future for the colony, the settlers found the poor sandstone soils around Sydney Cove scarcely conducive to successful harvests. In order to place the struggling colony on a firmer footing, there was great interest in discovering commodities which could be exploited. One of the first was the timber of *Toona australis*. Examples of the wood were dispatched to London in November 1791 and attracted the interest of the Admiralty, who ordered that convict ships on their return journey to England should carry as much of the timber as possible. At the time the similarity of the wood to mahogany was noted, indeed *Toona* is related to *Swietenia*, the South American mahogany. In the colony, however, the tree came to be known as cedar, or sometimes red cedar to distinguish it from the white cedar, *Melia azeredach*.

The wood of *Toona* is deep red in colour, easily worked, strong, and durable. Cedar, in addition to its attractive timber, had other advantages; the logs float, an important consideration when rivers provided the main means of access—many Australian species (including many eucalypts) as 'green' logs have specific gravities greater than 1—as a species it is easily identified and not readily confused with others, it is deciduous, and, leafless in winter or with a new flush of leaves in spring (Fig. 9.2), it was easily located. It soon became highly prized, both within the colony itself and for export. For nearly a century it was of major commerical importance and became known as red gold.

The first cedar to be exploited was from rainforests on the Hawkesbury River to the north-west of Sydney. In the early years of the nineteenth century, cedar was being cut in the Hunter Valley, shortly afterwards cedar-getters were active in the Illawarra region. Later, they moved to the north coast of New South Wales (Bolton 1981; Vader 1987; Baur 1989).

The cedar-getters have entered Australian mythology, heroes of a romantic phase of the nation's history. At the time their lot was one of great hardship and privation (Fig. 9.3). The history of cedar-getting has been reviewed by Vader (1987) and in local histories, such as Daley (1966). The cedar-getters were the pioneers along much of the east coast, although the process

Fig. 9.2 *Toona australis* (red cedar) in early spring, with the new flush of leaves, northern New South Wales.

Fig. 9.3 Cedar-getters in the New England region (the camp is within eucalypt forest) (photo: Forestry Commission of New South Wales)

was not a steady progression away from Sydney, but the opening up of country along major river valleys with links between catchments established later. Despite various attempts by successive Governors to control and regulate the cedar-getters, most operated outside the umbrella of authority.

Opening up 'new country' the first cedar-getters were dependent on rivers, not only for access, but for transport of their harvest. Logs were branded by their fellers, launched into rivers, and carried downstream to be collected in the estuaries and shipped to Sydney or other centres. Often logs were accumulated on river banks until periods of heavy rain when there was sufficient flow to carry them. On the bigger rivers, they were gathered into rafts for the journey downstream (Daley 1966; Vader 1987).

The last great rush of red gold took place in north-east Queensland in the last twenty years of the nineteenth century (Birtles 1982). At first, timber was provided for the first settlers and the mining industry but later the Atherton Tableland cedar formed the basis of a major export trade.

Some cedar was pit sawn in the bush to serve local uses or to permit easier transport when river transport was not possible. As exploitation moved away from the immediate vicinity of rivers, bullock teams were used for transport either to local sawmills or coastal ports.

Although the age of the cedar-getters probably ended around the turn of the last century, *Toona* continued to be felled and harvested, not only as an adjunct to land-clearing or routine forestry but also in selective logging operations in the tradition of the cedar-getters. Vader (1987) discussed what he called the third-wave cedar-getters, who were active from the late 1940s onwards, using modern machinery to obtain access to trees growing on the steep slopes of the Great Escarpment. The very high value of the timber made the costs of roading for even single trees worthwhile.

In the early 1980s extraction of red cedar by helicopter from otherwise inaccessible sites on private land in the Illawarra region of New South Wales was proposed (Mills 1987), but this practice was terminated by the use of conservation orders under the Heritage Act. Some limited quantities of cedar logs are still available for sale and realize high prices at auction; in 1987 at Wingham in New South Wales forty-one logs (with a total volume of 61 m^3) were auctioned for an average of \$443 m^{-3}, with a maximum price of \$1072 m^{-3} (Baur 1989).

Cedar was used for many purposes, not all of which did justice to its special properties. It was widely used in furniture and also architecturally in many of the grander earlier colonial houses (Vader 1987). In the bush, it was often used as a general-purpose timber, simply because it was readily available. Large quantities were used in some mining operations: Idriess (1959) illustrated fluming constructed of cedar for hydraulic tin-mining in north Queensland.

Extraction was very wasteful. In felling the largest trees, many smaller trees were often destroyed, and trees were sometimes felled in places from which the logs could not be extracted. Logs transported in rivers were often not recovered. On the Atherton Tableland in northern Queensland, the first cedar to be felled was used locally as there was no means of transporting it to the coast. In the early 1880s, it was decided to send cedar down the Barron River in flood, and it is reported that some fifteen million superfeet (45 000 m^3) of timber were lost, smashed over the Barron Falls (Birtles 1982).

Cedar-getting undoubtedly caused a substantial reduction in the number of mature trees and caused local disturbance to rainforest stands. During the late nineteenth century, numerous commentators criticized the profligacy of the cedar-getters—pointing out that the rate of extraction of large trees from accessible locations was not sustainable. However, the cedar-getters did not, by themselves, threaten the survival of either rainforest stands or red cedar, which is still a widespread species. In terms of long-term impacts on rainforest, the significance of the cedar-getters was that they identified rainforest stands for subsequent clearing for agriculture.

Although cedar-getters did not seriously

imperil *Toona australis*, they may have been largely responsible for reducing the onion cedar, *Owenia cepiodora*, a member of the same family, to its current endangered status. *O. cepiodora* is restricted to the Border Ranges region of northern New South Wales and southern Queensland. Much of its original habitat has been cleared for agriculture, but it was logged extensively because of the similarity of the timber to red cedar. Logs were soaked for some time in streams to remove the strong smell of onion from the wood before the sawn timber was sold as red cedar (Leigh *et al.* 1984). The population was reduced to a small number of regrowth trees, but in recent years the species has been used in tree-planting schemes in northern New South Wales (Baur 1989).

9.2.1 Agriculture

To the early European settlers, Australia was a harsh, inhospitable land. Sydney was largely surrounded by infertile soils and the sclerophyll vegetation yielded little of value; even the timber was heavy, hard, and difficult to work. The discovery of rainforest stands gave access to more easily worked timber and more fertile soils.

The luxuriance of the rainforest vegetation aroused hopes for very prosperous agriculture once the forest had been cleared. Atkinson (1826) suggested that:

The soil in many of these brushes is extremely rich, but the labour of clearing is immense; and very little land of this description has been hitherto brought into cultivation; it seems, however, well adapted for the production of vines and other fruits, being generally of a light friable texture, and lying in peculiarly warm sheltered situations.

While Hodgkinson (1845) wrote that:

The luxuriant and vigorous character of the brush, on alluvial land, in the northern part of the territory of New South Wales, cannot be surpassed in any tropical region. When this brushland is cleared, and cultivated, its fertility seems inexhaustible...

The equation of rainforest with soil fertility has been the downfall of many agricultural ventures throughout the tropics, but in eastern Australia rainforest was generally situated on more fertile soils with greater agricultural potential than the sclerophyll forest, woodland, or heath which usually occur on soils of (by world standards) very low nutrient status. Nevertheless, clearing of Australian rainforest did involve substantial loss of nutrients, and the original high expectations of the pioneers were frequently only briefly met. Agricultural utilization of land formerly under coachwood, *Ceratopetalum apetalum*, was particularly unsuccessful (Baur 1989).

The first rainforests cleared were to the northwest of Sydney on the alluvial flats of the Hawkesbury River, near the towns of Richmond and Windsor (Vader 1987). This area still supports highly productive market gardens. To the south of Sydney, the lowland rainforests of the Illawarra region were cleared relatively early in the nineteenth century, and dairy farming later became a major land use. Other small areas of rainforest were cleared, but the main assault in New South Wales was delayed until the 1860s, after the passing of the Land Act which promoted closer settlement. One area where advantage was taken of the new laws was the Big Scrub, the extensive area of lowland subtropical rainforest in north-eastern New South Wales, originally some 75 000–80 000 ha in extent. The clearing has been described by Daley (1966).

Clearing away the dense, semi-tropical vegetation on the lower Richmond was a long and tedious process. When a rough shack or tent was erected and the heavy timber was cut down, the selector had to wait for it to dry before he could burn it... Six months or more afterwards, when the hot summer sun had dried out the tangled mass of trees, vines and undergrowth, the selector returned to his block of land and put a running fire through the dry timber before using a hoe to plant corn and pumpkin seed among the charred stumps. Rare trees—rosewood, tulip, sandalwood and white beech, all prized by the cabinet-makers—were burnt in the great conflagrations which could be seen for miles. The smoke hung like a pall over the hills as the whole face of the country slowly changed and the land which Captain Cook had praised in its 'pure state of nature' was turned to other uses.

The early crops were not particularly profitable, but from the late 1870s the former Big Scrub became a major dairy farming region, and continued prosperously so until the 1950s (Frith 1977). The area is still an important dairy-farming centre (Fig. 9.4), but now there are also extensive tropical fruit orchards. On steep slopes and land of lower fertility, there has been vigorous growth by *Cinnamonum camphora* (camphor laurel), an introduced species whose fleshy fruit results in spread by birds, and which, because of its dense growth and possibly also because of allelochemical effects, may inhibit the re-establishment of native rainforest species.

In New South Wales, the higher-elevation rainforests were little affected by the first wave of clearance, but in the first two decades of the twentieth century, virtually all of the rainforests on the Comboyne and Bulga Plateaux were cleared (these two regions are outliers of the Tableland and the rainforests were part of the Hastings group—Adam 1987*b*). Little attempt was made to salvage timber, which was burnt, again to make way for dairy farms (Hannah 1979, 1981). Only remnants of rainforest now remain, and the small stands of *Nothofagus moorei* were totally destroyed. The Dorrigo region was one of the major centres of rainforest distribution in New South Wales, but suffered extensive clearance for agriculture from the end of the nineteenth century through to the 1920s: a range of

Fig. 9.4 North-east New South Wales (Mt. Warning on the horizon right of centre). All of this land was originally forested, much with rainforest prior to clearing in the late nineteenth century.

rainforest types were cleared, including some stands of *N. moorei* (Carter 1933).

The story in Queensland was much the same as in New South Wales. Clearing of lowland rainforest occurred around various centres. The extent of clearance of araucarian notophyll vine forest in south-east Queensland has been documented by Watson (1989), whose studies have shown that the original extent of these forests was far greater than previously recognized. The extensive rainforests of the Atherton Tableland in north Queensland were opened for selection in 1882, 'high prices were set as a deliberate policy to discourage speculators. Nevertheless, the quality of cedar on the land guaranteed a good sale' (Birtles 1982).

Early European selectors cleared the rainforest to expose the soil, usually confining their activities to five acres, the limit which one man could crop on his own. The task of clearing the 'scrub' was simplified by the practice of cutting a deep nick in all trees for an area of up to an acre and then chopping one tree to topple the rest as it fell. Since the 'drive' of fallen logs was oriented in the same direction, the procedure of 'burning off' the timber and other debris was easier. Usually only cedar was recognised as commercially valuable but in a few instances, logs of walnut and black bean were pit sawn for the construction of a small rough slab hut... Stumps were left in the ground to rot, a process which prevented the use of the plough for at least six years after clearing (Birtles 1982).

The staple crop was maize, although a range of fruit crops was also established. Considerable problems with insect pests occurred and, from a European viewpoint, depredations by the Aborigines were a major concern, prompting a variety of retaliatory action (Birtles 1982).

The first dairying on the Atherton Tableland commenced in the early 1880s, to supply mines on the goldfields (Birtles 1982); later, like the Big Scrub, it was to become a major dairying centre. Although extensive areas of rainforest were cleared in the 1880s and 1890s, the process continued well into this century. Ratcliffe (1938) writing of the region in 1930, described clearing for new dairy farms and observed that 'Australians revel in felling trees. It is the pulse of the pioneer still beating in their systems, I suppose. A man is leaving his mark on the earth's surface when he fells a hundred or two acres of forest to grow fruit or corn or green pastures for his cattle. The big trees have always been foes in the eyes of the Australian settler, and he has treated them with scant mercy'. Lamenting the destruction of potentially valuable timber, he questioned the economic wisdom of the exercise—'acre after acre irreplaceably lost, to make way for dairy farms which could probably not survive without the fairy wand of subsidy'.

The history of major rainforest clearance was not one of a steady progression from Sydney culminating in north Queensland, rather, as with the earlier cedar-getters, it was a series of leaps, reflecting the interaction of politics, economic circumstance, and opportunity.

These examples of total destruction of rainforest to make way for agriculture could be repeated many times over. Clearing continued for many years. The semi-evergreen vine thickets of the brigalow belt (see Chapter 3) were substantially cleared only after World War II (Donohue 1984; Webb 1984), and very little was conserved (Pulsford 1984; Sattler and Webster 1984). Even today, the remnants outside formal reserves remain vulnerable to clearing.

Loss of coastal rainforest to agriculture continued well after World War II. Much of the land devoted to sugar cultivation was originally under rainforest; the sugar industry has a long history but considerable expansion, particularly in north-east Queensland, took place after 1940, although King (1965) pointed out that clearing of rainforest was often associated with damage to soil structure and loss of productivity. As recently as 1964, about 7000 ha of lowland rainforest in the Tully River Valley was cleared for a pastoral venture (Winter *et al.* 1984; Keto *et al.* 1985; Frawley 1988).

The cumulative impact of alienation of rainforest for agriculture has been the loss of an estimated three-quarters of that present at the time of European settlement. As a proportional loss, this is greater than that so far suffered in most other countries with rainforest, but the loss has been much greater in some rainforest types than in others, for example the loss of lowland subtropical complex notophyll vine forest has been very high, while that of cool temperate microphyll moss/fern forest has been small. Further losses of public-owned land for agriculture are unlikely. However, stands of rainforest on private land remain vulnerable. Many of these stands (even though they are frequently very small) may be significant for the conservation of particular species, and private rainforest is an important constituent of the network of feeding sites necessary for the maintenance of nomadic fauna.

9.2.2 Forestry

To the first settlers, the properties of the timber were just another of the many disappointments to be endured. The main species around Sydney (*Eucalyptus* and *Angophora*) had heavy misshapen trunks, often with extensive gum veins, difficult to work and with little durability. Later, more useful eucalypt species were discovered, but these posed different problems; it was a long time before technologies appropriate for the utilization of the finest eucalypts were developed.

The discovery of rainforests provided welcome relief, and, in the form of red cedar, a valuable commodity. Cedar was only one of many components of the rainforest, many other species have wood with desirable properties, as was realized by at least some early settlers. Atkinson (1826) suggested that cedar was not the finest timber and extolled the virtues of rosewood (*Dysoxylum fraseranum*).

While a number of species may have been used locally, it was cedar which became a major commodity; whether other species could have supported an export trade if the markets had been aware of them, must remain a matter of speculation.

As settlement spread, other rainforest species were utilized; as land was cleared most of the timber was destroyed, but sufficient would have been salvaged to meet immediate local needs. Cedar continued to provide the bulk of timber needed for local furniture and cabinet-making but other species were used on occasion. Some became prized, white beech (*Gmelina leichhardtii*), rosewood, and rainforest conifers, on the mainland *Araucaria* spp. and *Agathis* spp., and in Tasmania the Huon pine *Lagarostrobos franklinii*. The piners in Tasmania have a history almost as long as cedar-getters on the mainland and endured possibly greater hardship in the remote, rugged, and cold west of the inland. Huon pine timber is easily worked but very durable and is prized both for furniture-making and boat-building. Pining commenced in about 1815 and the purpose of the convict settlement of Macquarie Harbour, in its brief existence between 1822 and 1833, was the exploitation of *Lagarostrobos* in the lower Gordon and King River Valleys. In later years, *Athrotaxis selaginoides* (King Billy pine) and *Phyllocladus aspleniifolius* (celery top pine) were also harvested.

For the first century of European settlement, there was little that could be called forestry. Despite extensive clearing, the resource appeared limitless. Various Governors had promulgated regulations to control timber-felling, commencing with Governor King in 1803, who had placed controls on clearing on the banks of the Hawkesbury River to limit erosion. In general, any rules were observed more in the breach. Various commentators lamented the profligate waste of resources and pointed to the likelihood of future shortages, but such concerns were unheeded. By the 1870s, however, the increasing rate of clearing and the increasing demands for timber were sufficient to impress even the most sceptical of the necessity for management of the forest resource. The first forest reserves were established in New South Wales in 1871; the next fifty years were marked by a series of reviews and interim administrative arrangements which eventually led in all states to the establishment of strong forestry services (Carron 1985).

In the early days of Australian forestry, the main interest was in eucalypt (hardwood) forests, and rainforest management did not loom large except in Queensland. Nevertheless, pining continued in Tasmania, and rainforest logging occurred in New South Wales as well as in Queensland. The logging industry was fragmented, with numerous small mills serving local communities. Rainforest timbers had become better known, and accounts such as Francis (1929), documented uses for different species. For many, the uses listed were unspecialized, for example, general-purpose joinery or the manufacture of packing-cases. Much of the rainforest timber entering the mills was salvaged from continuing clearance, but more highly valued species were sought out. In New South Wales and southern Queensland, *Araucaria cunninghamii* (hoop pine) was harvested for manufacture of butter boxes, as the timber did not cause tainting. Ratcliffe (1938) has a long description of hoop pine logging in the Border Ranges.

The first silviculture involving rainforest species was attempts to grow red cedar, dating from the 1880s. Red cedars were planted in many localities in both New South Wales and Queensland (Vader 1987; Baur 1989) but most of the plantings failed due to damage by the larvae of the cedar tip moth, *Hypsipyla robusta* (which does not kill the trees but, by destroying growing tips, results in poorly formed trees unlikely to yield commercially utilizable timber). Where plantings have been of isolated individual trees, growth has been more successful; such a planting pattern would be a closer approximation to natural regeneration than plantations, but does not readily lend itself to a long-term profitable venture. While large-scale planting of red cedar in commercial forestry is no longer practised, there is limited enrichment planting in snig tracks and log dumps, and individual trees have been widely planted on farmland. Growth rates of seedlings in open habitats can be increased by the provision of shelter (Applegate and Bragg 1989). Red cedar is now readily available from many nurseries in eastern Australia.

Rainforest management became an important part of the operations of the Queensland Forestry Commission in the 1920s. The then Commissioner, E. H. F. Swain, was of the view that the long-term future of forestry in that State would involve a major role for rainforests—in the south in the form of *Araucaria* plantations and in the north from a steady supply of high-value cabinet timbers (Carron 1985). In the south, existing rainforest stands were felled and replanted to *Araucaria cunninghamii* and to a lesser extent *A. bidwillii*; these plantations currently cover an area in excess of 40 000 ha, and support an important plywood industry. Largely for experimental purposes, a range of exotic conifers has also been planted in former rainforest sites (J. A. Dale 1983).

In north Queensland, some plantations were established, involving both conifers (*Araucaria* and *Agathis*) and *Flindersia brayleyana* (Queensland maple) (Baur 1968), but these were limited in area. Prior to World War II, relatively few rainforest species were utilized, but in the post-war period the number of merchantable species was substantially larger (around 160) (Nicholson 1985). Clearing for plantations would have involved the loss of small stems of economic species with the potential to grow to harvestable size, and establishment costs were high. Development of new plantations ceased, and silvicultural treatment aimed at stand improvement following logging. Improvement involved some loss of stand diversity, as the intention was to increase the proportion of desirable species while sharply reducing the presence of non-merchantable species.

Baur (1968) described in some detail the silviculture practices carried out in upland forests during the 1950s and 1960s. Prior to harvesting, climbers and undergrowth were cut; merchantable stem above certain girth limits (varying with species) were then removed (except where this would result in a very low stocking rate of smaller sizes of a species, in which case large individuals were left as seed sources). Subsequent to harvesting, non-merchantable stems, and possibly some small stems of harvestable species, were removed to give a fairly even spacing of stems. In some cases there was additional treatment to promote regeneration of particular species or even enrichment planting (which in some experimental trials included exotic species). An average figure for basal area in the virgin forest was about 12.9 m² ha⁻¹, after harvesting this was reduced to 9 m² ha⁻¹ and after treatment to 2 m² ha⁻¹. After initial harvesting, the felling cycle was expected to be of the order of 15–20 years.

Particular problems were experienced with dense growth of stinging trees, *Dendrocnide moroides* and *D. cordata*, after treatment, which was sufficient to totally prevent access to treated areas. Control measures, including herbicide use, were implemented to reduce this problem (van Altena and Harvey 1979).

There were some exceptions to the retention of seed trees; in the case of *Endiandra palmerstoni* (Queensland walnut), highly prized for cabinet work, all merchantable stems were removed. This was due to the slow growth rate, the large size needed before stems had commercial value, and its poor regeneration—'in effect, the forest service is prepared ultimately to lose this species in the treated forests, by preferring other more amenable and almost equally valuable cabinet timber species' (Baur 1968). The Department of Forestry, Queensland (1983), however, disputed this interpretation, pointing out that trees were not harvested until much larger than other merchantable species and 'thus, despite the poor reproductive capacity of the species, the logging rules increase the likelihood of its regeneration by extending its effective lifetime in logged stands'.

In the early 1970s, largely for economic reasons, the intensive post-logging treatment was discontinued, as was enrichment planting. Logging continued on a selective basis.

In New South Wales, the development of rainforest silviculture occurred somewhat later than in Queensland (see Baur 1968; 1977, 1989). Apart from logging of hoop pine (*Araucaria cunninghamii*) and the removal of a small number of other valuable species, there was no general rainforest harvesting in the 1920s and 1930s. In 1928, the Commissioner, N. W. Jolly, suggested that rainforest stands would require artificial regeneration, although natural regeneration of *Nothofagus moorei* and of *Araucaria cunninghamii* stands might be achieved.

In the early 1930s, some planting of *A. cunninghamii* in araucarian notophyll vine forest took place. In 1938, E. H. F. Swain, who became Commissioner in New South Wales following his dismissal from the position in Queensland (largely because of his opposition to the conversion of north Queensland rainforest to agriculture (see Humphreys 1983; Frawley 1988)), commenced a hoop pine plantation programme. This was initiated in response to pressure for the revocation of State Forest on the Queensland border for conversion of dairy farms; the forests were seen by the proponents as of little value. Swain, having appreciated the importance of the plantation programme in Queensland, proposed to create similar high-value stands in New South Wales. A number of plantations were established from Dorrigo northwards. The high costs of establishment and the low initial growth rates led to the programme being terminated in 1954, when about 1500 ha of plantation had been created.

Most of the lowland rainforest in New South Wales had been cleared for agriculture, much of what rainforest remained was difficult to access and only a small proportion of the species were regarded as merchantable. In the late 1930s, the technology for veneer production from New South Wales species was developed and, equally importantly, the methods for gluing veneers to produce plywood became usable on an industrial scale. In World War II, plywood became an extremely important commodity and coachwood, *Ceratopetalum apetalum*, from New South Wales rainforests was suddenly valuable. New machinery allowed access to previously inaccessible upland forests which were often very heavily logged. *C. apetalum* had, as its name coachwood suggests, been used for making the frames of coaches, but this was a relatively minor use; in the war it was harvested to provide rifle furniture while coachwood plywood became famous for its use in Mosquito aircraft and small boats.

After the war, the infrastructure of veneer mills remained and rainforest logging continued, mainly to supply them. Relatively few species were utilized.

Different management systems were developed for the various major rainforest types in the State (Baur 1977). In araucarian notophyll vine forest (dry rainforest) the major commercial species regenerate prolifically in some years, but the dry winters cause high seedling mortality so that natural regeneration is unreliable. Logging tracks and other openings were planted with hoop pine, which grew well.

Nothofagus moorei timber has few desirable properties. In addition, after logging, there is substantial die-back of the canopies of remaining trees. In consequence, *N. moorei* was logged for only a limited period. However, some stands have a lower storey including *Ceratopetalum apetalum* and *Doryphora sassafras*, which can be peeled for veneer. Removal of trees from this lower storey was practised in some localities. However, removal of understorey trees was associated with die-back of the *Nothofagus* canopy, possibly as a result of damage to the shallow root systems of *Nothofagus* during the snigging of logs (Floyd 1990).

Complex notophyll vine forest, the most species-rich of the New South Wales rainforest types, contains many species of low commercial value. After selective logging of commercially valuable species, canopy die-back occurred in some instances if canopy opening was excessive, so that, in general, logging prescriptions involved 50 per cent or greater canopy retention. Regeneration was generally good, but was in some cases supplemented by enrichment planting and treatment of tracks and log landings to promote seedling establishment.

A number of stands with a preponderance of low-value species were converted to eucalypt, predominantly *Eucalyptus grandis* (flooded gum). The rainforest was felled and after the debris was burnt, eucalypt seed was sown (Baur 1968). This practice was particularly applied to gully stands of rainforest (normally forest type 5 in the classification of the Forestry Commission of NSW), within large areas of tall open eucalypt forest.

The most important commercial rainforest type was simple notophyll vine forest dominated by *Ceratopetalum apetalum* (Fig. 9.5). Seedlings of the canopy species establish readily after logging. However, the stands were susceptible to crown die-back following disturbance. The incidence of die-back varies considerably and the cause has never been fully explained (see Baur 1989). Even in virgin forest, some trees exhibit partial die-back; the expression of die-back following logging was related both to the intensity

Fig. 9.5 Simple notophyll vine forest (*Ceratopetalum apetalum*) immediately after logging. Mt. Boss State Forest.

of logging and the pre-logging frequency of die-back.

Where *Ceratopetalum* stands were virtually clear-felled, dense seedling establishment occurred, but growth rates were slow, leading to estimates of a harvesting rotation of over 200 years. Given the experience of the much shorter rotations with *Pinus radiata* and eucalypts, Baur (1968) suggested that there was a psychological barrier against proposing any rotation longer than 100 years. The solution was to log selectively removing a small proportion of the standing crop, permitting a felling cycle of about 25 years. Setting a prescription for canopy retention was complicated by the difficulty of predicting die-back. Recovery from light die-back is possible; extensive die-back normally results in death. Selective canopy removal was thought possible in some areas, but in others the intensity of logging before die-back became a problem was too light to be economic. In these stands, despite the concerns raised by Baur (1968), the decision was made to permit heavier logging to maintain the industry, and then to close down for a long period until the next crop was available.

Much of the rainforest timber harvested in New South Wales was used for purposes which did not demand its special properties. For example, as shuttering for concrete-pours on building sites or for the tops of school desks. In addition, the return to the Forestry Commission from stumpage was generally lower for rainforest timbers than for eucalypts, despite the high royalties attracted by some species (Forestry Commission of NSW 1979). As community interest in the environment in general developed, there was increasing concern that rainforest resources were being put to inappropriate uses (Department of Arts, Heritage and the Environment 1986).

In 1976, the Forestry Commission of NSW developed an 'Indigenous Forest Policy'. In respect of rainforest, the policy stated that:

The broad objective for all rainforest areas is to reduce harvesting to selective fellings for speciality logs, at a level low enough to maintain canopy and rainforest structure. This would require the phasing out of general purpose timber harvesting in most rainforest areas. The rate of selective logging of speciality timbers would generally be too low to support mills now primarily dependent on rainforests.

Where selection logging is successfully carried out without destroying the ecological viability of the rainforest, this may be continued to meet current market commitments. However, these commitments should be reduced, where necessary, in time to avoid the need for logging above the sustained yield level after the first cutting cycle.

Where market commitments or the nature of the forest type force a continuation of intensive logging in rainforests, rehabilitation should be carried out by planting openings at a stocking sufficient to provide an acceptable tree cover. In types which originally carried Hoop Pine, this species may be used, otherwise eucalypts suitable to the site should be planted.

As the Commission observed (Forestry Commission of NSW 1979), the policy reflected changes in attitudes evolved over 40 years from 'actively promoting the use of rainforest timbers during the 1940s and 1950s, to a cautious holding of the *status quo* in the 1960s, and finally in the 1970s to an expressed policy of phasing down routine rainforest logging operations wherever possible'. However, in implementing the policy, 'the Forestry Commission has been caught in the bind of having to meet commitments entered into many decades ago, commitments made with the strong support of, or under direction from the Government of the day—commitments which subsequent Governments have been loath to terminate or reduce because of their important role in providing local employment and decentralised economic activity' (Forestry Commission of NSW 1979).

Hurditch (1985) pointed out that points of the policy were subject to various interpretations. The timber industry regarded veneer and plywood production as a speciality use, and thus may have entertained hopes of continued supplies to veneer mills. However, the indication in the policy that mills previously dependent on rainforest sources could not be supported and the projections of future harvesting indicated that the

Forestry Commission had a much narrower concept of speciality use.

The decline in harvesting from rainforest is illustrated in Fig. 9.6, which also demonstrates the very heavy logging during and immediately after the war. During the 1950s and 1960s, the proportion of rainforest timbers to eucalypts in the total harvest was much greater than the relative areas occupied by rainforest and sclerophyll forest (Baur 1989). The species composition of the harvest is shown in Table 9.1. The decline in utilization of rainforest timbers reflects the implementation of the phasing down policy and diminution of the resource; the Government

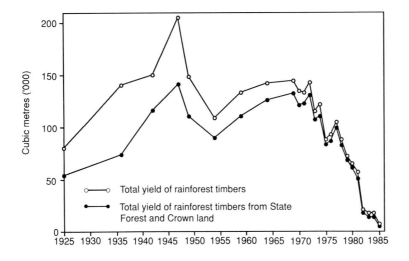

Fig. 9.6 Yield of rainforest timbers in New South Wales (compiled from data in Forestry Commission of NSW annual reports).

Table 9.1 Composition (per cent total volume) of the rainforest harvest in New South Wales, 1979–86 (from Forestry Commission of NSW annual reports).

Species	1979–80	1980–1	1981–2	1982–3	1983–4	1984–5	1985–6
Ceratopetalum apetalum	21.88	21.08	24.04	15.24	19.58	13.82	10.26
Doryphora sassafras	12.46	13.71	16.60	17.39	15.84	10.83	18.96
Sloanea woollsii	14.48	16.33	11.51	10.08	17.89	21.56	21.83
Argyrodendron actinophyllum	6.34	6.29	6.45	10.02	8.62	10.74	5.45
Araucaria cunninghamii	6.15	4.51	6.10	9.04	3.61	2.78	.20
Schizomeria ovata	4.24	3.77	5.83	2.81	6.80	14.30	18.72
Argyrodendron trifoliolatum	7.21	5.42	5.15	—	—	—	—
Geissois benthami	7.29	7.40	5.03	4.51	5.33	3.90	—
Caldcluvia paniculosa	4.03	4.82	4.40	4.32	5.94	4.19	5.87
Cryptocarya erythroxylon	3.39	4.21	2.87	5.72	1.50	3.38	0.43
Orites excelsa	1.77	1.90	2.23	1.19	1.63	1.66	2.21
Dysoxylum fraseranum	2.05	1.91	1.87	3.89	2.03	3.44	0.55
Litsea reticulata	1.53	1.40	1.48	1.38	1.55	2.32	1.90
Cryptocarya glaucescens	0.85	0.37	0.74	—	—	—	—
Alphitonia excelsa	—	—	—	0.35	0.92	1.44	7.83

decision, made in 1982, drastically to curtail rainforest logging only pre-empted the inevitable by a few years. Baur (1989) has pointed out that this situation arose from exploitation above the sustainable annual cut, because politicians and senior forestry administrators failed to recognize that sufficient was known about rainforest regeneration to manage the forests on a sustained yield basis. Nevertheless, although the harvest taken from New South Wales rainforest was far in excess of sustainable yields for the species concerned, the logged sites are regenerating vegetation which is still recognizably rainforest.

The history of forestry in Tasmania differs from that on the mainland in that from the 1920s onwards, harvesting for pulp and paper manufacture has predominated (Dargavel 1988). In recent decades, as well as the domestic industry, there has been a major export of wood-chips to Japan. The rainforest timbers of Tasmania have long been favoured for boat-building and for fine cabinet work, particularly Huon pine (*Lagarostrobos franklinii*), blackwood (*Acacia melanoxylon*), myrtle (*Nothofagus cunninghamii*), and sassafras (*Atherosperma moschatum*). As well as occurring in 'pure' rainforest, there are extensive areas of mixed forest with rainforest taxa under an open eucalypt canopy. When these stands were logged for pulpwood, much of rainforest timber was harvested as well, and regeneration was managed so as to favour eucalypts. Some 'pure' rainforest was also havested for pulpwood and the stands were converted to eucalypt forest.

In 1982, a moratorium on logging in large rainforest stands was instituted to allow a review of conservation status and also of regeneration. However, the demand for rainforest logs continued to be met from mixed forest. In addition, non-sawlog quality rainforest timber continues to be salvaged as pulpwood from mixed forest and in areas to be flooded by hydroelectricity schemes. In the mid 1980s, about 100 000 tonnes a year of rainforest timber was pulped (Boyer and Hickey 1987). The sawlog harvest in recent years is documented in Table 9.2.

9.2.3 Other forest products

Before the arrival of the Europeans, with their narrow perception of forests as being primarily a source of timber, the Aborigines, without the technology (or the need) to fell trees had valued rainforests as a source of numerous resources.

Only one Australian rainforest food has so far entered the list of international crops. The macadamia nut (*Macadamia tetraphylla*, *M. ternifolia*) was first cultivated on a commercial scale in Hawaii, but extensive orchards have now been

Table 9.2 Average annual sawlog harvest of non-eucalypt native species in Tasmania between 1981–2 and 1985–6 (data from Boyer and Hickey 1987).

Species	*Volume (m^3/pa)*
Acacia melanoxylon (blackwood)	12 500
Phyllocladus aspleniifolius (celery top pine)	6500
Nothofagus cunninghamii (myrtle)	5000
Lagarostrobos franklinii (Huon pine)	500*
Athrotaxis selaginoides (King Billy pine)	200*
Atherosperma moschatum (sassafras)	
Eucryphia lucida (leatherwood)	1800
Acacia dealbata (silverwattle)	

* Mostly salvaged from sites to be flooded for hydroelectric power generation.

established in Australia, although most of these have not yet reached full production. Many other rainforest fruits were utilized by Aborigines. Some of these were also sampled by European settlers, for example, fruits of lilly-pilly, *Acmena* spp., and *Syzygium* spp. have been used in jam-making; numerous other examples are given by Cribb and Cribb (1974), but no attempt has been made to commercialize them. At the present time there is increasing interest in 'bush tucker' both for its novelty and because it is perceived to be 'health food'. It is possible that this interest may be translated into domestication of more species.

Aborigines made extensive use of rainforest species for medicinal purposes. Until recently, there has been little serious attention given to the Aboriginal pharmacopaeia. Research on the chemistry and possible uses of Australian plants has a long history (Rae 1988), and systematic survey began during World War II, with the establishment of what became the Australian Phytochemical Survey (Webb 1969). During the course of the survey, a large number of species was screened and many novel compounds discovered, including a large number with promising pharmacological properties (Webb 1969). Many of the more interesting compounds are from rainforest species. There is clearly potential for further investigation of these compounds and of additional species. Recently a compound extracted from *Castanospermum australe* has shown potential for the treatment of AIDS patients (Beckmann 1988).

One species which has been commercially exploited is *Duboisia myoporoides*, a small tree of subtropical rainforest margins and understorey. The leaves are a source of the alkaloids hyoscine and hyoscyamine; in 1982–3 some 35 000 kg of *Duboisia* leaves were harvested in New South Wales (*Annual Report of the Forestry Commission of New South Wales 1982–3*).

Honey, including that from rainforest, was highly valued by Aborigines (e.g. Lumholtz 1889). Europeans introduced the honey-bee, *Apis mellifera*, into Australia, and honey is now exported in quantity. Beehives are largely situated in heathland and open forest on the mainland; while there is some honey production from mangroves, rainforest in general is not regarded as profitable habitat by apiarists. In Tasmania, however, one of the most distinctive honeys is produced from leatherwood (*Eucryphia lucida*). In good seasons up to 500 tonnes of leatherwood honey are produced (Boyer and Hickey 1987). Concern has been expressed that there will be a reduction in the availability of leatherwood nectar because of the conversion of mixed forest to eucalypt forest following logging. Leatherwood regenerates well after logging if there is a good seed source, seed bed, and no fire. Conversion of mixed forest to eucalypt is promoted by a post-logging burn detrimental to leatherwood (Neyland and Hickey 1990).

There are numerous other minor products of rainforest, such as the stems of lawyer cane (the climbing palm *Calamus*) used in basket-making and cane furniture, and various seeds and fruits used in ornaments and souvenirs.

A major horticultural industry based on rainforest species is developing in Australia. For many years there was a minor trade based on tree ferns and epiphytic orchids salvaged during logging operations. In addition, there was widespread poaching of these plants, a practice which, unfortunately, is still prevalent in some areas. A number of rainforest trees have been widely planted as ornamental and street trees, including silky oak (*Grevillea robusta*), umbrella tree (*Schefflera actinophylla*), fire wheel tree (*Stenocarpus sinuatus*), lilly-pillies (*Syzygium* spp.), and figs (*Ficus* spp.). Many of these species are surprisingly hardy and have been grown successfully in localities far removed from rainforest. In the last two decades, an increased interest in growing native species, and the long debate over the conservation of rainforest, have created a horticultural demand for rainforest species and several hundred are now regularly available from a network of specialist nurseries. Although some species are difficult to grow, providing a challenge for the keenest gardeners, others are relatively hardy and can be grown in a wide range of situations (see Jones 1986; Nicholson and Nicholson 1985).

This planting of rainforest plants serves not only to increase the diversity of the garden landscape but also to arouse wider community interest in rainforests. A number of species now well established in cultivation are relatively rare in the wild; horticulture may play an important conservation role in maintaining and increasing the stock of threatened species, although care will have to be taken to maintain the genetic diversity of species in cultivation.

9.2.4 Mining industry

The development of European Australia received a great impetus with the discovery of gold in both New South Wales and Victoria in the 1850s. The next fifty years saw a series of rushes and booms, not just for gold but for a whole range of minerals. Mining continues to be one of the mainstays of the economy and is likely to remain so.

Despite the importance of mining, and the occurrence of mines across the continent, the total area directly impacted is very small. Nevertheless, the wider area affected is much larger, and the role that mining has played in opening up regions for settlement should not be ignored. A number of rainforests were affected. In north Queensland, gold-rushes on the Gilbert and Palmer Rivers first brought white settlement to the Atherton Tableland (Bolton 1963; Birtles 1982), where substantial amounts of rainforest timber were consumed by the gold-miners. Later tin-mining became a major activity in north Queensland rainforests (Idriess 1959; Bolton 1963; Birtles 1982), and locally caused intense disturbance. Unwin et al. (1988b) reported that in the Rossville area (30–35 km south of Cooktown) there was vigorous recovery of both rainforest and eucalypt forest following tin-mining. Changes in fire regime following settlement associated with mining resulted in a spread of rainforest into previously open forest.

At the other extremity of the continent, the copper mines at Queenstown on the west coast of Tasmania consumed considerable quantities of timber, including rainforest species, and, in consequence of the pollution from the smelters, the death of large areas of vegetation (Blainey 1954). The denuded landscape around Queenstown, today a tourist attraction, is stark testimony to the need for controls to achieve environmental protection.

Between these two geographic extremes, there have been numerous mining operations, at various scales and for a great variety of minerals, a number of which have had impacts on rainforests. Operations, which economically were of relatively minor significance, could nevertheless have major impacts on particular stands of vegetation. A well-documented example is Mt. Dromedary in southern New South Wales (Forestry Commission of NSW 1987), which supports cool temperate rainforest dominated by *Eucryphia moorei* and depauperate subtropical (complex notophyll) rainforest, almost at its southern limit. Goldmining was carried out for nearly seventy years from 1852. Vegetation was cleared for houses and for mine sites, and surrounding areas were heavily burnt. Large quantities of wood, both eucalypt and rainforest, were cut for fuel in the boilers for the ore-crushing plant and for construction. As well as mining reef gold near the mountain summit, alluvial gold was won from the creeks by hydraulic sluicing. This has had the most lasting effect on the environments; creek beds which before mining were rich in ferns are now entrenched with exposed rock walls. Many of the other mining sites are now hidden by vegetation regrowth and tunnels and shafts have collapsed. In general, the rainforest has recovered well, although much of the regrowth is by coppicing from stumps. In the most heavily disturbed area, rainforest has been converted to tangles of vine scrub, which may slowly regenerate to forest. Comparison of photographs taken during the mining phase with the present vegetation of the mountain provides clear evidence of the regenerative potential of both eucalypt and rainforest communities (Forestry Commision of NSW 1987).

These examples of rainforest recovery belie the common public perception that rainforest is particularly fragile. Westoby (1986) argued that:

> We should stop wittering about ecosystems being fragile. Rainforests are fragile, deserts are fragile, sclerophyll forests are fragile, the open ocean is fragile; I have yet to hear the media describe any ecosystem as anything but fragile. It is perfectly true that some of the world's ecosystems are being destroyed at an unprecedented rate—but it is dangerously misleading to suggest that this is happening because the ecosystems are fragile. The tropical forests are contracting in area not because they are susceptible to subtle ecological side-effects which our scientific knowledge has been unable to anticipate, but because huge amounts of labour power and fossil energy are being systematically devoted to felling and removing the structuring species which make these forests what they are.

Frequently the failure of rainforest to recover from disturbance is not the direct result of the initial disturbance, but a consequence of the repeated disturbance of the cleared site. The rate of recovery after disturbance ceases may depend on the intensity of disturbance and the nature of the initial community. Within any community, some species may be less resilient than others so that permanent changes may be initiated by disturbance.

Stands of littoral rainforest, in both New South Wales and Queensland, have been subject to total clearance during mining for the mineral sands, rutile, zircon, and illmenite, with little, if any, post-mining regeneration. As the environment in littoral rainforest is marginal for rainforest, this particular community type may be much less resilient than others.

In western Tasmania, rainforests have been inundated by hydroelectricity generation impoundments. These schemes have been the subject of great controversy (Gee and Fenton 1978). The most valuable timbers, particularly Huon pine (*Lagarostrobos franklinii*), have been salvaged from the rising waters.

9.3 EXTINCTIONS AND INTRODUCTIONS

Despite the considerable loss of rainforest habitat over the last 200 years, there is little evidence that there has been an accompanying extinction of species.

No mammal associated with rainforest is known to have become extinct. The parma wallaby (*Macropus parma*), a species of rainforest margins, was thought to be extinct in Australia (but survived as an introduced feral population on Kawau Island, New Zealand). However, it was rediscovered in New South Wales in 1967 and, although not common, is now thought to be widely distributed between the Watagan Mountains and the Gibraltar Range (Maynes 1983). The richest rainforest mammal faunas are found in north Queensland (Chapter 7), where a number of species are rare and endangered (Strahan 1983). Continued fragmentation and disturbance of rainforests in north Queensland pose a threat to the survival of these species in Australia, although several also occur in Papua New Guinea.

The position with birds is similar, with no known extinctions on the mainland, although a number of species have declined and may now be endangered. On Lord Howe Island, however, nine of the fifteen species present at the time of European contact are now extinct. Some species of rainforest birds were subject to hunting, notably the brush turkey *Alectura lathami* and various pigeons, but while this may have caused local reductions, it does not seem to have threatened the survival of the species.

There is similarly little evidence for loss of reptile or amphibian species, although there is concern that two of the most remarkable frogs in the world, the gastric brooding species which were found in rainforest streams in south-east Queensland, may have disappeared in recent years. *Rheobatrachus silus* was recorded from the Blackall and Conondale Ranges in south-east Queensland, and *R. vitellinus* from Clarke Range near the boundary of the Eungella National Park

(Tyler 1985). In both species, the stomach is converted to a brood sac and the young are born through the mouth. So little is known about these species that it is difficult to know whether their apparent rapid decline should be cause for concern or simply be a part of the natural demographic cycle. Gastric brooding frogs are of great interest in their own right but also because an understanding of how gastric secretions are controlled so as not to digest the offspring may assist medical research into the treatment of ulcers.

Among the invertebrates, so little is known about distributions and population sizes that it is impossible to make any statements about extinctions.

Among the higher plants of rainforest, there is little conclusive evidence of recent extinctions, but a number of species are known only from type collections or are believed to have very small populations (Leigh et al. 1984; Floyd 1987). A number of the rarest species occur in small rainforest stands in private ownership, so that their chances of future survival must be regarded as precarious.

With the total clearance of many rainforest stands, there must have been numerous examples of local extinctions and consequently a reduction in genetic diversity in widespread species. There may also have been unrecorded total extinctions but at least in the case of the relatively well-studied vertebrates this seems unlikely. The rainforest fauna has certainly survived the European invasion more successfully than that of the arid zone, where a large number of extinctions have occurred (Strahan 1983). The fossil record indicates substantial loss of rainforest fauna during the late Tertiary contraction of closed forest in the face of advancing aridity (Chapter 7). The fauna that survived this period, and the climatic fluctuations of the Quaternary, may be a 'robust' selection equipped to tolerate a range of disturbances. Nevertheless, the concentration of much of the specialist rainforest fauna in northern Queensland raises concern for its continued survival if further habitat loss occurs.

Humans have certainly been responsible for introducing numerous species to Australia. The longest-established introduction, the dingo *Canis familiaris dingo*, includes rainforest amongst its wide range of habitats. More recently arrived carnivores, the fox and the feral cats, may be found in rainforest, particularly in small isolated stands. Feral pigs are widespread in northern and eastern Australia (Pavlov 1983) and are found in rainforest; rooting and wallowing may cause localised disturbance. Asian water buffalo (*Bubalus bubalis*) have been feral in the Northern Territory since the 1820s. There is evidence that use of rainforest stands by buffalo causes major changes to the structure and subsequently to floristics. These changes seem to be initiated by compaction of the soil which reduces infiltration of water to recharge the groundwater causing death of the largest trees (Braithwaite *et al.* 1984).

Large stands of rainforest are generally devoid of introduced plants except around the perimeter, but small stands and disturbed areas may become weed infested. In Queensland and New South Wales, the most widespread introduced species is probably *Solanum mauritianum*, a fast-growing, softwooded small tree with large leaves. This species is now almost ubiquitous as an early colonist of disturbed areas but does not persist (except along roads and the edges of clearings) as the canopy is re-established. A species which can form very dense stands in disturbed areas, either blocking or deflecting the natural course of succession is *Lantana camara* (see Webb *et al.* 1972; Hopkins 1981). Although biological control measures have been effective in some circumstances (Haseler 1981) *Lantana* is still a major problem to land managers in the eastern coastal zone.

Many coastal habitats in New South Wales have been invaded by *Chrysanthamoides monilifera* spp. *rotundata*, bitou bush. This composite is unusual in its family in that the fruit is drupaceous, and part of the reason for its success in Australia is the spread of fruit by birds. Bitou bush scrambles over small trees, smothering the canopy. It may form dense growth around the edge of littoral rainforest (for example, the Iluka rainforest—Adam 1987b). To date, there has been little expansion into Queensland but bioclimatic predictions would suggest that bitou bush would flourish along almost all the coast, with the potential to invade numerous other rainforest stands.

A number of introduced vines are aggressive invaders of disturbed rainforest. These may enshroud trees, eventually causing collapse of the canopy. The remnant lowland complex notophyll rainforest stands of northern New South Wales have been particularly vulnerable to this type of invasion (Floyd 1984). Three species of particular concern are *Macfadyena unguis-cati*, *Anredera cordifolia*, and *Cardiospermum halicacabum*.

Small disturbed isolated stands of rainforest in northern New South Wales are also liable to invasion by *Tradescantia albiflora*, which can form a dense carpet over large areas of the forest floor, inhibiting regeneration of canopy species.

Few species of exotic trees or shrubs have become established. The most successful are the bird-dispersed *Cinnamomum camphora* and privets (*Ligustrum sinense* and *L. lucidum*) (Floyd 1984). These are mainly a problem in isolated stands rather than in disturbed stands within large contiguous blocks of forest. In parts of the former Big Scrub in northern New South Wales, *C. camphora* has been an aggressive invader of abandoned agricultural land; once established, it suckers freely. These dense groves may be resistant to invasion by other species, so that opportunities to re-establish rainforest may be constrained. However, *C. camphora* stands may provide corridors for faunal movement between isolated rainforest remnants, so that eradication may not be desirable.

9.4 CONSERVATION

Formal conservation in Australia has a long history compared with other parts of the world (Frawley 1989; Hall 1989). The second-oldest area designated as national park in the world was the National Park (now Royal National Park) immediately to the south of Sydney, which was dedicated in 1879. Justification was seen in terms of opportunities for public recreation and enjoyment of the bush rather than for the protection of particular communities and species. Nevertheless, Royal National Park contains a great range of vegetation types, including a number of stands of rainforest. For many Sydneysiders, it provides the first encounter with rainforest.

Other colonies followed suit and by Federation in 1901, national parks had been dedicated in various localities. The expansion of the national parks system during the first half of the twentieth century was relatively slow, despite increased public interest generated by the growth of the bush-walking movement after the 1920s. Management of national parks within the States was relatively uncoordinated until the late 1960s–early 1970s when professional management authorities came into being in all States. In some cases, the Park Service is independent, in others it remains part of a larger land management department, and administrative arrangements tend to vary with changes in government. The 1970s and 1980s saw a great expansion in the national park system in all States.

Some important areas of rainforest were included in early national parks. In Queensland there had been a campaign in the late nineteenth century to declare a national park covering the rainforests along the New South Wales border. The proposal originated with Robert Collins, who had been inspired by the concept of national parks on a visit to the USA. The proposal was taken up later by Romeo Lahey, a sawmiller, and despite strong local opposition, the Lamington National Park was dedicated in 1915 (Groom 1949). In New South Wales, a series of reserves, dedicated for various purposes, protected rainforest at Dorrigo. The oldest of these areas was gazetted in 1901, after various other parcels of land were reserved the protected area became known as Dorrigo Mountain National Park. The New England National Park was dedicated in

1935. Other park proposals in this period were less successful. In north-east New South Wales, Mt. Warning became a Park in 1920 but despite strong local support, plans for a park in the Nightcap Range did not reach fruition.

The pressure for conservation came from a variety of groups; local landholders and councils, the bushwalkers who became an increasingly important lobby from the 1920s onwards, and natural history societies. Interest in the natural history of rainforests goes back to the earliest European collectors; organized natural history societies were important in documenting many sites (see, for example, Carter 1933) but natural historians did not play as important a role in conservation as did naturalists in Britain (Adam 1988).

Since the 1960s, the conservation movement has been an important influence on national and state politics. A number of large and active groups were formed, the first being the Australian Conservation Foundation, more recent societies include The Wilderness Society, and the Australian chapters of Greenpeace and Friends of the Earth as well as many more specialized groups. The reasons for this upsurge in interest have yet to be fully explained. In part, it is a reflection of the growth of the environmental movement seen throughout the western world in the 1960s. It may also reflect a declining colonial influence and the development of increased interest in, and respect for, things uniquely Australian. Although Australia is one of the most highly urbanized nations, images (and myths) of the bush play a major part in defining a national identity. The 1960s also saw greater mobility and leisure time for many people so that areas once remote now became accessible.

As the conservation groups developed, conflicts between conservationists and those wishing to exploit natural resources achieved greater publicity, and the views of conservationists gained credibility in policy debates. Although a number of issues have achieved prominence, the management of forests has been at the centre of some of the most intense controversies even though, or perhaps because, forested land is a small portion of the continent. However, initially, rainforest logging was not regarded as a major national issue, although there were a number of locally based campaigns aimed at securing the transfer of particular forests to national parks. The public suddenly became aware of rainforest in 1979, when clashes between conservationists and loggers occurred at Terania Creek in the Nightcap Range in north-east New South Wales (Watson 1990). Although this was one of the smallest rainforest logging proposals, it was the catalyst which led, in a few years, to massive changes in official policy towards rainforest management.

In the early 1970s, there had been an influx of new residents into rural northern New South Wales, mainly of former city-dwellers intent on leading an 'alternative life style'. Terania Creek was just one of the areas where new settlement occurred. The head of Terania Creek is a steep-sided basin supporting rainforest. The southern part of the basin had been logged, and parts of the rainforest totally cleared, but the head of the basin was largely intact. Between 1974 to 1979 opposition from local residents had led to the deferral of plans by the Forestry Commission of NSW to log in the basin. In 1979, the Commission determined to proceed. The logging proposal had been scaled down from earlier plans and was restricted to a selective logging of brush box (*Lophostemon confertus*) and blackbutt (*Eucalyptus pilularis*) growing in the rainforest margin. In all, less than two thousand trees were to be felled.

The proposal was strongly opposed, for, although the Commission argued that they were not proposing to log rainforest, the residents argued that the marginal communities should be regarded as an integral part of the rainforest and that, in addition, roads and snig tracks would traverse 'true' rainforest.

In August 1979 when logging was to commence, the foresters and tree-fellers were confronted by large numbers of protesters who lay in front of bulldozers, climbed trees to be felled, and generally hindered operations. There was some vandalism (disowned by the organizers of the protest), in the form of inserting steel spikes in trees to be felled and damage to sawn logs. These activities attracted considerable media attention and caused concern to a government which was fearful that injury or death to protesters might occur. After three weeks of confrontation, logging was suspended, and subsequently State Cabinet appointed a retired judge, Justice Isaacs, 'to conduct an inquiry into the proposed logging of Terania Creek and recommend whether logging should or should not proceed'.

The Inquiry lasted almost two years and cost about $1 million (on top of a bill for police at the protests of $500 000) (Kendell and Buivids 1987). The proceedings were conducted in an adversarial manner with little chance of agreement between parties. Much time was spent in attempting to resolve whether or not the area to be logged was, in fact, rainforest. The eventual report of the Inquiry recommended that logging be permitted.

While the Isaacs' Inquiry debated the future of one small stand of rainforest, the public, newly aware of the occurrence of rainforest, began to question the Forestry Commission's policies elsewhere in New South Wales. Two regions in particular attracted attention—Washpool and the Hastings. Washpool contained the largest remaining intact stand of *Ceratopetalum apetalum*-dominated forest, while the Hastings region contained a diversity of floristic types. Under new planning legislation, the New South Wales Environmental Planning and Assessment Act 1979, the Forestry Commission was required to prepare Environmental Impact Statements (EIS) for its operations in both regions (Forestry Commission of NSW 1980, 1981). Both documents attracted strong criticism from environmental groups. In the case of Washpool, the Department of Environment and Planning carried out a detailed assessment of the Commission's EIS, including an independent analysis of ecological issues (Fox 1983) and a study of possible alternative sources of supply to the mills which would draw from Washpool. The adequacy of the EIS for the Hastings forests was challenged in the Land and Environment Court by the National Parks Association.

By the early 1980s, it was clear that the New South Wales Government was faced with one controversy after another over rainforest logging (Colong Committee 1983). On 26 October 1982, after an all-day Cabinet meeting, the New South Wales Government decided upon a rainforest policy which involved, in summary:

Conservation of certain rainforest areas in national parks and nature reserves; maintenance of employment levels consistent with that existing and predicted from the Forestry Commission's current management proposals;
Identification of alternative timber sources, the availability of which will be guaranteed.
A 'Rainforest Fund' be established with an initial sum of $1 million to promote and encourage the development of new technologies and to assist affected industries during the transition period in which these changes of policies will be implemented.
In the management of hardwood resources in the Forestry management areas affected by the new rainforest parks and reserves, the Government has reaffirmed the principle of sustained yield.

The Premier, Neville Wran, subsequently said of this decision:

I know it is not everyone who thought it was a great thing to save the rainforests, but I make this predication here today: when we are all dead and buried and our children's children are reflecting on what was the best thing the Labor Government in New South Wales did in the 20th century, they will come up with the answer that we saved the rainforests.

Implementation of the Government's decision was achieved through the Forestry Revocation and National Parks Reservation Acts 1983, 1984. Implementation of these acts led to the removal of 118 528 ha from the control of the Forestry Commission and transfer to the National Parks and Wildlife Service (NPWS) as national parks and nature reserves. In addition, 1825 ha were set aside as flora reserves under the control of the Forestry Commission (Department of Environment and Planning 1983, 1984). While rainforest was a major vegetation type on the areas transferred to NPWS there were also considerable stands of other forest types (Fig. 9.7), conservation of larger areas being necessary to permit continuing dynamic fluctuation between rainforest and tall open forest, as well as adding to the national park system some examples of communities otherwise poorly conserved.

In order to meet commitments to mills utilizing rainforest timbers during the phase-down period, supplies were made available by relogging some forests not included in the transfers. The loss of eucalypt areas, in addition to rainforest, resulted in some adjustment in quotas, although part of the short fall was made up by the purchase of private *Eucalyptus grandis* plantations near Coffs Harbour. In order to provide alternative supplies to veneer mills, the Forestry Commission embarked on an experimental programme investigating plywood manufacture using high-grade eucalypt timber (Forestry Commission of NSW Annual Report 1986-7), and limited commercial utilization of eucalypts in plywood now occurs.

In New South Wales, the conservation of rainforests ceased to be a major issue as a result of these actions of the State Government. North of the border, the conservation of Queensland rainforest continued to be one of the major issues for environmental groups for the rest of the 1980s. Although there had been various local campaigns for particular sites, the conservation of the rainforests of the wet tropics of north-east Queensland became a national issue in the early 1980s, with the construction by the local shire council of a road in the Cape Tribulation National Park, from Cape Tribulation to Bloomfield through the most extensive coastal lowland rainforest in Australia. Construction started in December 1983, to be met by a blockade of protesters. Work ceased in less than a month, with the onset of the wet season. In August 1984, work resumed, and despite a further blockade, the road was completed (Kendell and Buivids 1987). The unsealed road is vulnerable to erosion in the wet season, and eroded sediment has caused pollution of adjacent coastal waters.

While the construction of the Cape Tribulation road became a focus of attention, the broader issue of conservation of rainforests in north-east Queensland was increasingly under debate. The Australian Heritage Commission (a Commonwealth agency) commissioned a study of the area which strongly argued that the wet tropics were of World Heritage status (Keto and Scott 1986), while a number of publications brought the region to national prominence (Borschmann 1984; Keto 1985; Russell 1985).

9.4.1 The role of the federal government in conservation and land-use decisions

Before considering north Queensland further, it is necessary to digress and to discuss the role of the Commonwealth Government in land-use and conservation. On 1 January 1901, the Commonwealth of Australia came into being as a Federation of what had previously been six colonies but which then became six states. The governance of the Commonwealth was defined by a written constitution, which established among other things the different roles of the State and Federal Governments. The Constitution makes no specific mention of the role of the Federal Government in environmental matters. Control and management of land-use were State responsibilities; forestry and national park declaration and management were therefore under State control. The Commonwealth controls national parks and reserves in Commonwealth territory, and also Uluru and Kakadu National Parks in the Northern Territory, which has yet to achieve statehood.

The Constitution does, however, give the Commonwealth considerable indirect powers to influence land-use in the States. The Commonwealth controls economic policy and, through taxation policy (both in the form of penalties and incentives) and control of foreign investment, this can influence land-use practices. The Commonwealth Government can regulate international

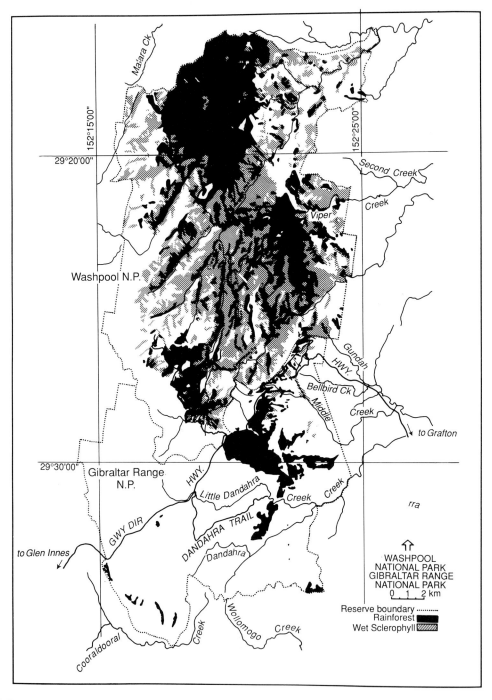

Fig. 9.7 Distribution of rainforest and tall open forest (wet sclerophyll forest) in the Washpool and Gibraltar Range National Parks (from Adam 1987b).

trade, and powers originally intended merely as tools of economic or strategic policy were used in 1976 to prevent mineral sands mining on Fraser Island, the world's largest sand island off the southern Queensland coast. The Federal Government refused to grant an export licence, an action which was subsequently confirmed as a valid use of the Commonwealth's powers by the High Court of Australia. The Federal Government also provides funds to the States. Some of these take the form of special-purpose grants related to specific activities which may include conservation.

In the 1970s, responding to the increased public interest in the environment, the Commonwealth began to seek a greater involvement in conservation and environmental issues. The establishment of the Australian Heritage Commission in 1975 enshrined the concept of the National Estate in legislation—the National Estate being 'those places, being components of the natural environments of Australia or the cultural environment of Australia that have aesthetic, historic, scientific or social significance or other special value for future generations as well as for the present community'. A prime function of the Australian Heritage Commission is the maintenance of a Register of the National Estate. Inclusion of sites on the Register does not ensure their protection, as the States are not bound to take action, but the Register has high public recognition and strong opposition is virtually guaranteed for any proposal adversely affecting National Estate sites. Many rainforest stands are included upon the Register.

The Constitution states that only the Commonwealth can enter into treaties or conventions with foreign powers, but where the Commonwealth has entered into a treaty then it acquires legislative power to implement its provisions.

The Commonwealth Government is party to a number of treaties and conventions which deal with environmental issues. In the context of the rainforest debate, the most important such agreement is the Convention Concerning the Protection of the World Cultural and Natural Heritage (The World Heritage Convention) which was adopted by the General Conference of UNESCO in 1972. Australia was one of the first countries to ratify this Convention, to which some one hundred countries are now party. The Convention provides for the nomination of properties for inscription on the World Heritage List. The decision as to what is included is made by the World Heritage Committee, based in Paris, and comprising representatives of twenty-one nations.

Properties inscribed on the List are judged to be of 'outstanding universal value'. The World Heritage Committee has established a set of criteria by which the value of both natural and cultural properties can be judged. For natural properties, the criteria are:

(i) be outstanding examples representing the major stages of the earth's evolutionary history. This category would include sites which represent the major 'eras' of geological history such as 'the age of reptiles' where the development of the planet's natural diversity can well be demonstrated and such as the 'ice age' where early man and his environment underwent major changes; or

(ii) be outstanding examples representing significant ongoing geological processes, biological evolution and man's interaction with his natural environment. As distinct from the periods of the earth's development, this focuses upon ongoing processes in the development of communities of plants and animals, landforms and marine and fresh water bodies. This category would include for example (a) as geological processes, glaciation and volcanism, (b) as biological evolution, examples of biomes such as tropical rainforests, deserts and tundra (c) as interaction between man and his natural environment, terraced agricultural landscapes; or

(iii) contain unique, rare or superlative natural phenomena, formations or features or areas of exceptional natural beauty, such as superlative examples of the most important ecosystems to man, natural features, (for instance, rivers, mountains, waterfalls), spectacles presented by great concentrations of animals, sweeping vistas covered by natural vegetation and exceptional combinations of natural and cultural elements; or

(iv) be habitats where populations of rare or endangered species of plants and animals still survive. This category would include those ecosystems in which concentrations of plants and animals of universal interest and significance are found.

It should be realized that individual sites may not possess the most spectacular or outstanding single example of the above, but when sites are viewed in a broader perspective with a complex of many surrounding features of significance, the entire area may qualify to demonstrate an array of features of global significance.

In addition to the above criteria, the sites should also meet the conditions of integrity.

- The areas described in (i) above should contain all or most of the key interrelated and interdependent elements in their natural relationships; for example, an 'ice age' area would be expected to include the snow field, the glacier itself and samples of cutting patterns, deposition and colonization (striations, moraines, pioneer stages of plant succession, etc.).

- The areas described in (ii) above should have sufficient size and contain the necessary elements to demonstrate the key aspects of the process and to be self-perpetuating. For example, an area of 'tropical rainforest' may be expected to include some variation in elevation above sea level, changes in topography and soil types, river banks or oxbow lakes, to demonstrate the diversity and complexity of the system.

- The areas described in (iii) above should contain those ecosystem components required for the continuity of the species or of the objects to be conserved. This will vary according to individual cases; for example, the protected area for a waterfall would include all or as much as possible, of the supporting upstream watershed; or a coral reef area would be provided with control over siltation or pollution through the stream flow or ocean currents which provide its nutrients.

The areas described in (iv) above should be of sufficient size and contain the necessary habitat requirements for the survival of the species.

Nominations to the List are made by parties to the Convention: in the case of Australia all nominations must be made by the Commonwealth and not by the States. After nominations are made to the World Heritage Committee they are subject to rigorous independent evaluation, which for natural properties' assessment is by the International Union for the Conservation of Nature and Natural Resources (IUCN).

The first properties were inscribed on the World Heritage List in 1978. Eight Australian properties have been placed upon the List.

(1) Kakadu National Park in the Northern Territory (Stage 1 inscribed in 1981, Stage 2, in 1987);

(2) The Great Barrier Reef (inscribed in 1981);

(3) The Willandra Lakes Region in New South Wales (inscribed in 1981);

(4) The Lord Howe Island Group (inscribed in 1982);

(5) The Western Tasmania Wilderness National Parks (inscribed in 1982);

(6) The Australian East Coast Temperate and Sub-Tropical Rainforest Parks (the New South Wales rainforests, inscribed in 1986);

(7) Uluru National Park in the Northern Territory (inscribed in 1987);

(8) The Wet Tropical Rainforests of North-East Australia (inscribed in 1988).

Rainforest is an important component of the vegetation of Kakadu, Lord Howe Island, and western Tasmania, and a minor feature of the Great Barrier Reef. In the case of the New South Wales rainforests and the wet tropics, rainforest was the *raison d'être* of the nominations.

Prior to 1987, nominations were made with the consent, and normally on the initiative, of the State concerned. The nomination of south-west Tasmania was originally proposed by the Tasmanian Government, but following an election and change of government was strongly opposed, nevertheless the Federal Government refused to withdraw the nomination. The Tasmanian Government sent a delegation to Paris to lobby against listing, but the decision in favour of the nomination was unanimous. The inscription of Kakadu, Stage 2, was likewise vehemently opposed by the Northern Territory Government which similarly, unsuccessfully, lobbied the

World Heritage Committee (see Adam 1987a; Mellanby 1987).

The opposition to the listing of south-west Tasmania arose because the Tasmanian Government had determined to construct a hydroelectric dam on the Gordon River, downstream of its junction with the Franklin River. If this had gone ahead then it would have resulted in the flooding of an area which includes some notable archaeological sites and stands of vegetation including rainforest. In addition, the dam, roads, and associated works, would have been an intrusion in an otherwise near pristine area.

The World Heritage Committee, when accepting this nomination, expressed concern 'at the likely effect of dam construction in the area on those natural and cultural characteristics which make the property of outstanding universal value' and recommended 'that the Australian authorities take all possible measures to protect the integrity of the property'.

Early in 1983, the Federal Government changed from the Liberal/National Party coalition led by Mr Fraser to the Labor administration of Mr Hawke. The new Government moved to prevent the construction of the dam: firstly by use of regulations under the National Parks and Wildlife Conservation Act 1975 and subsequently by new legislation—the World Heritage Properties Conservation Act 1983—an act to strengthen the Commonwealth's powers to give effect to the Convention.

The validity of the Commonwealth's action was challenged in the High Court of Australia by the Tasmanian Government. The Court, by a majority verdict, upheld the validity of sections of the World Heritage Properties Conservation Act, so confirming the Commonwealth's ability to prevent the construction of the dam (Coper 1983). The Court did not consider the merits of dam construction; rather the question was whether the Commonwealth had the ability to override the State and to order that construction be halted.

The Northern Territory Government also took legal action in an attempt to prevent Stage 2 of Kakadu National Park being added to the World Heritage List. In this case, the Commonwealth's powers were also ultimately upheld.

The World Heritage Properties Conservation Act is not limited to protection of sites actually inscribed on the World Heritage List. The Act is also applicable to areas which are declared, in regulations issued by the Commonwealth Government, as meeting the criteria for World Heritage areas. The Commonwealth Government was urged by conservation groups to use these powers to protect rainforest in north Queensland but was reluctant to do so (Cohen 1984)—partly because of a desire to achieve a co-operative solution involving the State Government, but also because of doubts that the validity of such regulations would stand challenge.

Application of the criteria laid down in the Convention must involve judgement. Arguments could be made that many sites satisfy one or more of the criteria. Nevertheless, it is clearly the intention of the Convention that the World Heritage List include not just important sites but only those of 'outstanding universal value'. The World Heritage List recognizes the value that people of many nations may place on the protection of various natural and cultural phenomena (protection which, in a sense, forms a symbol for certain concepts of civilization), inscription of sites on the List involves recognition by the countries, in whose territories the sites occur, of stewardship of the properties for the whole world, not just for present generations but for those of the future. If the symbolic value of the World Heritage List is to be maintained, then the number of sites included on it must be limited or the criteria will become devalued. However, it will be difficult to give absolute definition to the criteria, interpretation must have regard to the precedent set by earlier successful nominations. Nomination does not ensure inscription, the assessment process provides for a critical examination of the evidence supporting the case for properties meeting the criteria.

Recognizing that World Heritage Sites are of 'universal' significance, the Convention provides for the establishment of a World Heritage Fund to support the management of World Heritage sites. Australia, as a developed nation, is unlikely to draw upon the fund.

Against this background, we can now look specifically at the New South Wales and Queensland rainforest nominations.

9.4.2 The New South Wales rainforest nomination

The decision to include substantial areas of New South Wales rainforest within conservation reserves was made in 1982. In the 1984 State election, the rainforest controversy was reopened. Although the official policy of the opposition parties was that existing national park boundaries would be retained, there was a strong perception that elements in the National Party would seek to reverse at least part of the 1982 decision. The Labor Premier, Neville Wran, made a commitment that if re-elected he would seek World Heritage Listing for New South Wales rainforests. After the election, his government moved to give effect to this promise.

The preparation of the nomination faced a number of problems. The New South Wales rainforests exist as a series of discrete stands differing in their species and community composition. No one site could be chosen as either representative or as, in some sense, the best. It was therefore decided to nominate a series of sites, each individually with outstanding features and which, together, illustrate the range and diversity of rainforests in the State.

Rainforest is not the only vegetation type represented in the nominated sites. It was decided to nominate the whole sites and not simply to delineate the rainforest stands (see Fig. 9.7). This was done so as to include a wide range of rainforest—sclerophyll forest transitions; not only are these stands of very tall eucalypts over rainforest spectacular, but they illustrate the dynamic nature of the rainforest boundary, which is, in itself, an important feature of the Australian rainforest (see Chapter 4). In addition, non-rainforest communities form a buffer around the rainforest stands and so help the nominated sites satisfy the integrity criteria adopted by the World Heritage Committee. In order to provide a theme linking the sixteen nominated sites (Fig. 9.8), the documentation stressed the relationship between the existing diversity of New South Wales rainforests and their development over geological time.

At the time, the New South Wales Government announced its intention to proceed to nomination, the future of the north Queensland rainforest had become a major issue. It was suggested by some conservation groups that New South Wales rainforests would be an inferior substitute for a nomination of the Queensland forests. As the New South Wales nomination was developed, it became increasingly clear that this was not a valid argument (Hitchcock 1986*a*, *b*). The arc of eastern Australian rainforests exhibits considerable variation with latitude, each segment exhibits a particular range of diversity reflecting selection from the originally widespread rainforest cover (Chapter 6). The rainforests of New South Wales are different from those in northern Queensland, but are no less interesting or important for that.

It was argued (Adam 1987*b*) that in aggregate the nominated sites in New South Wales amply satisfy all four criteria for natural properties to be included on the World Heritage List. Four of the sites are on the New South Wales/Queensland border (Mt. Nothofagus Flora Reserve, Border Ranges National Park, Limpinwood Nature Reserve, Numinbah Nature Reserve), with contiguous rainforest in Queensland. There would clearly have been merit in nominating areas of rainforest regardless of State boundaries, however, nomination was an initiative of the New South Wales Government and it was unlikely that the then Queensland Government would have cooperated.

New South Wales is not (and cannot be) a party to the World Heritage Convention so the nomination, although prepared by the State, was submitted by the Commonwealth. The World Heritage Committee accepted the concept of inclusion of a number of sites in single nomination

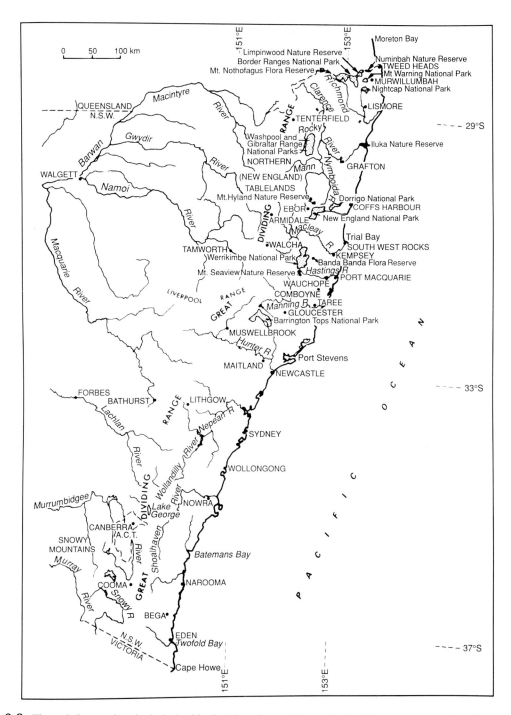

Fig. 9.8 The rainforest sites included with the New South Wales World Heritage nomination (from Adam 1987b).

and at its 1986 meeting the New South Wales rainforests were inscribed on the World Heritage List.

9.4.3 The north Queensland rainforests

The nomination of the New South Wales rainforests was the result of a State Government initiative, the role of the Commonwealth was simply to facilitate its presentation to the World Heritage Committee. The nomination of the north Queensland rainforests by the Commonwealth was strongly opposed by the Queensland Government.

The diversity and complexity of the rainforests of north-east Queensland have long been recognized. Calls for the conservation of particular areas became more frequent in the 1970s, and received at least partial response in the form of declaration of a number of national parks. The Greater Daintree region of some 350 000 ha was nominated for inclusion on the Register of the National Estate in 1976 and was listed in 1980. The area was described as containing the most extensive relatively untouched tropical rainforest left in Australia and the largest extent of coastal rainforest. This listing by the Commonwealth was largely symbolic, as it imposed no obligations upon the State Government.

In January 1984, the Australian Heritage Commission engaged the Rainforest Conservation Society of Queensland to report on the claims for international conservation significance of the wet tropical rainforests of north-east Queensland between Townsville and Cooktown. The report was completed in June 1984 and was published in 1986 (Keto and Scott 1986). The report summarized available information on the rainforests of the region and their geological history. The report concluded that there was a strong case for the region satisfying the criteria for World Heritage Listing.

In 1984, the Standing Committee on Environment and Conservation in the Federal House of Representatives had reported on the Daintree region, and had suggested that if the Australian Heritage Commission report concluded that the region was worthy of World Heritage nomination, the Commonwealth should invoke the World Heritage Properties Conservation Act to protect the area, although the then Minister for Home Affairs and Environment expressed doubt about the wisdom of such action (Cohen 1984).

The Commonwealth Government continued to urge that the region be nominated for World Heritage Listing, and was strongly supported by the environmental movement. The Queensland Government remained adamantly opposed.

In June 1987, the Commonwealth Government announced that it would proceed towards nomination, and in August launched a public consultation process to help decide the boundary of the area to be nominated and to prepare strategies for the social and economic development of the region. In December 1987, the nomination (DASETT 1987) was lodged with the World Heritage Committee. During 1988, following assessment of the nomination, some minor revisions of the boundaries were made, and in December the World Heritage Committee inscribed the Wet Tropical Rainforest of north-east Australia on the World Heritage List. The nomination, encompassed over 900 000 ha (see Fig. 9.9), the majority in public ownership (State Forests, Timber Reserves, and National Parks, vacant Crown Land and leasehold areas), less than one per cent of the area is private freehold.

The area listed:

spans part of the tablelands of the Great Dividing Range, the lower coastal belt and the intermediate Great Escarpment. It is an area of diverse environments rising from coastal plains to undulating tablelands and to highlands with peaks above 1600 m. Much of the topography is rugged, with numerous fast-flowing rivers, gorges and waterfalls ... the values on which the World Heritage nomination were based are:

(a) outstanding examples representing eight major stages in the evolutionary history of the earth;

(b) numerous outstanding examples of ongoing geological processes and biological evolution;

(c) many features of exceptional natural beauty;

THE HUMAN INFLUENCE

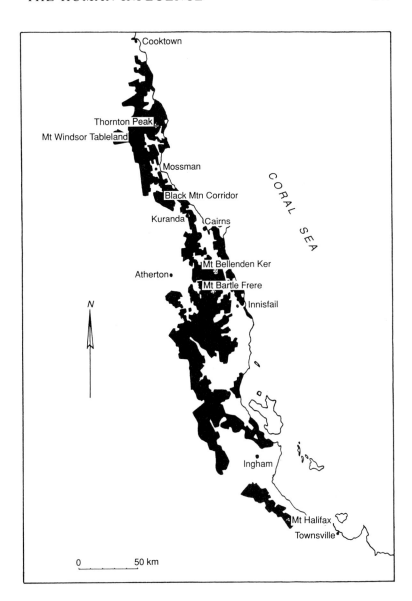

Fig. 9.9 The wet tropics of north-east Queensland World Heritage area

(d) the only remaining habitat for numerous plant and animal species of universal significance that are considered rare, vulnerable or endangered;

(e) the only recognised extant Aboriginal rainforest culture. (DASETT 1988).

The nomination argued that the areas concerned met all four criteria for the Listing of Natural Properties, and that, in addition, there was important cultural value associated with the survival of the only recognized extant Aboriginal rainforest culture.

Following the lodging of the nomination, the Commonwealth acquired a responsibility under the World Heritage Convention to protect those values of the region which were the basis of the nomination. The Federal Government took the view that continued logging of rainforest was incompatible with the nomination and made regulations under the World Heritage Properties Conservation Act to terminate commercial forestry operations. The Federal Government also announced a substantial package of financial support for the region. Although it is now clear that rainforest logging will not occur in the world heritage listed area, the position regarding other activities is less certain. There is currently (in 1991) a major controversy over proposals by the Queensland government to develop a hydro-electricity generating scheme (the Tully–Millstream Dam) which would involve flooding an area of rainforest within the listed area.

These decisions were strongly opposed by the Queensland Government, which lobbied the World Heritage Committee to reject the nomination and mounted a legal challenge to the Commonwealth's action. However, in 1989, there was a change of government in Queensland. The incoming Labor Administration dropped legal action and agreed to co-operate with the Commonwealth over the management of the listed areas.

Unlike the New South Wales' nomination where all the sites were existing National Parks, Nature Reserves, or Flora Reserves, only part of the Queensland nomination is in pre-existing conservation reserves. Accordingly, substantial changes to the management of the area will be required.

In April 1988, the Commonwealth Government released proposals for the management of the region (DASETT 1988). This document established the context and objectives of management and suggested mechanisms by which management would be implemented. Central to the scheme for implementation of management objectives is collaboration between the Commonwealth and the Queensland State Governments through a joint management authority. There is precedent for such a model in the Great Barrier Reef Marine Park Authority which is responsible for management of the Great Barrier Reef, also a World Heritage Site. Co-operation between Tasmania and the Commonwealth occurs over the management of the Western Tasmania World Heritage Area.

9.5 THE FORESTRY VERSUS CONSERVATION DEBATE

Having looked briefly at the controversies of the past twenty years, it is appropriate to examine some of the underlying issues.

In part, the rainforest debate in Australia can be seen as just one aspect of a broader questioning of the aims, methods, and outcomes of forestry by environmental groups. This questioning is not unique to Australia but has been particularly intense there.

The issues raised by environmentalists question not just the immediate environmental

impacts of logging, but wider concerns over land allocation and use, the economics of forestry and timber utilization, and the social structure of rural communities dependent on timber industries (Routley and Routley 1975; Cameron and Penna 1988).

The debate between foresters and environmentalists has been marked more by heat than light. The forestry profession, at least at the highest levels, has often taken the view that foresters are the experts in all aspects of forest management, and has not been prepared to concede equal status to others. Part of the cause of the divide between foresters and many environmental groups is that the 'forestry profession, with some honorable exceptions, has been remarkably insensitive to the changing needs and values of society' (Westoby 1978). Leslie (1987) has suggested that a particular problem arises from differing interpretations of the term 'multiple-use management'. Leslie suggested that for foresters 'In the Australian context multiple-use management is considered to be timber production plus other uses. Management which excludes timber production is dismissed as single use'. To many environmental groups the multiple-use concept applies across the whole forest estate—different parts of which may have different uses and meet different needs. Under this interpretation, it would not be expected that every forest stand be regarded as a timber supply, nor stands reserved from logging represent locked-up resources (the frequent assertion of forest industry).

In coming to terms with these issues, it has to be recognized that the forestry services labour under some difficulties not of their own making. Governments may have entered into agreements to supply mills, without knowledge of the availability of resources, but in the expectation that supplies would be found. This was a factor in the continued logging of New South Wales' rainforests into the 1980s (Forestry Commission of NSW 1979). The sawmilling industry was dominated by small firms with a poor record of investment and innovation, making it difficult to change the nature of the timber supplied to particular mills.

In considering the argument in favour of continued rainforest logging, the different situations in New South Wales and Queensland need to be remembered. In New South Wales, the debate took place when the availability of timber for veneer mills was already limited. The logging practices then in effect had a fairly short predicted life. Disruption and re-organization of the industry was inevitable, and the decision to curtail logging only advanced these changes by a few years.

In Queensland, the availability of timber for the mills was projected over a much longer period, and, unlike the NSW Forestry Commission, which had already adopted a policy of phasing-down general-purpose rainforest logging, the Queensland Forestry Department proposed a continuation of existing selective logging on a sustained yield basis.

Were it not for the dedication of timber reserves and the establishment of forestry in the late nineteenth century, there is little doubt that much more forest, both sclerophyll and rainforest, would have been cleared for agriculture. The very existence of rainforest as a topic for debate in the late twentieth century is a measure of the success of foresters in halting alienation. Nevertheless, it cannot be denied that forestry has been responsible for some loss of rainforest habitat through establishment of plantations on former rainforest or through conversion of rainforest to eucalypt-dominated stands.

During the rainforest debate of the early 1980s, it was frequently claimed that rainforest was already adequately conserved, either in national parks and nature reserves or in areas within productive forests set aside from logging. The environmental lobby argued that the then existing conservation reserves did not adequately sample the diversity of rainforests. The communities well represented in reserves were those unlikely to be under threat because of environmental constraints or species composition. On the other hand, of the area of unlogged rainforest in State Forest, some would never have been logged because of steepness of slope, proximity to watercourses, difficulty of access, or low merchantable value. In addition, some areas would have been set aside because of particular scientific values or as representative stands of particular communities. The environmental lobby argued that the areas which would be excluded from logging would be fragmented and possibly not viable as reserves in the long term. More importantly, the setting aside of areas was seen merely as a concession which did not challenge the supremacy of timber production; in terms of proportional representation in reserves, communities of high commercial value were likely to be significantly underrepresented.

The particular properties and beauty of rainforest timbers give them a cultural significance as the raw material for craft workers producing works of art and high-value speciality furniture (Figs. 9.10, 9.11). In addition, some of the major public buildings constructed in recent years (for example the Sydney Opera House and the Federal High Court), have made extensive use of rainforest timbers. However, while these speciality uses contribute to cultural heritage, they represented only a small part of the harvest from rainforest logging, much of which was used for purposes which did not require the special qualities of the timbers.

The forestry lobby argued that their operations were not incompatible with the objectives of the World Conservation Strategy, objectives also embodied in the National Conservation Strategy for Australia, and that forestry and conservation should not be regarded as antithetical. Despite this, forestry found itself in the dock before a jury of public opinion, and was found wanting; despite apparent victories in the battles (for example, the Terania Creek Inquiry and the challenge to the Hastings EIS) the war was lost—why?

Throughout the debate, it has been difficult to obtain agreement as to the facts, let alone their interpretation. Much of the confusion can be related to the tendency of the different sides to seek answers to different questions.

One set of questions relates to the practicalities of timber production:

1. Is it possible to conduct rainforest logging on a sustainable yield basis?
2. If yes, then can harvesting continuously support industry or will it by a cyclic operation?
3. For publicly owned and managed forests, will such operations be economically viable (for private operations then it would also be legitimate for the public to enquire whether viability was dependent on government subsidy)?

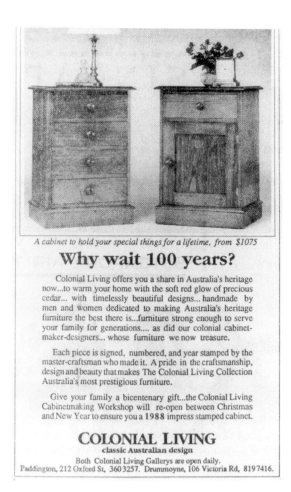

Fig. 9.10 Advertisement for furniture made from rainforest timbers (courtesy of Colonial Living)

Lighting Studios (Aust) Pty. Ltd.

COLONIAL LIVING

classic Australian design

THE COLLECTION: We proudly present COLONIAL LIVING. A collection of reproduction Australian Colonial Furniture, crafted from famous Cedar. These are authentic recreations of antique pieces originally made by such master craftsmen as Joseph Sly, Andrew Lenehan, James Whitesides and later John Hill to name a few.

This chapter in Australia's history (1840 - 1865), the golden era in cabinetmaking; not only did we have notable craftsmen, but Australian Cedar (TOONA AUSTRALIS), the most beautiful of all timbers.

THE FURNITURE: Each piece is an authentic replica, faithfully made by the same time honoured methods as the original. All our joints are morticed and tennoned or double dowelled, all our drawers are dove tailed, guaranteed for strength and durability.

Every piece is made to order and bears the impressed stamp of COLONIAL LIVING, the Cabinetmaker, the YEAR and IT'S REGISTERED NUMBER.

Fig. 9.11 Advertisement for furniture made from rainforest timbers (courtesy of Colonial Living)

The general consensus among foresters is that the answer to the first question is—yes (although perhaps with some qualifications) (see Baur 1989; Horne and Hickey 1991). Opponents of logging would query the meaning of 'sustainable' and would argue that in any case, the database available for analysis is small and suffers serious limitations, being based on small plots which are not necessarily representative of larger areas. These concerns are valid, but could equally be raised about the information available in many other land-use issues. In an ideal world there should be a moratorium on exploitation until such deficiencies had been addressed. Nevertheless it is worth examining the information on which foresters base their response. The information base is a mixture of empirical field evidence and modelling based on that evidence. While it is appropriate to approach models with a degree of scepticisim, particularly in extrapolating to the future, model results have to be considered seriously. Simply to dismiss models because they are models is to reject one of the few tools available for environmental prediction.

The available data were discussed by Horne and Hickey (1991). Once-only selective logging models have been prepared for complex notophyll vine forest (Horne and Gwalter 1982) and simple notophyll vine forest (*Ceratopetalum apetalum*) (Horne and Gwalter 1987) in New South Wales. Both models suggest that stand basal area will recover after logging, although the rate of recovery will depend on the proportion of canopy removed.

Hickey and Felton (1987) suggested that in Tasmanian rainforests basal area would similarly recover from a single selective logging, but the rates of recovery would be much slower than those reported from the mainland (it takes at least 500 years to reach a commercial diameter of 60 cm in the case of *Lagarostrobos franklinii*).

There are far fewer estimates of the effects of multiple cycles of selective logging. Shugart *et al.* (1980) modelled repeated 30-year logging cycles in complex notophyll vine forest; the results suggest that over time the volume of available commercially utilizable timber would decline. (In general the harvest from the first cut is likely to be greater than the long-term sustainable yield—Department of Forestry, Queensland 1981). Modelling of the more species-rich rainforests of north-east Queensland suggests that log production from selection logging can be maintained over long periods, although again with an indication of declining volume in later cycles (Preston and Vanclay 1988; Vanclay 1989).

Estimates by foresters of projected yields and length of logging cycles have, over the years, been subject to revision, sometimes drastically. In the case of the north Queensland rainforests, predictions of sustainable annual yield (on a 40-year cycle) were subject to reduction. Between 1948 and 1978 the annual log quota was set at 207 000 m^3. This was not established as a sustainable yield but came to be accepted as such. There was no public suggestion of this figure not being sustainable, and neither of any intention in the future to reduce it (Frawley 1988). By the 1970s, it was clearly evident that the quota was not sustainable, and in 1978 a series of annual reductions in quota were instituted, leading to an allocation in 1986 of 80 000 m^3, which was estimated as the long-term sustainable yield (Department of Forestry, Queensland 1981). From 1987, however, the allocation was further reduced to less than 60 000 m^3. This reflects the accumulation of more and better data on which to base predictions, and demonstrates that research results were being used to refine management prescriptions. In the rainforest debate, doubt was cast on the projections then being advanced—the figures had been inaccurate in the past, what guarantee was there that they were now right? In the debate over Washpool, the Forestry Commission's claims that alternative supplies were not available were disputed by independent consultants (Department of Environment and Planning 1982). While the merits of some of the proposed alternatives could be questioned, the challenge to the figures was an important factor in reducing the Commission's political credibility.

Whether sustainable yield management could support a continuously operating industry would depend on the area available for logging and the annual yield. In the case of the New South Wales rainforests, by the 1970s the resource available could not have sustained industry on the scale then operating. In the case of Huon pine in Tasmania it would clearly be impractical to contemplate sustainable high-volume production, but a very low-intensity harvesting for high-value craft uses might be sustainable.

The economics of forestry are a source of considerable debate, conventional accounting often rendering it difficult to assess the long-term viability of operations. However, if the models discussed above are regarded as indicators of future resources, then even if multiple logging cycles are feasible the volume yields will be relatively small. Given the infrastructure (roads, fire protection, etc.) required for forest management, the true costs of production will be high; whether the market would be prepared to pay 'true' prices may be doubtful.

However, even if those opposed to logging conceded that sustained yield harvesting were feasible, they would argue that there is more to forest management than timber production, and they ask whether selective logging results in any significant changes to the forest environment (a question which would generate a secondary debate as to the meaning of 'significant change').

Under a selective logging regime, both the species composition and the size class of the harvest can be controlled. Nevertheless, it is likely that in the first cycle a greater proportion of large trees will be removed and, inevitably, over time, that the frequency of the largest-diameter classes will be greatly reduced (Winter et al. 1987c). From a wood production point of view, this is desirable, as the largest trees have passed their productivity peak (Vanclay 1989), but from the point of view of public enjoyment of forest the loss of forest 'giants' is regarded as a serious impact. Even if there is only a one-off logging, the time required to regrow very large trees may be considerable.

Both sides of the debate could agree on the effect of logging on the size/age structure of individual stands, even if they disagreed as to its significance. In the case of other possible impacts of logging, there were either different interpretations of the data or a total lack of relevant information.

Does selective logging result in changes in stand composition and/or diversity?

There is no evidence that logging has caused the extinction of any rainforest species in Australia. However, little is known about the occurrence of many taxonomic groups, so that the effect of logging on the total biota must also be unknown; extrapolations from the few groups which have been investigated are not necessarily justified.

In the case of trees, considerable controversy has been engendered by claims that logging leads to an increase in diversity. The Department of Forestry, Queensland (1983) concuded that:

- typical virgin stands show negative increments, high mortality and low recruitment rates compared to logged or treated stands. This indicates that one function of logging is to remove timber in trees which are becoming senescent;

- the response of stands to logging or treatment by increased increments and rapid re-establishment of physical structure indicates that rainforest stands are adapted to periodic disturbance;
- selective logging does not cause the systematic removal of particular species from the stands and evidence exists that species diversity is better maintained in logged than in virgin stands.

This argument that not only did logging have no long-term adverse effects but that it may have positive benefit was one which environmental groups found particularly difficult to accept (Webb 1987), and its basis was severely criticized by Lowe (1990). The implication that logging can be regarded as euthanasia for senile trees was contested by the environmental lobby, who pointed to the important role of dying and dead trees in the ecosystem. In addition, although logging removes some senescent trees, many of the harvested stems would not fall into this category. Nicholson *et al.* (1988) discussed the monitoring of a number of rainforest stands, virgin and logged, in north Queensland. Their data show that, in the case of some of the virgin rainforest stands, there was a decline in species richness and diversity over time. On the other hand, 'logging tends to reverse this natural decline in diversity, though not always completely'. Nicholson *et al.* (1988) argued that their data support Connell's model (Connell 1978, 1979), which postulates that a certain level of disturbance is necessary to maintain high species diversity. The role of disturbance in regeneration is now well recognized (see Chapter 5). However, to leave an impression that, left unlogged, diversity in rainforest will decline, would be incorrect. The factors promoting disturbance in virgin rainforest operate intermittently in time and disjunctly in space. Any particular research plot will have a low probability of disturbance (although two of Nicholson *et al.*'s (1988) plots were affected by Cyclone Winifred in 1986), but over a large enough area these will be a mosaic of patches at different stages since the last disturbance, and overall species richness and diversity will be maintained. There are few data on the effects of imposing logging on long-term natural perturbation cycles.

The interpretation of Nicholson *et al.*'s (1988) data has been debated by Saxon (1990) and Nicholson *et al.* (1990). Saxon (1990) contended that Nicholson *et al.*'s (1988) conclusions were untenable and that logging was not to be equated with natural disturbance, a conclusion strongly contested by Nicholson *et al.* (1990).

Horne and Gwalter's (1982, 1987) modelling suggested that, in both complex and simple notophyll vine forest in New South Wales, logging would not result in a loss in species richness, but diversity would change as the proportion of faster-growing species would increase.

The effects of selective logging on fauna are reviewed by Horne and Hickey (1991) (see also Chapter 5). There are insufficient data to reach any general conclusions. However, particular studies have resulted in suggestions for modifications to management practices in order to protect certain species. Crome and Moore (1989) examined the impact of logging on bowerbirds in rainforest on the Mount Windsor Tableland in north-east Queensland. Three species of bowerbirds in the area have bowers or courts, the golden bowerbird *Prionodura newtoniana*, the satin bowerbird *Ptilonorhynchus violaceus*, and the tooth-billed catbird *Ailuroedus dentirostris*. Bowerbird activity was concentrated along ridge systems, which are also the preferred locations for roads during logging. The satin bowerbird is predominantly a forest-edge species and was rare in the study area. During Crome and Moore's (1989) study, the one golden bowerbird bower in a logged area was destroyed by a felled tree, but another bower was built nearby and was used during the logging operation. There was a loss of a third of the tooth-billed catbird courts on the logged area, but the same number of courts were built on the adjoining unlogged area, so that in total there was no loss. What would have been the fate of these birds if there had not been adjacent unlogged areas is unknown. In this study, logging did not adversely affect the bowerbird populations, however, Crome and Moore

(1989) suggested that continued damage and reduction to vegetation cover on ridges and upper slopes could reduce bowerbird populations, and suggest that during road construction extra clearing should be minimized, continued maintenance clearing of verges should be limited, and walking-tracks should be routed to avoid bowers and courts. Implementation of these suggestions would probably assist conservation of other organisms, but ideally, management regimes should be based on ecosystem studies and not be developed in an *ad hoc* fashion.

Selective logging involves more than the removal of the harvested trees. Roads, log dumps, and snig tracks are constructed, requiring clearing of vegetation and localized disturbance to the soil. During felling of the selected trees, others may be knocked over or damaged. (In an area studied by Crome and Moore (1989) on the Mt. Windsor Tableland in north-east Queensland, of 151 trees greater than 10 cm diameter lost during logging only seven were harvested, the remainder were pushed over by bulldozers or knocked over by other falling trees).

While admitting that, in the past, logging may have sometimes caused extensive disturbance, foresters argue that modern controls on road and snig track construction and on logging near creeks reduce the possibility of erosion and maintain catchment water quality. Damage during felling could be minimized by greater training of forest workers. Forestry practices in both New South Wales and Queensland rainforest changed over time (Nicholson 1985), and logging in north Queensland in the 1980s was subject to a high degree of control (Ward and Kanowski 1985) by trained personnel (Dale 1985). Australian foresters point to these controls as a model for rainforest management elsewhere. Nevertheless the ability of the forestry services to police environmental controls on logging operations was widely questioned—publicity was given to examples where it was said that erosion control and catchment protection requirements were not being met. Regardless of the actual extent of breaches of environmental controls, the adverse publicity influenced community opinion.

The networks of roads and disturbed areas resulting from logging provide sites for the establishment of exotic species (for example *Lantana camara*). Although regeneration may occur on smaller roads or snig tracks, the larger roads established during logging are often retained for public access. For many of the public, their major opportunities for visiting rainforest are provided by the existence of these roads, nevertheless the networks of roads may permanently fragment the habitat, and result in significant road deaths of the larger fauna, and, in some instances, create management problems from overuse of small areas.

Immediately after a logging operation, there is inevitably an increase in the amount of potentially flammable fuel on the forest floor. If this material burns, then regeneration may be affected and fire may spread into adjacent rainforest. Horne and Hickey (1991) argued that because of the tendency to summer drought, post-logging fire was a much greater threat to Tasmanian rainforest than on the mainland.

Fires may be an important mechanism by which nutrients could be lost from the ecosystem, jeopardizing the chances for long-term maintenance of rainforest. Timber harvesting must also represent a direct loss of nutrients (in addition, any erosion associated with logging will also be responsible for nutrient losses). Whether logging-losses of nutrients adversely affect the ecosystem will depend on their magnitude and on the rates of additions to the system from weathering and deposition. Plowman (1990) and Congdon and Lamb (1990) have emphasized that so little is known about decomposition processes and nutrient cycles in Australian rainforest that there is no sound basis for predicting the long-term effects of timber harvesting on nutrient availability and forest growth.

Empirical evidence demonstrates that, following logging in rainforest, the regenerating vegetation would still be classified as rainforest (Fig. 9.12, 9.13) (unless management intervention

'In the final analysis, the real debate on rainforest management represents a clash of fundamentally opposed viewpoints' (Department of Forestry, Queensland 1983). Central to the disagreement were different values placed on the non-timber production roles of forests. Quantification and evaluation of often intangible concepts are, of course, difficult, but these difficulties should not be used to deny the importance to many people of non-utilitarian features of the landscape.

In attempting to present the debate to the public, there was a tendency for the forestry lobby to represent itself as rational and responsible and the environmental lobby as irrational and emotional; a deliberately promotes conversion to a different vegetation type.) Although various short-term studies have demonstrated that certain features of the regenerating rainforest are not significantly different from the pre-logging condition (Horne and Hickey 1991) the data are insufficient to conclude that logging has no long-term effects and the effects of repeated harvesting are even less certain.

Disagreement about technicalities was not, however, the underlying basis of the argument—

Fig. 9.12 Regeneration several years after logging in complex notophyll vine forest (cool subtropical rainforest), Mt. Boss State Forest, New South Wales. This stand is adjacent to cool temperate rainforest and emergent *Nothofagus moorei* are visible in the background

Fig. 9.13 Regeneration in simple notophyll vine forest (dominated by *Ceratopetalum apetalum*), following selective logging. Billilimbra State Forest, New South Wales

statement from Richards (1952) being quoted. 'Few writers on the rain forest seem able to resist the temptation of the "purple passage", and in the rush of superlatives they are apt to describe things they never saw or to misrepresent what was really there. In attempting to paint an accurate picture of a rain forest it will be necessary to point out and correct some of these common errors' (see, for example, Department of Forestry, Queensland 1983). While in any discussion of scientific issues data must be presented fairly and debated without bias, the sense of wonder inspired by rainforest, and its tendency to encourage the 'purple passage' is in itself a significant value to be recognized.

The rejection of the case advanced by environmental groups is encapsulated in the final paragraphs of the 1983 Queensland Position Paper (Department of Forestry, Queensland 1983).

The lexicon of the conservation movement provides some clues to its viewpoint. Metaphors relating the rainforest to 'museums' and 'cathedrals' abound, while forest management is described, at best, as 'pilfering' and, at worst, as 'destruction'. The ethos of the conservation movement presents an extremely myopic view of rainforest ecosystems: that they are unchanging yet fragile systems which depend for their survival on an absence of perturbation. In the light of the demonstrable climatic changes on a recent geologic timescale and the signally impressive recuperative abilities of rainforests, this must be seen as analogous to the proposition that the earth stands still because it appears thus to the earthbound observer.

Probably as a result of observing slow-growing organisms, foresters are inured to the idea of long-term dynamism in forest communities: that what we have today will surely be different tomorrow. This perspective is often lacking in commentary on forest management. There are many examples in Australia to show that the application of correct logging techniques produces vigorous, healthy forest. In Europe, where sound forest management has been practised in many forests for centuries, the examples are legion. Although logged many times, management of these forests scarcely warrants the comments of 'desecrating a cathedral' or 'vandalizing a museum'. While rainforests are much more complex systems, the research evidence supports the long-held belief of foresters that perpetuation, and not destruction, is the result of the judicious application of selective logging.

Most members of the conservation lobby would acknowledge the resilience of rainforests in the face of environmental change: indeed, this is one of the features of Australian rainforests stressed in the World Heritage nominations. The fact that many European conservation reserves have been subject to logging over many centuries, and that to protect conservation values, former logging practices may be reintroduced, would also be recognized. However, the forestry case ignores the context in which the rainforest debate took place and also the unique conservation opportunities afforded by Australian rainforests.

The rainforest debate took place after nearly two hundred years of European impact, during which perhaps three-quarters of the original rainforest has been lost by clearance; much of what remains has been logged. The remaining area of virgin rainforest is small and, if forestry strategies had been implemented, most of the loggable virgin rainforest would eventually have been harvested. In New South Wales this would have occurred relatively quickly. The concept of virgin forest is one which has great appeal to many people. Foresters argue that faced with a virgin stand and one logged some decades earlier, few members of the general public could detect a difference. This may well be true, but it is equally true that knowing that a forest has been logged lowers the value of the rainforest experience for many people. The opportunities to conserve substantial stands of virgin rainforest were rapidly diminishing, and many agreed with the proposition that the opportunity should be taken before it was lost. Most of the people supporting conservation would never visit the forests in question, but this does not diminish their case. Not only was there the opportunity to conserve virgin rainforest, but also to protect wilderness areas. The concept of wilderness is one which has engendered considerable controversy in Australia (Hall 1989), and its definition is contentious as it clearly cannot be an area totally devoid of human influences; after at least 40 000 years of occupation, no area in Australia would meet this criterion. The generally accepted working rule in Australia is that a wilderness is a large area (at least 25 000 ha) devoid of roads or other obvious constructions. Logging, which is inevitably associated with road construction, is clearly incompatible with the maintenance of wilderness values. The arguments in favour of wilderness are difficult to quantify and thus tend to be dismissed by 'practical' land-users. Nevertheless, the case advanced by Stanton (1982) strikes a chord with many people:

I have not chosen to give compelling scientific evidence of the value of Cape York Peninsula wilderness. I have, instead, preferred to dwell on the ways it can enrich the inner life of those who take the trouble to experience it. It may be desirable that rationality and science should rule the world, but there is little evidence that the behaviour of human beings was ever guided for long by their precepts. It is the heart that still rules, and if the morality of the arguments for caring for one of nature's last realms in this devastated continent (which Xavier Herbert so feelingly calls 'my poor destructed country') is not enough, the cold arguments of science will not suffice.

The decision in the rainforest debate lay with neither the forestry nor the environmental lobby but with Government, a decision complicated in the case of Queensland by the opposed verdicts of Commonwealth and State Governments. It was the task of government to assess the merits of the cases advanced by the conflicting parties, an assessment which also had to consider public opinion, and the economic and employment implications of any decision. While forestry and related industries are of major national economic significance, this is not the case for the rainforest segment, although by virtue of the scale of the operations, rainforest logging in Queensland was more valuable than that in New South Wales. Nevertheless, no government would willingly expose itself to a charge of destroying small rural communities. In both New South Wales and Queensland, it was alleged that phasing-out of rainforest logging would lead to massive unemployment and social disruption. In New South Wales, the Government, in agreeing to the transfer of State Forest to National Park, also established a 'Rainforest Fund' to promote and encourage the development of new technologies and to assist industries affected by the decision. In the case of north-east Queensland, the Commonwealth Government proposed a substantial financial package to provide assistance to the communities affected by the termination of logging. The effectiveness of the Commonwealth proposals remains to be proven; in New South Wales, the dire predictions of long-term unemployment and destruction of communities made by opponents of the conservation option did not eventuate (but see Watson 1990), although some jobs were lost.

In arguing for the transfer of extensive areas of forest to national parks, environmental groups have stressed the long term benefits from increased tourism. A number of projections suggest that in economic terms alone tourism will be of greater benefit to local communities than continued logging. Tourism, both domestic and international, is one of the most rapidly growing sectors of the Australian economy and visits to national parks have increased considerably. Increased usage of rainforest national parks will require careful management to prevent visitor pressure from jeopardizing the very values for which the parks were established. Concerns have been expressed that existing management of tourism in the wet tropics of north-east Queensland is inadequate to prevent habitat degradation (McIntyre *et al.* 1989).

9.5.1 The importance of the World Heritage List

The debate about rainforest conservation in Australia has, to the general public, become very much tied up with a debate about World Heritage status. To some, it would seem that rainforest can only be conserved if it is placed on the World Heritage List. This is clearly not the case, but the impression arises because the only way that the conservation and management of the north Queensland rainforests could have been achieved in the manner desired by the Federal Government, prior to the 1989 State election, was through invocation of the powers of the World Heritage Properties Conservation Act. That this is so, arises out of the particular nature of the Australian Constitution (and its interpretation by the High Court).

The major political battles between the Commonwealth and certain States and Territories over World Heritage nominations have, in part, been about different approaches to conservation and land-use, but from the point of view of the States they have been about 'States' rights' and this issue has overriden any rational consideration of the other matters involved. Disagreement between the States and the Commonwealth over the extent of the Commonwealth's powers goes back to Federation, and such tensions between different levels of government are not unique to Australia. They do raise the question as to the extent to which Australia is one nation. If Australians in general feel that western Tasmania or north Queensland are part of their heritage, and that they are prepared to accept certain responsibilities towards those areas, should these sentiments override local opinion? Local community opposition to World Heritage nominations has centred on the evils of interference by Canberra and on an 'us versus them' divide between a small misunderstood rural population and an urban population, mainly in Sydney and Melbourne, which might be well meaning but is ignorant of the 'real' issues involved.

Inclusion of sites on the World Heritage List affords them the strongest possible protection, in that much of the major change in land-use would require concurrence of both State and Federal Governments, rather than the State Government alone. In urging that other countries conserve outstanding examples of rainforest, Australia can demonstrate that she has taken exceptional steps to guarantee the protection of her rainforest.

The Australian Government has, in the nominations for the New South Wales and Queensland rainforests, provided two of the most detailed and comprehensive nomination documents yet submitted. This is further evidence of the seriousness with which responsibilities under the Convention are addressed, and also provides models for future nominations.

9.6 AUSTRALIAN RAINFORESTS IN A WORLD CONTEXT

The destruction of rainforests is of major international concern. Rainforests are the greatest centres of genetic diversity on earth and the accelerated loss of diversity is viewed with alarm. The rate of clearance is such as potentially to have significant impacts on the composition of the atmosphere (McElroy and Wofsy 1986); through its contribution to the greenhouse effect, rainforest clearance may be a factor in changing world climate. After clearance there is frequently rapid loss of fertility and land degradation. The case for rainforest conservation is being advanced with increasing urgency. Concern is normally concentrated on the great but declining expanses of rainforest in Amazonia, West Africa, and South-east Asia. However, Myers (1980, 1986) included Australia as one of the areas undergoing broad-scale depletion of rainforest at a rapid rate, suggesting that little would remain by 1990; a claim which even in 1980 was surprising. The threat of conversion of rainforest to agricultural land was small; in 1980 there was a strong possibility that much of the rainforest in Queensland, not in national parks, would be logged, but while this would have destroyed primary forest, the regeneration would still have been rainforest.

Even before European settlement, the area of rainforest in Australia was a very small proportion of the world's total. However, the contribution of Australia to the genetic diversity of global rainforest biota is far greater than the small area of the forests might suggest. This species diversity has two causes: firstly, the long period of isolation of Australia from other continents permitted evolution of numerous endemic taxa; secondly, the great diversity of rainforest structural types—from microphyll mossy forest to complex mesophyll vine forest—is correlated with a comparable diversity in biota. Conservation of Australian rainforests is thus important because of the contribution these forests make to world genetic resources, and also because they provide a textbook example of the effects of isolation on evolution. So far we can only trace parts of the evolutionary process through the fossil record. A fuller understanding of the development of Australian rainforests will require greater study of the fossil record, but will also demand more investigation of intact rainforest stands.

Australian rainforests are unique in not facing great pressures from a growing population desperate for land and fuelwood. Australia is one of the few countries where the long-term conservation of relatively large tracts of rainforest can be reasonably assured. Australia still has the opportunity of guaranteeing the conservation of a larger range of rainforest types than most other nations. In addition, although the rainforests have been subject to human influences for thousands of years, they have never been subject to slash-and-burn shifting cultivation. The long history of such practices elsewhere makes it difficult to identify, with any degree of certainty, primary rainforest. In this respect, Australia is unique, which adds further value to conserving the remaining virgin rainforest, not only in its own right but for the opportunities for study and research that it offers.

Australia is one of the few scientifically advanced nations with indigenous rainforest. It is well equipped to conduct a wide range of research on rainforest ecology and management. Much research, including basic survey and inventory, as well as studies on ecosystem processes and dynamics, will be required if Australia's rainforests are to be properly managed. In addition, there are opportunities to provide training of management and research personnel from neighbouring countries (Cassells *et al.* 1988).

If the rainforest cover of many tropical countries is to be maintained then it will be largely on the basis of conservative management for sustainable yield of forest resources (including, but not exclusively, timber). Nevertheless there is a very strong case for a considerable increase in the area reserved for conservation. It is frequently argued that national parks can be developed as tourist attractions and so contribute to economic development. This argument has been used in Australia to support many of the rainforest conservation measures of the 1980s. However, it is not certain that all conservation objectives are compatible with tourism as there has been little experience in the integration of tourism with natural resource management. It will be necessary to develop expertise in the management of rainforest national parks and Australia's developing experience in this regard may well prove valuable to other countries.

9.7 THE FUTURE OF AUSTRALIA'S RAINFORESTS IN A 'GREENHOUSE' WORLD

There has been considerable publicity given to the possibility of rapid climatic change as a result of changes in atmospheric composition, in particular increases in carbon dioxide, methane, nitrous oxide, ozone, and chlorofluorocarbons (Pearman 1988*b*). There is uncertainty as to the changes which may occur, but the scenario for Australia used as the basis for discussion in Pearman (1988*a*) suggests that over the next fifty years there may be:

(1) a rise of 2–4 °C in mean annual temperature, with greatest warming in the south and in winter;

(2) higher spring, summer, and autumn rainfall by up to 50 per cent in those regions which receive rain from the southern penetration of tropical/subtropical air during the monsoon season. The increase may be greatest at the southern limits of the summer rainfall zone.

Winters in the winter rainfall regions may be 20 per cent or more drier;

(3) a general rise in sea-level between 20–140 cm;

(4) a shift of 200–400 km south in the southern limit of tropical cyclones, and an increase in maximum intensity of 30–60 per cent. The frequency of occurrence of cyclones may increase;

(5) a decrease in windspeed by 20 per cent north of 36 °S and an increase in the south.

Australian rainforests have experienced great environmental changes over their history, and their present nature and distribution reflects such changes. The response of the rainforests to the greenhouse effect may differ from previous accommodation of climatic change for two reasons. Firstly, the postulated rate of change is probably faster than those experienced in the past; secondly, the continuity of natural vegetation over the landscape has been fragmented by agricultural clearance and urban development, breaking migration corridors, and possibly limiting the ability of species to respond to change.

Accepting these provisos, possible changes to rainforest can be speculated upon.

The increased temperature and greater summer rainfall north of Sydney will probably be advantageous for rainforest growth. Expansion of rainforest into adjacent eucalypt forest might be predicted. Cool temperate rainforest might be adversely affected. Busby (1988) suggested a contraction in the range of *Nothofagus cunninghamii* in Victoria and Tasmania; a similar contraction in range for *N. moorei* in New South Wales would also be predicted. However, such predictions do not allow for the inertia of established populations. Although conditions might not permit regeneration of a species, established trees might endure the changes, although stands would not recover after disturbance.

It is difficult to predict changes in the cyclone regime (Holland *et al.* 1988). However, cyclones are a major disturbance factor affecting rainforests, and changes in frequency, intensity, and extent of cyclones could have major impacts on rainforests. It is also difficult to predict the effect of the 'greenhouse' changes on fire regimes, but Beer *et al.* (1988) suggest the possibility of a considerable increase in Forest-fire Danger Index in south-east Australia. This could have important repercussions on small rainforest stands in southern New South Wales and Victoria, which are already vulnerable to damage in bush fires.

On the coast, the rise in sea-level might cause a regression of the seaward margin of mangroves. On the other hand, the increased temperature might permit a southwards extension of some species.

Changes of the magnitude predicted would have substantial impacts on land use, human activities, and population distribution. There may be considerable changes in the patterns of visitation to rainforests, with important implications for management.

9.8 MANAGEMENT ISSUES

Even without considering the greenhouse effect, there are many issues facing those responsible for management of Australian rainforests.

In both New South Wales and Queensland there are extensive areas of rainforest which are not included in conservation reserves, but remain under the control of the forestry services and have been subject to various logging regimes. Will these forests ever be subject to further cutting cycles? In New South Wales the closure of some mills and the re-equipping of others could mean that, even if harvesting were permitted, the industry would not have the capacity to process the resource for veneer production. The possibility of sale of high-value speciality timber for craft use and furniture manufacture exists: this currently occurs on a small scale and if the appropriate infrastructure were developed it could be expanded. Use could also be made of rainforest logs from wet sclerophyll forest. However, public opinion is likely to be very strongly opposed to anything which could be portrayed as a recommencement of rainforest logging. Baur (1989) has made suggestions as to how a low, but sustainable, yield of rainforest timbers, with realistic royalties, could be harvested in New South Wales. Hopkins (1990) has suggested that if rainforest logging continues, managers should change from 'exploiting' to 'farming' ('harvesting at lower intensities or putting something back into improving the system of growing timber trees rather than cutting them from natural forests'). In order to reduce disturbance he suggested that non-traditional methods of extraction should be considered, such as using airships or helicopters. (If the resource is given sufficient value such methods would be viable—helicopter extraction of red cedar was contemplated in the Illawarra region in NSW in the early 1980s—Mills 1987).

Contrary to the prevailing view in the media and general public, rainforest logging has not been banned in public forests in Australia. However, transfers of land to national parks and other controls on forest operations elsewhere have reduced rainforest harvesting very considerably. This situation has been attacked by the forestry profession and has also been criticized from other quarters. Flannery and Conlon (1989) have argued that the present policies are short-sighted and have suggested an inevitable reversal in the future. ('Are we simply locking away these resources so that they can be pillaged by our resource starved children?') They also argued that a demonstration of sustained yield rainforest harvesting could provide a model for neighbouring countries. Environmental groups, on the other hand, argue that the better way of emphasizing the importance of rainforest is to maximize protection.

With the exception of *Araucaria cunninghamii* and *A. bidwillii* plantations, and possibly some small plantations of *Acacia melanoxylon* in Tasmania, there are unlikely to be commercial plantings of rainforest species for harvesting. Even when species can be established in monoculture, for example Queensland maple, *Flindersia brayleyana*, growth rates are very slow (Department of Forestry, Queensland 1983). In south-east Queensland, continuing clearance of notophyll vine forest for establishment of *A. cunninghamii* plantations has raised opposition from environmental groups. Over 4000 hectares have been cleared since 1980. (On the other hand, establishment of plantations on marginal farming land, which would have supported rainforest, would be generally welcomed.)

However, when economic considerations are not paramount, there is considerable interest in re-establishing rainforest on sites which had been cleared for agriculture (Clark *et al.* 1980). Rainforest will naturally re-establish in a few decades when agriculture is abandoned on sites adjacent to surviving rainforest (Horne *et al.* 1981). At sites now distant from rainforest, planting is necessary. Most success has been achieved by repeating the pattern of natural succession following disturbance (see Chapter 5), starting with fast-growing shrubs, followed by underplanting with short-lived trees. As birds are attracted to the growing stand, propagules are likely to be introduced so that planting of these species is unnecessary (Clark 1980).

Many stands of rainforest are small. Protection of these stands is essential to maintain the geographic spread of rainforest which may be essential for nomadic or migratory fauna species, and to conserve many rare species which are restricted to isolated stands. As stands become smaller the diversity of the resident fauna is likely to decline, as has been demonstrated by Howe *et al.* (1981) for bird species. A particular feature of these small stands is that they have large perimeter to area ratios and are, therefore, vulnerable to edge disturbance. One of the most severe threats is posed by fire. Rainforest is unlikely to burn in most years, but in dry years fire may sweep through whole stands, as occurred with some rainforests in southern New South Wales and Victoria in the summer of 1983. Even if the whole stand does not burn, then repeated burning (either in wildfires or fuel-reduction burns) of adjacent areas will cause attrition of the rainforest boundary. In some areas there has been a tendency to regard rainforest stands, particularly those in gullies, as natural fire-breaks, and in management plans a tendency to treat the rainforest as the protection for sclerophyll forest, rather than specifying measures to protect the rainforest. Over the long term, such a strategy will be deleterious to rainforest survival.

In Victoria, rainforest today occurs as numerous small stands which are mostly on public land (see Chapter 3, Fig. 3.26). This has afforded the State Government the opportunity of developing management guidelines, which, if successfully implemented, should ensure the conservation of rainforest in that State. The guidelines also apply to lands adjacent to rainforest, so as to provide buffer zones. A major thrust of the guidelines is to ensure protection of rainforest stands from damage by fire, particularly in fuel-reduction burning.

Similar guidelines would be a useful aid to the management of public land in other States, but would be difficult to apply to private land. Although it is difficult to enforce management on private land through legislation, proposed changes in use can be controlled through planning regulations. In New South Wales, special controls exist on littoral rainforest (State Environmental Planning Policy 26, Department of Environment and Planning 1988), including those in private ownership. Any development in some hundred mapped littoral rainforest stands requires approval from both Local and State Governments, and any proposal must be accompanied by an Environmental Impact Statement. Buffer zones are designated around each stand, and restrictions apply to development in these zones.

These policies of the Victorian and New South Wales Governments are attempts to widen the scope of rainforest conservation and management. However, particularly in the case of privately owned rainforest, there is a need to provide positive incentives for management (Floyd 1987). These could take a number of forms, assistance with developing and implementing management plans, grants or loans for management works or, from the Federal Government, tax incentives. Given the significance of privately owned stands for the conservation of many rare species, developing appropriate co-operative management strategies with private landholders is likely to be an important component of any overall rainforest conservation strategy.

The management of small rainforest stands surrounded by eucalypt forest poses fewer problems than for isolated stands adjacent to agricultural or urban areas. In northern New South Wales there are a number of examples of this type (Fig. 9.14), which are either nature reserves or are under some other form of tenure which favours conservation. Many of these tiny stands are remnants of once larger stands. Their survival, with still recognizable rainforest structure and composition, more than a century after clearing of their surrounds, is of considerable interest.

Nevertheless, these stands often display considerable degradation. The most serious problems are caused by weeds which have become established from deliberate, if misguided, planting, from dumped garden rubbish, by transport by birds, or, in the case of riverine stands, in floods. Programmes of weed control have been commenced at a number of sites. The best-documented example is Wingham Brush, a small reserve on the Manning River, subject to regular floods. The most serious weeds at this site are vines: *Macfadyena unguis-cati*, *Anredera cordifolia*, and *Cardiospermum halicacabum*. These vigorous climbers smother the canopy, eventually killing edge trees and expanding gaps (Fig. 9.15). The most successful treatment of the vines has been to cut the stems at about 1 m height and to paint the cut ends with glyphosate herbicide. The details of the treatment programme are discussed by Floyd (1984, 1990), Fox (1988), and Stockard and Hoye (1990).

One management option for very small isolated stands is to extend their area by planting. This has been attempted in Victoria Park Nature Reserve in north-east New South Wales, one of the few surviving fragments of the Big Scrub, the vast stand of lowland subtropical rainforest which once covered some 75 000 ha. The present area of rainforest at Victoria Park is only about 8.5 ha. The rest of the reserve consists of grasslands, dominated by *Pennisetum clandestinum*, or *Stenotaphrum secundatum*, both alien species which are very vigorous and create dense vegetation mats.

The reforestation of these grassland areas has been under way since 1978 (Floyd 1984). After the grass had been killed with herbicide, initial plantings were of the introduced *Solanum mauritianum*, mixed with smaller numbers of the native *Commersonia bartramia*. After about three years, the *Solanum* began to thin out, but by this time a large number of other seedlings, both fast-growing species and longer-lived slow-growing species, were present as a result of natural spread of propagules by either wind or birds.

Management of small areas of rainforest is

Fig. 9.14 An example of a rainforest remnant in New South Wales. Maguires Creek (photo: courtesy P. Eby, National Parks and Wildlife Service).

Fig. 9.15 Vine towers in Wingham Brush (photo: M. D. Fox).

clearly far more expensive, per unit area, than management of larger stands. Re-establishment, at least in the initial stages, is also expensive. It is not clear, therefore, whether Australian experience with small rainforest stands offers a practical model for other countries. It would suggest that even if rainforest is very much reduced and degraded, it is not necessarily lost, but the costs of rehabilitation may be beyond the means of developing countries. In terms of whole ecosystems, there may be an important difference between rainforest in New South Wales and that in, say, South-east Asia. In New South Wales, there is not a specialist vertebrate fauna restricted to rainforest (the same cannot be said of the invertebrates and some small rainforest stands are protected to conserve particular invertebrate species: for example, Hayters Hill Nature Reserve of only 4.5 ha). If there were a specialist fauna, as there is in South-east Asia and north Queensland (see Chapter 7), it is doubtful whether it could survive in totality in very small stands. (The decline in fauna but the maintenance of flora has been documented for the small Bukit Timah reserve on Singapore by Corlett 1988.) The extent to which the invertebrate fauna is impoverished, as a result of habitat fragmentation, is unknown. In the long term, doubt must be expressed as to whether some nature reserves are large enough to maintain their floristic diversity through the process of gap replacement regeneration, when the fall of one major tree could open up a large proportion of the whole stand. Australia is well equipped to monitor the dynamics of small stands, and the results of such studies will be important to future management, and also of wider interest to the conservation of rainforest in general.

9.8.1 Threats from development

Problems seem likely to arise from real-estate developers seeking to benefit from the increased popular interest in rainforest, by sale either of rainforest or of adjacent sites (Figs. 9.16, 9.17, 9.18). Purchasers of such land may be destined to prove the truth of Oscar Wilde's statement 'Yet each man kills the thing he loves'. Destruction and fragmentation of the forest through road-building and house construction, soil erosion, introduction of feral plants and animals, and continuing disturbance, are all likely results of real-estate development. It is particularly regrettable that land sales have been occurring in the lowland tropical rainforest of north-east Queensland, threatening one of the richest and rarest rainforest types in Australia. Unfortunately, the World Heritage Listing of the north-east Queensland rainforests does not provide an avenue for curtailing these activities.

THE HUMAN INFLUENCE

9.9 AUSTRALIAN RAINFOREST— A PARADIGM FOR WORLD RAINFOREST?

It is now accepted that tropical rainforest is not unchanging, and neither has it enjoyed a stable environment over millions of years. Rainforest has experienced environmental changes as great as those in temperate latitudes (Whitmore and Prance 1987; Flenley 1979). The history of environmental change and its effects on vegetation, although still imperfectly understood, have

Fig. 9.16 Advertisement for the sale of rainforested land in north-east Queensland (courtesy of Quaid Real Estate).

Fig. 9.17 A dense stand of *Livistona australis* (cabbage tree palm) at Pacific Palms, New South Wales.

Fig. 9.18 Land near that illustrated in Fig. 9.17 following clearing for property development—prior to clearing, it was dominated by *Livistona*, some individuals have been left to form part of the landscaping.

been better established in Australia than elsewhere, and the picture of differentiation in response to variations in climate, soil type, and fire would, if painted with a broad enough brush, serve as more general account. What is particularly striking in Australia is that it is possible to read the present variation in rainforest, not only as an expression of the interactions between the contemporary environment and the flora, but as a series of chapters which recapitulate the essential features of some 70 million years of rainforest history.

There are some features of Australian rainforest which are particularly distinctive—most notably the relationships, both evolutionary and ecological (particularly the role of fire), between rainforest and the currently more widespread sclerophyll vegetation. Although these relationships are fundamentally not very different from those between rainforest and other vegetation types elsewhere in the world, what makes them so striking are the special features of the sclerophyll vegetation. Why have so few elements in the rainforest contributed to the development of the sclerophyll vegetation, and what factors led to the development of the ecological characteristics of this quintessentially Australian vegetation type? These questions are, as yet, only partially answered.

One of the striking features of the most recent history of Australian rainforests has been the strength of public interest in their conservation. Public opinion has led to political action, overturning long-established land-use practices. Environmental issues are now perceived by politicians across the whole spectrum of politics as being high on the agenda of government. Will the issue of rainforest conservation be the catalyst for change in policies and attitudes of governments elsewhere?

REFERENCES

Abdulhadi, R. and Lamb, D. (1988). Soil seed stores in a rainforest succession. *Proceedings of the Ecological Society of Australia* **15**, 81–7.

Adam, P. (1987*a*). Australian deserts defended. *Nature* **325**, 570.

Adam, P. (1987*b*). *New South Wales rainforests. The nomination for the World Heritage List*. National Parks and Wildlife Service, Sydney.

Adam, P. (1988). *A sunburnt country. Europeans and the Australian environment*. University of Durham, Durham.

Adam, P. (1990). *Saltmarsh ecology*. Cambridge University Press, Cambridge.

Adam, P. and King, R. J. (1990). Ecology of unconsolidated shores. In *Biology of marine plants* (ed. M. N. Clayton and R. J. King), pp. 296–309. Longman Cheshire, Sydney.

Adam, P., Stricker, P., and Anderson, D. J. (1989). Species richness and soil phosphorus in plant communities in coastal New South Wales. *Australian Journal of Ecology* **14**, 189–98.

Adam, P., Urwin, N., Weiner, P., and Sim, I. (1985). *Coastal wetlands of New South Wales*. Coastal Council of New South Wales, Sydney.

Adam, P., Wilson, N. C., and Huntley, B. (1988). The phytosociology of coastal saltmarsh vegetation in New South Wales. *Wetlands (Australia)* **7**, 35–85.

Adamson, D., Selkirk, P. M., and Mitchell, P. (1983). The role of fire and lyre birds in the sandstone landscape of the Sydney Basin. In *Aspects of Australian sandstone landscapes* (eds. R. W. Young and G. C. Nanson), pp. 81–93. Australian and New Zealand Geomorphology Group Special Publication.

Allaway, W. G. (1987). Exploitation and destruction of mangroves in Australia. In *Mangrove ecosystems of Asia and the Pacific. Status, exploitation and management* (eds. C. D. Field and A. J. Dartnall), pp. 183–92. AIMS, Townsville, Queensland.

Allaway, W. G. and Ashford, A. E. (1984). Nutrient inputs by seabirds to the forest on a coral island off the Great Barrier Reef. *Marine Ecology—Progress Series* **19**, 297–8.

Allen, A. D. (1966). *Geology of the Montague Sound 1:250,000 sheet area (SD/51-12), Western Australia*. Bureau of Mineral Resources, Australian Records 1966/20.

Allen, A. D. (1971). *1:250 000 Geological Series—explanatory notes. Montague Sound Western Australia Sheet SD/51-12*. Bureau of Mineral Resources, Geology and Geophysics, Canberra.

Altena, A. C. van and Harvey, A. M. (1979). Control of stinging trees (*Dendrocnide* spp.: Urticaceae) in north Queensland by high-volume foliar spraying. *Research Note. Department of Forestry Queensland* **27**, 1–7.

Andrews, E. C. (1916). The geological history of the Australian flowering plants. *American Journal of Science* **42**, 171–232.

Aplin, K. P. and Archer, M. (1987). Recent advances in marsupial systematics with a new syncretic classification. In *Possums and opossums: studies in evolution* (ed. M. Archer), pp. xv–lxxii, Surrey Beatty, Sydney.

Applegate, G. B. and Bragg, A. L. (1989). Improved growth rates of red cedar (*Toona australis* (F. Muell.) Harms) seedlings in growth tubes in north Queensland. *Australian Forestry* **52**, 293–7.

Archer, M. and Hand, S. J. (1987). Evolutionary considerations. In *Koala, Australia's endearing marsupial* (ed. L. Cronin), pp. 79–106. Reed, Sydney.

Archer, M., Every, R. G., Godthelp, H., Hand, S. J., and Scally, K. B. (1990). *Yingabalanaridae*, a new family of enigmatic mammals from Tertiary deposits of Riversleigh, northwestern Queensland. *Memoirs of the Queensland Museum*, **25**, 193–202.

Archer, M., Flannery, T. F., Ritchie, A., and Molnar, R. E. (1985). First Mesozoic mammal from Australia—an early Cretaceous monotreme. *Nature* **318**, 363–6.

Archer, M., Godthelp, H., Hand, S. J., and Megirian, D. (1989*a*). Fossil mammals of Riversleigh, northwestern Queensland: preliminary overview of biostratigraphy, correlation and environmental change. *Australian Zoologist* **25**, 29–65.

Archer, M., Hand, S. J., and Godthelp, H. (1988). A new order of Tertiary zalambdodont marsupials. *Science* **239**, 1528–31.

Archer, M., Hand, S. J., and Godthelp, H. (1989*b*). Ghosts from green gardens. *Riversleigh Notes* **7**, 4–7.

Armesto, J. J. and Rozzi, R. (1989). Seed dispersal syndromes in the rainforest of Chiloé: evidence for the importance of biotic dispersal in a temperate rain forest. *Journal of Biogeography* **16**, 219–26.

Armstrong, J. A. (1979). Biotic pollination mechanisms in the Australian flora—a review. *New Zealand Journal of Botany* **17**, 467–508.

Armstrong, P. (1979). Biogeography: themes and case-studies. In *Western landscapes* (ed. J. Gentilli), pp. 88–105, University of W. A. Press, Perth.

Ash, J. (1983a). Growth rings in *Agathis robusta* and *Araucaria cunninghamii* from tropical Australia. *Australian Journal of Botany* **31**, 269–75.

Ash, J. (1983b). Tree rings in tropical *Callitris macleayana* F. Muell. *Australian Journal of Botany* **31**, 277–81.

Ash, J. (1988). The location and stability of rainforest boundaries in north-eastern Queensland, Australia. *Journal of Biogeography* **15**, 619–30.

Ashton, D. H. (1981a). Tall open-forest. In *Australian vegetation* (ed. R. H. Groves), pp. 121–51. Cambridge University Press, Cambridge.

Ashton, D. H. (1981b). Fire in tall open-forests (wet sclerophyll forests). In *Fire and the Australian biota* (eds. A. M. Gill, R. H. Groves, and I. R. Noble), pp. 339–66. Australian Academy of Science, Canberra.

Ashton, D. H. and Frankenberg, J. (1976). Ecological studies of *Acmena smithii* (Poir.) Merrill & Perry with special reference to Wilson's Promontory. *Australian Journal of Botany* **24**, 453–87.

Ashton, P. S. (1969). Speciation among tropical forest trees: some deductions in the light of recent evidence. *Biological Journal of the Linnean Society, London* **1**, 155–96.

Atkinson, J. (1826). *An account of the state of agriculture and grazing in New South Wales*. J. Cross, London.

Audley-Charles, M. G. (1987). Dispersal of Gondwanaland: relevance to evolution of the angiosperms. In *Biogeographical evolution of the Malay archipelago* (ed. T. C. Whitmore), pp. 5–25, Oxford University Press, Oxford.

Augspurger, C. K. (1984). Seedling survival of tropical tree species: interactions of dispersal distance, light-gaps and pathogens. *Ecology* **65**, 1705–12.

Augspurger, C. K. and Kelly, C. K. (1984). Pathogen mortality of tropical tree seedlings: experimental studies of the effects of dispersal distances, seedling density and light conditions. *Oecologia, (Berlin)* **61**, 211–17.

Bailey, I. W. and Sinnott, F. W. (1916). The climatic distribution of certain types of angiosperm leaves. *American Journal of Botany* **3**, 24–39.

Balgooy, M. M. J. von (1987). A plant geographical analysis of Sulawesi. In *Biogeographical evolution of the Malay Archipelago* (ed. T. C. Whitmore) pp. 94–102. Oxford University Press, Oxford.

Ball, M. C. (1988). Ecophysiology of mangroves. *Trees*, **2**, 129–42.

Bardsley, K. (1985). The effects of Cyclone Kathy on mangrove vegetation. In *Coasts and tidal wetlands of the Australian monsoon region* (eds. K. N. Bardsley, J. D. S. Davie and C. D. Woodroffe), pp. 167–85. ANU Northern Australia Research Unit, Darwin.

Barlow, B. A. (1981). The Australian flora: its origin and evolution. In *Flora of Australia. Vol. I. Introduction.* (ed. A. S. George), pp. 25–75. Australian Government Publishing Service, Canberra.

Barlow, B. A. (1988). Patterns of differentiation in tropical species of *Melaleuca* L. (Myrtaceae). *Proceedings of the Ecological Society of Australia* **15**, 239–47.

Barlow, B. A. and Hyland, B. P. M. (1988). The origins of the flora of Australia's wet tropics. *Proceedings of the Ecological Society of Australia* **15**, 1–17.

Baur, G. N. (1957). Nature and distribution of rainforests in New South Wales. *Australian Journal of Botany* **5**, 190–233.

Baur, G. N. (1962a). *Forest vegetation in north-eastern New South Wales*. Forestry Commission of NSW Research Note 8. Forestry Commission of NSW, Sydney.

Baur, G. N. (1962b). *Silvicultural practices in rainforests of northern New South Wales*. Forestry Commission of NSW Research Note 9. Forestry Commission of NSW, Sydney.

Baur, G. N. (1965). *Forest types in New South Wales. Forestry Commission of NSW Research Note 17.* Forestry Commission of NSW, Sydney.

Baur, G. N. (1968). *The ecological basis of rainforest management*. Forestry Commission of NSW, Sydney.

Baur, G. N. (1977). Rainforests in the service of man. *Parks and Wildlife* **2**, 18–24.

Baur, G. N. (1989). *Notes on the silviculture of major N.S.W. types 11. Rainforest types*. Forestry Commission of NSW, Sydney.

Bawa, K. S. (1979). Breeding systems of trees in a tropical wet forest. *New Zealand Journal of Botany* **17**, 521–4.

Beadle, N. C. W. (1954). Soil phosphate and the delimitation of plant communities in eastern Australia. I. *Ecology* **35**, 370–5.

Beadle, N. C. W. (1962). Soil phosphate and the delimitation of plant communities in eastern Australia. II. *Ecology* **43**, 281–8.

Beadle, N. C. W. (1966). Soil phosphate and its role in moulding segments of the Australian flora and vegetation with special reference to xeromorphy and sclerophylly. *Ecology* **47**, 991–1007.

Beadle, N. C. W. (1981). *The vegetation of Australia*. Fischer, Stuttgart.

Beadle, N. C. W. and Costin, A. B. (1952). Ecological classification and nomenclature. *Proceedings of the Linnean Society of NSW* **77**, 61–82.

Beard, J. S. (1955). The classification of tropical American vegetation-types. *Ecology* **36**, 89–100.

Beard, J. S. (1967). An inland occurrence of mangrove. *West Australian Naturalist* **10**, 113–15.

Beard, J. S. (1976). The monsoon forests of the Admiralty Gulf, Western Australia. *Vegetatio* **31**, 177–92.

Beard, J. S., Clayton-Greene, K. A., and Kenneally, K. F. (1984). Notes on the vegetation of the Bougainville Peninsula, Osborn and Institut Islands, North Kimberley district, Western Australia. *Vegetatio* **57**, 3–13.

Beckmann, R. (1987a). Myrtles, *Platypus* and fungi. *Ecos* **51**, 18–20.

Beckmann, R. (1987b). Birds—the Gondwanan connection. *Ecos* **52**, 23–5.

Beckmann, R. (1988). Trees may aid the AIDS fight. *Ecos* **56**, 20–2.

Beer, T., Gill, A. M., and Moore, P. H. R. (1988). Australian bushfire danger under changing climatic regimes. In *Greenhouse—planning for climate change*. (ed. G. I. Pearman) pp. 421–7. CSIRO, Melbourne.

Bell, A. (1983). Fire and rainforest in Tasmania. *Ecos* **37**, 3–8.

Bell, F. C. (1981). Conserving the total resource potential of Australian rainforests. In *Proceedings of the rainforest conference* (eds. J. H. Shaw and S. J. Riley), pp. 38–45. Geographical Society of NSW, Sydney.

Bell, F. C., Winter, J. W., Pahl, L. I., and Atherton, R. B. (1987). Distribution, area and tenure of rainforest in northeastern Australia. *Proceedings of the Royal Society of Queensland* **98**, 27–39.

Bell, H. L. (1984). A bird community of lowland rainforest in New Guinea. 6. Foraging ecology and community structure of the avifauna. *Emu* **84**, 142–58.

Bell, J. D., Pollard, D. A., Burchmore, J. J., Pease, B. C., and Middleton, M. J. (1984). Structure of a fish community in a temperate tidal mangrove creek in Botany Bay, New South Wales. *Australian Journal of Marine and Freshwater Research* **35**, 33–46.

Belshaw, J. (1978). Population distribution and the pattern of seasonal movement in northern New South Wales. In *Records of times past. Ethnohistorical essays on the culture and ecology of the New England tribes* (ed. I. McBryde) pp. 65–81. Australian Institute of Aboriginal Studies, Canberra.

Bevege, D. I. (1968). Inhibition of seedling hoop pine (*Araucaria cunninghamii* Ait.) on forest soils by phytotoxic substances from the root zones of *Pinus*, *Araucaria* and *Flindersia*. *Plant and Soil* **24**, 263–73.

Bigwood, A. J. and Hill, R. S. (1985). Tertiary araucarian macrofossils from Tasmania. *Australian Journal of Botany* **33**, 645–56.

Bird, J. (1981). Barilla production in Australia. In *Plants and man in Australia*. (eds. D. J. Carr and S. M. Carr), pp. 274–80. Academic Press, Sydney.

Birtles, T. G. (1982). Trees to burn: settlement in the Atherton–Evelyn rainforest, 1880–1900. *North Australia Research Bulletin* **8**, 31–86.

Bishop, M. J. (1981). The biogeography and evolution of Australian land snails. In *Ecological biogeography of Australia* (ed. A. Keast), pp. 923–54. Junk, The Hague.

Bishop, P. and Bamber, R. K. (1985). Silicified wood of early Miocene *Nothofagus*, *Acacia* and Myrtaceae (aff. *Eucalyptus* B) from the Upper Lachlan Valley, New South Wales. *Alcheringa* **9**, 221–8.

Blainey, G. (1954). *The hills of Lyell*. Melbourne University Press, Melbourne.

Boles, W. (1991). Riversleigh researchers, Part 1—Walter Boles discusses his work on the Riversleigh birds. *Riversleigh Notes* **13**, 2–4.

Bolton, G. C. (1963). *A thousand miles away. A history of north Queensland to 1920*. ANU Press, Canberra.

Bolton, G. C. (1981). *Spoils and spoilers. Australians make their environment 1788–1980*. Allen and Unwin, Sydney.

Borschmann, G. (1984). *Greater Daintree. World Heritage tropical rainforest at risk*. ACF, Melbourne.

Boutland, A. (1989). Review paper: forests and aboriginal society. In *Australia's ever changing forests* (ed. K. J. Frawley and N. Semple), pp. 143–68. Australian Defence Force Academy, Canberra.

Bowdler, S. (1983). Rainforest: colonised or coloniser? *Australian Archaeology* **17**, 59–66.

Bowler, J. M. (1982). Aridity in the late Tertiary and Quaternary of Australia. In *Evolution of the flora and fauna of arid Australia*. (eds. W. R. Barker and P. J. M. Greenslade), pp. 35–45. Peacock Press, Frewville, South Australia.

Bowman, D. M. J. S. (1988). Stability amidst turmoil?: towards an ecology of north Australian eucalypt forests. *Proceedings of the Ecological Society of Australia* **15**, 149–58.

Bowman, D. M. J. S. and Dunlop, C. R. (1986). Vegetation patterns and environmental correlates in coastal forests of the Australian monsoon tropics. *Vegetatio* **65**, 99–104.

Bowman, D. M. J. S. and Wilson, B. A. (1988). Fuel characteristics of coastal monsoon forests, Northern Territory, Australia. *Journal of Biogeography* **15**, 807–17.

Bowman, D. M. J. S., Wilson, B. A., and Fensham, R. J. (1990). Sandstone vegetation patterns in the Jim Jim Falls region, Northern Territory, Australia. *Journal of Ecology*, **15**, 163–74.

Boyer, P. and Hickey, J. (1987). *The rainforest of Tasmania*. Tasmanian Government Printer, Hobart.

Braithwaite, R. W., Dubzinski, M. L., Ridpath, M. G., and Parker B. S. (1984). The impact of water buffalo on the monsoon forest ecosystem in Kakadu National Park. *Australian Journal of Ecology* **9**, 309–22.

Breeden, S. and Breeden, K. (1970). *Tropical Queensland*. Collins, Sydney.

Brennan, K. (1986). *Wildflowers of Kakadu*. Published by the author, Jabiru, Northern Territory.

Brennan, P. F. (1988). Forest history of the continental islands, Great Barrier Reef. In *Changing tropical forests. Historical perspectives on today's challenges in Asia, Australasia and Oceania*. (eds. J. Dargavel, K. Dixon, and N. Semple), pp. 327–40, CRES, Canberra.

Bridgewater, P. B. (1985). Variation in the mangal along the west coastline of Australia. *Proceedings of the Ecological Society of Australia* **13**, 243–56.

Bridgewater, P. B. (1987). The present Australian environment—terrestrial and freshwater. In *Fauna of Australia* Vol. 1A. (eds. G. R. Dyne and D. W. Walton), pp. 69–100. Australian Government Publishing Service, Canberra.

Briggs, D. and Walters, S. M. (1984). *Plant variation and evolution*, (2nd edn.). Cambridge University Press, Cambridge.

Broadbent, J. and Clark, S. (1976). *A faunal survey of East Australian rainforests: interim report*. Australian Museum, Sydney.

Brock, J. (1988). *Top End native plants*. Published by the author, Winnellie, Northern Territory.

Brown, B. N. (1976). *Phytophthora cinnamomi* associated with patch death in tropical rainforests in Queensland. *Australian Plant Pathology Society Newsletter* **5**, 1–4.

Brown, M. J. and Podger, F. D. (1982). Floristics and fire regimes of a vegetation sequence from sedgeland-heath to rainforest at Bathurst Harbour, Tasmania. *Australian Journal of Botany* **30**, 659–76.

Buckley, R. C. (1982). Patterns in north Queensland mangrove vegetation. *Australian Journal of Ecology* **7**, 103–6.

Bundock, M. (1898). Notes on the Richmond River blacks. Unpublished MS (Bundock Papers A6939), Mitchell Library, Sydney.

Bunt, J. S. and Williams, W. T. (1980). Studies in the analysis of data from Australian tidal forests ('Mangroves'). I. Vegetational sequences and their graphic representation. *Australian Journal of Ecology* **5**, 385–90.

Bunt, J. S., Williams, W. T., and Clay, H. J. (1982*a*). River water salinity and the distribution of mangrove species along several rivers in North Queensland. *Australian Journal of Botany* **30**, 401–12.

Bunt, J. S., Williams, W. T., and Duke, N. C. (1982*b*). Mangrove distributions in north-east Australia. *Journal of Biogeography* **9**, 111–20.

Burbidge, N. T. (1960). The phytogeography of the Australian region. *Australian Journal of Botany* **8**, 75–212.

Bureau of Meteorology (1977). *Climatic Atlas of Australia; Map Set 5, Rainfall*. Australian Government Publishing Service, Canberra.

Bureau of Meteorology (1989). *Climate of Australia*. Australian Government Publishing Service, Canberra.

Burges, A. and Johnston, R. D. (1953). The structure of a New South Wales rain forest. *Journal of Ecology* **41**, 72–83.

Burgess, I. P., Floyd, A., Kikkawa, J., and Pattemore, V. (1975). Recent developments in the silviculture and management of sub-tropical rainforest in New South Wales. *Proceedings of the Ecological Society of Australia* **9**, 74–84.

Busby, J. R. (1984). *Nothofagus cunninghamii* (Southern Beech) vegetation in Australia. *Australian Flora and Fauna Series* **1**, 1–69.

Busby, J. R. (1987). Floristic communities and bioclimates of *Nothofagus cunninghamii* rainforest in south-eastern Australia. In *The rainforest legacy. Australian national rainforests study.* Vol. 1. (eds. G. Werren and A. P. Kershaw), pp. 23–31. Australian Government Publishing Service, Canberra.

Busby, J. R. (1988). Potential impacts of climate change on Australia's flora and fauna. In *Greenhouse—planning for climate change.* (ed. G. I. Pearman) pp. 387–98. CSIRO, Melbourne.

Busby, J. R. and Bridgewater, P. B. (1977). Studies in Victorian vegetation II. A floristic survey of the vegetation associated with *Nothofagus cunninghamii* (Hook.) Oerst. in Victoria and Tasmania. *Proceedings of the Royal Society of Victoria* **89**, 173–82.

Byrne, D. (1987). *The Aboriginal and archaeological significance of the New South Wales rainforests*. Forestry Commission of New South Wales, Sydney.

Calaby, J. H. (1966). *Mammals of the upper Richmond and Clarence Rivers, New South Wales.* CSIRO Division of Wildlife Research Technical Paper 10.

Cambage, R. H. (1914). The mountains of eastern Australia and their effect on the native vegetation. *Proceedings of the Royal Society of New South Wales* **48**, 267–80.

Cameron, D. G. (1987). Temperate rainforests of East Gippsland. In *The rainforest legacy. Australian national rainforests study. Vol. 1* (ed. G. Werren and A. P. Kershaw), pp. 33–46. AGPS, Canberra.

Cameron, J. I. and Penna, I. W. (1988). *The wood and the trees*. ACF, Melbourne.

Carron, L. T. (1985). *A history of forestry in Australia*. ANU Press, Canberra.

Carter, H. J. (1933). *Gulliver in the bush. Wanderings of an Australian entomologist*. Angus and Robertson, Sydney.

Cassells, D. S., Bonell, M., Gilmour, D. A., and Valentine, P. S. (1988). Conservation and management of Australia's tropical rainforests: local realities and global responsibilities. *Proceedings of the Ecological Society of Australia* **15**, 313–26.

Chapman, V. J. (1976). *Mangrove vegetation*. Cramer, Vaduz, Liechtenstein.

Chenery, E. M. and Sporne, K. R. (1976). A note on the evolutionary status of aluminium accumulators among dicotyledons. *New Phytologist* **76**, 551–4.

Chesterfield, E. A., Taylor, S. J., and Molnar, C. D. (1990). Recovery after wildfire: warm temperate rainforest at Jones Creek. *Arthur Rylah Institute for Environmental Research. Technical Report Series* **101**.

Christian, C. S. and Aldrick, J. M. (1977). *Alligator Rivers study*. Australian Government Publishing Service, Canberra.

Christidis, L. (1987). Phylogeny and systematics of estrildine finches and their relationships to other seed-eating passerines. *Emu* **87**, 119–23.

Christophel, D. C. (1981). Tertiary megafossil floras of Australia as indicators of floristic associations and palaeoclimate. In *Ecological Biogeography of Australia* (ed. A. Keast), pp. 377–90. Junk, The Hague.

Christophel, D. C. and Greenwood, D. R. (1988). A comparison of Australian tropical rainforest and Tertiary fossil leaf-beds. *Proceedings of the Ecological Society of Australia* **15**, 139–48.

Churchill, D. M. (1973). The ecological significance of tropical mangroves in the early Tertiary flora of Southern Australia. *Geological Society of Australia Special Publication* **4**, 79–86.

Clark, R. L. and Guppy, J. C. (1988). A transition from mangrove forest to freshwater wetland in the monsoon tropics of Australia. *Journal of Biogeography* **15**, 665–84.

Clark, R. V. (1980). Rainforest maintenance and regeneration for landholders. In *Reforestation*. (eds. R. V. Clark, D. Hume, and J. Seed) pp. 42–9, Northern Rivers reforestation seminar committee, Ballina.

Clark, R. V., Hume, D., and Seed, J. (eds.) (1980). *Reforestation*. Northern Rivers reforestation seminar committee, Ballina, New South Wales.

Clarkson, J. R. and Kenneally, K. F. (1988). The floras of Cape York and the Kimberley: a preliminary comparative analysis. *Proceedings of the Ecological Society of Australia* **15**, 259–66.

Clayton-Greene, K. A. and Beard, J. S. (1985). The fire factor in vine thicket and woodland vegetation of the Admiralty Gulf region, north-west Kimberley, Western Australia. *Proceedings of the Ecological Society of Australia* **13**, 225–30.

Clements, A. M. (1983). Suburban development and resultant changes in the vegetation of the bushland of the northern Sydney Region. *Australian Journal of Ecology* **8**, 307–19.

Clough, B. F. (ed.) (1982). *Mangrove ecosystems in Australia. Structure, function and management.* AIMS/ANU Press, Canberra.

Cogger, H. G. (1977). Reptiles and frogs. *Parks and Wildlife* **2**, 65–6.

Cogger, H. G. and Heatwole, H. (1981). The Australian reptiles: origins, biogeography, distribution patterns and island evolution. In *Ecological biogeography of Australia* (ed. A. Keast), pp. 1331–71. Junk, The Hague.

Cohen, B. (1984). Letter. *Habitat* **12**, (4), 29–30.

Collins, N. M. (1988). Termites. In *Malaysia* (ed. Earl of Cranbrook), pp. 196–211. Pergamon Press, Oxford.

Colong Committee (1983). *How the rainforest was saved*. The Colong Committee, Sydney.

Congdon, R. A. and Lamb, D. (1990). Essential nutrient cycles. In *Australian Tropical Rainforests Science—values—meaning* (ed. L. J. Webb and J. Kikkawa), pp. 105–13. CSIRO, Melbourne.

Connell, J. H. (1978). Diversity in tropical rainforests and coral reefs. *Science* **199**, 1302–10.

Connell, J. H. (1979). Tropical rain forests and coral reefs as open non-equilibrium systems. In *Population dynamics* (eds. R. Anderson, B. Turner, and L. Taylor), pp. 149–63. Blackwell, Oxford.

Connell, J. H., Tracey, J. G., and Webb, L. J. (1984). Compensatory recruitment, growth and mortality as factors maintaining rain forest tree diversity. *Ecological Monographs* **54**, 141–64.

Coper, M. (1983). *The Franklin Dam case. Commentary and full text of the decision in Commonwealth of Australia v State of Tasmania*. Butterworths, Sydney.

Corlett, R. T. (1988). Bukit Timah: the history and significance of a small rain-forest reserve. *Environmental Conservation* **15**, 37–44.

Corner, E. J. H. (1954). The evolution of tropical forest. In *Evolution as a process* (eds. J. S. Huxley, A. C. Hardy, and E. B. Ford), pp. 34–46. Allen and Unwin, London.

Costin, A. B. (1983). Mountain lands in the Australian region: some principles of use and management. *Proceedings of the Ecological Society of Australia* **12**, 1–13.

Cremer, K. W. (1960). Eucalypts in rain forest. *Australian Forestry* **24**, 120–6.

Cribb, A. (1987). Eating up the past. *Landscope* **3**, (1), 44–8.

Cribb, A. B. and Cribb, J. W. (1974). *Wild food in Australia*. Collins, Sydney.

Cribb, A. B. and Cribb, J. W. (1985). *Plant life of the Great Barrier Reef and adjacent shores*. University of Queensland Press, St Lucia.

Crome, F. H. J. (1975). The ecology of fruit pigeons in tropical northern Queensland. *Australian Wildlife Research* **2**, 155–85.

Crome, F. H. J. (1978). Foraging ecology of an assemblage of birds in lowland rainforest in northern Queensland. *Australian Journal of Ecology* **3**, 195–212.

Crome, F. H. J. (1990). Vertebrates and successions. In *Australian tropical rainforests. Science—values—meaning*. (eds. L. J. Webb, and J. Kikkawa), pp. 53–54, CSIRO, Melbourne.

Crome, F. H. J. and Irvine, A. K. (1986). 'Two bob each way': the pollination and breeding system of the Australian rain forest tree *Syzygium cormiflorum* (Myrtaceae). *Biotropica* **18**, 115–25.

Crome, F. H. J. and Moore, L. A. (1989). Display site constancy of bowerbirds and the effects of logging on Mt. Windsor Tableland, north Queensland. *Emu* **89**, 47–52.

Crome, F. H. J. and Moore, L. A. (1990). Cassowaries in north-eastern Queensland: report of a survey and a review and assessment of their status and conservation and management needs. *Australian Wildlife Research* **17**, 369–85.

Crome, F. H. J. and Richards, G. C. (1988). Bats and gaps: microchiropteran community structure in a Queensland rain forest. *Ecology* **69**, 1960–9.

Cullen, P. J. (1987). Regeneration patterns in populations of *Athrotaxis selaginoides* D.Don. from Tasmania. *Journal of Biogeography* **14**, 39–51.

Dale, J. A. (1983). Management studies in the escarpment rainforests of south east Queensland. *Research Paper. Department of Forestry Queensland* **14**, 1–90.

Dale, J. A. (1985). Training of forestry personnel for the control and management of harvesting in Queensland. In *Managing the tropical forest* (eds. K. R. Shepherd and H. V. Richter), pp. 187–96. Development Studies Centre, Australian National University, Canberra.

Dale, M. (1981). Response to Felton (1981). *Bulletin of the Ecological Society of Australia Inc.* **11**, (1), 3–4.

Dale, M., Kershaw, P., Kikkawa, J., Parsons, P., and Webb, L. (1980). Resolution by the Ecological Society of Australia on Australian rainforest conservation. *Bulletin of the Ecological Society of Australia Inc.* **10**, (3), 6–7.

Dale, M. B. and Webb, L. J. (1975). Numerical methods for the establishment of associations. *Vegetatio* **30**, 77–87.

Dale, P., Hulsman, K., Jahnke, B. R., and Dale, M. (1984). Vegetation and nesting preferences of black noddies at Masthead Island, Great Barrier Reef. I. Patterns at the macro-scale. *Australian Journal of Ecology* **9**, 335–41.

Daley, L. T. (1966). *Men and a river. Richmond River district 1828–1895*. Melbourne University Press, Melbourne.

Dargavel, J. (1988). Changing capital structure, the State, and Tasmanian forestry. In *World deforestation in the twentieth century* (ed. J. F. Richards and R. P. Tucker), pp. 189–210. Duke University Press, Durham, North Carolina.

Dawson, J. W. (1986). Floristic relationships of lowland rainforest phanerograms of New Zealand. *Telopea* **2**, 681–95.

Department of Arts, Heritage and the Environment (1986). Forests and woodlands. In *State of the environment in Australia: source book*. pp. 103–35. Australian Government Publishing Service, Canberra.

Department of Conservation, Forests and Lands (1987). *Victoria's rainforests: an overview*. Department of Conservation Forests and Lands, Melbourne.

Department of Environment and Planning (1982). *Proposed forestry operations in the Washpool area. Environmental Impact Assessment*. Department of Environment and Planning, Sydney.

Department of Environment and Planning (1983). *New South Wales Government rainforest policy 1982*. Department of Environment and Planning, Sydney.

Department of Environment and Planning (1984). *New South Wales Government rainforest policy—further progress 1984*. Department of Environment and Planning, Sydney.

Department of Environment and Planning (1988). *State Environmental Planning Policy 26—Littoral Rainforests*. Department of Environment and Planning, Sydney.

Department of Forestry, Queensland (1981). *Timber production from North Queensland's rainforests. A position paper*. Department of Forestry, Queensland, Brisbane.

Department of Forestry, Queensland (1983). *Rainforest research in north Queensland*. Department of Forestry, Queensland, Brisbane.

Department of the Arts, Sport, the Environment, Tourism and Territories (DASETT) (1987). *Nomination of wet tropical rainforests of north-east Australia by the Government of Australia for inclusion in the World Heritage List*. DASETT, Canberra.

Department of the Arts, Sport, the Environment, Tourism and Territories (DASETT) (1988). *Wet tropical rainforests of north-east Australia. Future management arrangements: a discussion paper*. DASETT, Canberra.

Dettmann, M. E. (1989). Antarctica: Cretaceous cradle of austral temperate rainforests? In *Origins and evolution of the Antarctic biota* (ed. J. A. Crame), pp. 89–105. Geological Society, Bath.

Dettmann, M. E., Pocknall, D. T., Romero, E. J., and Zamaloa, M. de C. (1990). *Nothofagidites* Erdtman ex Potonié, 1960; a catalogue of species with notes on the palaeogeographic distribution of *Nothofagus* Bl. (Southern Beech). *New Zealand Geological Survey Palaeontological Bulletin* **60**, 1–79.

Diels, L. (1906). Die Pflanzenwelt von West Australien. *Die Vegetation der Erde VII*. Engleman, Leipzig, Germany.

Division of National Mapping (1975). *Australia 1:250,000 Map series Gazetter*. Australian Government Publishing Service, Canberra.

Donohue, J. T. (1984). Land administration in the brigalow belt. In *The brigalow belt of Australia* (ed. A. Bailey), pp. 257–61. Royal Society of Queensland, Brisbane.

Duke, N. C. (1988). An endemic mangrove species, *Avicennia integra* sp. nov. (Avicenniaceae), in northern Australia. *Australian Systematic Botany* **1**, 177–88.

Elliott, H. J., Bashford, R., and Palzer, C. (1981). Defoliation of dry sclerophyll eucalypt forest by *Stathmorrhopa aphotista* Turner (Lepidoptera: Geometridae) in south Tasmania. In *Eucalypt dieback in forests and woodlands*. (eds. K. M. Old, G. A. Kile, and C. P. Ohmart), pp. 134–9. CSIRO, Melbourne.

Elliot, H. J., Kile, G. A., Candy, S. G., and Ratkowsky, D. A. (1987). The incidence and spatial pattern of *Nothofagus cunninghamii* (Hook.) Oerst. attacked by *Platypus subgranosus* Schedl. in Tasmania's cool temperate rainforest. *Australian Journal of Ecology* **12**, 125–38.

Ellis, R. C. (1971). Rainfall, fog drip and evaporation in a mountainous area of southern Australia. *Australian Forestry* **35**, 99–106.

Ellis, R. C. (1985). The relationships among eucalypt forest, grassland and rainforest in a highland area in north-eastern Tasmania. *Australian Journal of Ecology* **10**, 297–314.

Embleton, B. J. J. (1984). Past global settings. (a) Continental palaeomagnetism. In *Phanerozoic earth history of Australia* (ed. J. J. Veevers), pp. 11–16, Oxford University Press, Oxford.

Evans, J. W. (1981). A review of the present knowledge of the family Peloridae and new genera and new species from New Zealand and New Caledonia (Hemiptera: Insecta). *Records of the Australian Museum* **34**, 381–406.

Fedorov, A. A. (1966). The structure of the tropical rain forest and speciation in the humid tropics. *Journal of Ecology* **54**, 1–11.

Felton, K. (1981). ESA Rainforest resolution. Some feedback. *Bulletin of the Ecological Society of Australia Inc*. **11**, (1), 2–3.

Field, C. D. and Dartnall, A. J. (eds.) (1987). *Mangrove ecosystems of Asia and the Pacific. Status exploitation and management*. AIMS, Townsville, Australia.

Figgis, P. (ed.) (1985). *Rainforests of Australia*. Weldon, Sydney.

Figgis, P. (1988*a*). The wet tropics. In *Australia's wilderness heritage*, Vol. 1 (eds. P. Figgis and G. Mosley), pp. 166–85. Weldon, Sydney.

Figgis, P. (1988*b*). The Great Sandy Region. In *Australia's Wilderness Heritage*, Vol. 1 (eds. P. Figgis and G. Mosley), pp. 188–205. Weldon, Sydney.

Fisher, H. J. (1985). The structure and floristic composition of the rainforest of the Liverpool Range, New South Wales and its relationships with other Australian rainforests. *Australian Journal of Ecology* **10**, 315–23.

Flannery, T. (1987). The mountain diprotodontid—*Hulitherium thomasettii*. In *The antipodean ark* (ed. M. Archer and S. Hand), pp. 65–7. Angus and Robertson, North Ryde, Australia.

Flannery, T. and Conlon, T. (1989). A vaccine for the plague. *Australian Natural History* **23**, 148–55.

Flannery, T. F. and Plane, M. D. (1986). A new late Pleistocene diprotodontid (Marsupialia) from Pureni, Southern Highlands Province, Papua New Guinea. *BMR Journal of Geology and Geophysics, Australasia* **10**, 65–76.

Flenley, J. R. (1979). *The equatorial rain forest: a geological history*. Butterworths, London.

Flood, J. (1983). *Archaeology of the Dreamtime*. Collins, Sydney.

Florence, R. G. (1963). Vegetational pattern in east coast forests. *Proceedings of the Linnean Society of New South Wales* **88**, 164–79.

Florence, R. G. (1964). Edaphic control of vegetational pattern in east coast forests. *Proceedings of the Linnean Society of New South Wales* **89**, 171–90.

Floyd, A. G. (1976). Effect of burning on regeneration from seeds in wet sclerophyll forest. *Australian Forestry* **39**, 210–20.

Floyd, A. G. (1977). Regeneration. In *Rain forests* (ed. W. Goldstein), pp. 50–2. National Parks and Wildlife Service, Sydney.

Floyd, A. G. (1984). Management of small rainforest areas. In *Small natural areas—their conservation and management*, pp. 191–9. The National Trust of Australia (NSW), Sydney.

Floyd, A. G. (1987). Status of rainforest in northern New South Wales. In *The rainforest legacy. Australian national rainforests study*, Vol. 1 (eds. G. Werren and P. Kershaw), pp. 95–117. Australian Government Publishing Service, Canberra.

Floyd, A. G. (1989). *Rainforest trees of mainland south-eastern Australia*. Inkata Press, Melbourne.

Floyd, A. G. (1990). *Australian rainforests in New South Wales*. 2 vols. Surrey Beatty, Sydney.

Ford, H. A. (1989). *Ecology of birds. An Australian perspective*. Surrey Beatty, Sydney.

Ford, H. A., Paton, D. C., and Forde, N. (1979). Birds as pollinators of Australian plants. *New Zealand Journal of Botany* **17**, 509–19.

Ford, J. (1982). Origin, evolution and speciation of birds specialised to mangroves in Australia. *Emu* **82**, 12–23.

Forestry Commission of NSW (1976). *Indigenous forest policy*. Forestry Commission of NSW, Sydney.

Forestry Commission of NSW (1979). *Background paper—rainforest logging*. Forestry Commission of NSW, Sydney.

Forestry Commission of NSW (1980). *Proposed forest operations in the Washpool area. Environmental Impact Statement*. Forestry Commission of NSW, Sydney.

Forestry Commission of NSW (1981). *Proposed rainforest logging operation Hastings Catchment, Environmental Impact Statement*. Forestry Commission of NSW, Sydney.

Forestry Commission of NSW (1987). *Mount Dromedary. A pretty high mountain*. Forestry Commission of NSW, Sydney.

Forestry Commission of New South Wales (1989). Forest types in New South Wales. *Research note* **17**. Forestry Commission of New South Wales, Sydney.

Fox, A. (1985). Monsoon forests. In *Rainforests of Australia* (ed. P. Figgis), pp. 81–105. Weldon, Sydney.

REFERENCES

Fox, M. D. (1983). *A vegetation survey of the Washpool area, northern New South Wales*. Department of Environment and Planning, Sydney.

Fox, M. D. (1988). The ecological status of alien plant species. In *Weeds on public land—an action plan for today* (ed. R. G. Richardson), pp. 42–5. Weed Science Society of Victoria, Melbourne.

Frakes, L. A., McGowran, B., and Bowler, J. M. (1987). Evolution of Australian environments. In *Fauna of Australia. Volume 1A. General articles*. (eds. G. R. Dyne and D. W. Walton), pp. 1–16, Australian Government Publishing Service, Canberra.

Francis, W. D. (1929). *Australian rain-forest trees*. Government Printer, Brisbane.

Fraser, L. and Vickery, J. W. (1938). The ecology of the Upper Williams River and Barrington Tops Districts II. The rain-forest formations. *Proceedings of the Linnean Society of NSW* **63**, 139–84.

Frawley, K. (1988). Exploitation to preservation: appraisals of the rainforest lands of north-east Queensland, Australia. In *Changing Tropical Forests. Historical perspectives on today's challenges in Asia, Australasia and Oceania* (ed. J. Dargavel, K. Dixon, and N. Semple), pp. 181–96. CRES, Canberra.

Frawley, K. (1989). The history of conservation and the National Park in Australia: a state of knowledge review. In *Australia's ever changing forests* (eds. K. J. Frawley and N. Semple), pp. 395–417, Australian Defence Force Academy, Canberra.

Frith, D. W. (1984). Foraging ecology of birds in an upland tropical rainforest in North Queensland. *Australian Wildlife Research* **11**, 325–47.

Frith, H. J. (1977). The destruction of the Big Scrub. In *Rain Forests* (ed. W. Goldstein), pp. 7–12. National Parks and Wildlife Service, Sydney.

Frith, H. J. (1982). *Pigeons and doves of Australia*. Rigby, Adelaide.

Frodin, D. (1987). The mangrove ecosystem in Papua New Guinea. In *Mangrove ecosystem of Asia and the Pacific. Status, exploitation and management* (eds. C. D. Field and A. J. Dartnall), pp. 53–63. AIMS, Townsville, Queensland.

Fuller, L. (1980). *Wollongong's native trees*. Weston, Kiama, New South Wales.

Galloway, P. J. (1988). Plate tectonics and the distribution of cool temperate southern hemisphere macrolichens. *Botanical Journal of the Linnean Society* **96**, 45–55.

Gardner, C. A. (1923). Botanical notes, Kimberley Division of Western Australia. *Forests Department Bulletin* **32**, 1–105.

Gardner, C. A. (1944). The vegetation of Western Australia with special reference to the climate and soils. *Journal of the Royal Society of Western Australia* **28**, xi–lxxxvii.

Garwood, N. C., Janos, D. P., and Brokaw, N. (1979). Earthquake—caused landslides: a major disturbance to tropical forests. *Science* **205**, 997–9.

Gasteen, W. J. (1987). The brigalow lands of eastern Australia: agricultural impact and landuse potential versus biological representation and stability. In *The Rainforest legacy. Australian national rainforests study*, Vol. 1 (eds. G. Werren and A. P. Kershaw), pp. 143–52. Australian Government Publishing Service, Canberra.

Gay, F. D. (1970). Isoptera. In *Insects of Australia*. pp. 275–93, CSIRO, Melbourne.

Gee, H. and Fenton, J. (eds.) (1978). *The South West book. A Tasmanian Wilderness*. ACF, Melbourne.

Gentilli, J. (1986). Climates. In *Australia—a geography*. Vol. 1. *The natural environment*, (2nd edn) (ed. D. N. Jeans), pp. 14–48. Sydney University Press, Sydney.

George, A. S. (1978). Notes on the vegetation and flora of the Cape Londonderry peninsula, north Kimberley, Western Australia. *Western Australian Herbarium Research Note* **1**, 1–15.

Gilbert, J. M. (1959). Forest succession in the Florentine Valley, Tasmania. *Papers and Proceedings of the Royal Society of Tasmania* **93**, 129–51.

Gillison, A. N. (1981). Towards a functional vegetation classification. In *Vegetation classification in Australia* (eds. A. N. Gillison and D. J. Anderson), pp. 30–41. CSIRO/ANU Press, Canberra.

Gillison, A. N. (1985). Dry rainforests and other unique rainforests. In *Rainforests of Australia* (ed. P. Figgis), pp. 233–59. Weldon, Sydney.

Gillison, A. N. (1987). The 'dry' rainforests of *Terra Australis*. In *The rainforest legacy. Australian national rainforests study* (ed. G. Werren and P. Kershaw), pp. 305–21. Australian Government Publishing Service, Canberra.

Gillman, G. F., Sinclair, D. F., Knowlton, R., and Keys, M. G. (1985). The effect on some soil chemical properties of the selective logging of a north Queensland rainforest. *Forest Ecology and Management* **12**, 195–214.

Graham, A. W. and Hopkins, M. S. (1990). Soil seed banks of adjacent unlogged rainforest types in North

Queensland. *Australian Journal of Botany* **38**, 261–8.

Groom, A. (1949). *One mountain after another*. Angus and Robertson, Sydney.

Guillaumet, J.-L. (1984). The vegetation: an extraordinary diversity. In *Madagascar* (eds. A. Jolly, P. Oberlé, and R. Albignac), pp. 27–54. Pergamon, Oxford.

Hall, C. M. (1989). The 'worthless lands hypothesis' and Australia's national parks and reserves. In *Australia's ever changing forests* (ed. K. J. Frawley and N. Semple), pp. 441–56. Australian Defence Force Academy, Canberra.

Hamilton, A. G. (1897). On the fertilization of *Eupomatia laurina*, R. Br. *Proceedings of the Linnean Society of NSW* **22**, 48–57.

Hand, S. (1989). On the winds of fortune. *Australian Natural History* **23**, 130–8.

Hand, S., Godthelp, H., Novacek, M. J., and Archer, M. (1991). An early Tertiary bat from the Tingamarra local fauna of Southeastern Queensland. In *Abstracts 4th Australian Bat Research Conference*, University of Queensland, Brisbane.

Hanley, J. R. (1985). Why are there so few polychaetes (Annelida) in northern Australian mangroves? In *Coasts and tidal wetlands of the Australian monsoon region* (eds. K. N. Bardsley, J. D. S. Davie, and C. D. Woodroffe), pp. 239–50, ANU. North Australian Research Unit, Darwin.

Hannah, H. (1979). *The mountain speaks. A folk history of the Bulga Plateau*. Helen Hannah, Elands, NSW.

Hannah, H. (1981). *Together in this jungle scrub. A folk history of the Comboyne Plateau*. Helen Hannah, Elands, NSW.

Harper, J. L., Lovell, P. H., and Moore, K. G. (1970). The shapes and sizes of seeds. *Annual Reviews of Ecology and Systematics* **1**, 327–56.

Harris, J. H. (1984). Impoundment of coastal drainages of south-eastern Australia, and a review of its relevance to fish migrations. *Australian Zoologist* **21**, 235–50.

Harrison, J. L. (1962*a*). Mammals of Innisfail. I. Species and distribution. *Australian Journal of Zoology* **10**, 45–83.

Harrison, J. L. (1962*b*). The distribution of feeding habits among animals in a tropical rain forest. *Journal of Animal Ecology* **31**, 53–63.

Hartley, T. G. (1986). Floristic relationships of the rainforest flora of New Guinea. *Telopea* **2**, 619–30.

Haseler, W. H. (1981). Lessons from early attempts at biological control of weeds in Queensland. In *Proceedings of the Fifth International Symposium on Biological Controls of Weeds*. (ed. E. S. Del Fosse), pp. 3–9. CSIRO, Melbourne.

Haynes, C. D. (1985). The pattern and ecology of *munwag*: traditional Aboriginal fire regimes in north central Arnhemland. *Proceedings of the Ecological Society of Australia* **13**, 203–14.

Heatwole, H. (1981). Herpetofaunal assemblages. In *Proceedings of the Melbourne herpetological symposium* (eds. C. B. Banks and A. A. Martin), pp. 156–8. The Dominion Press, Blackburn, Victoria.

Heatwole, H. (1987). Major components and distributions of the terrestrial fauna. In *Fauna of Australia Vol. 1A. General articles* (eds. G. R. Dyne and D. W. Walton), pp. 101–35. Australian Government Publishing Service, Canberra.

Helman, C. (1987). Rainforest in southern New South Wales. In *The rainforest legacy. Australian national rainforests study*, Vol. 1. (eds. G. Werren and A. P. Kershaw), pp. 47–70. Australian Government Publishing Service, Canberra.

Herbert, D. A. (1932). The relationships of the Queensland flora. *Proceedings of the Royal Society of Queensland* **44**, 2–22.

Herbert, D. A. (1950). Present day distribution and the geological past. *Victorian Naturalist* **66**, 227–32.

Herbert, D. A. (1960). Tropical and sub-tropical rain forest in Australia. *Australian Journal of Science* **22**, 283–90.

Herbert, D. A. (1967). Ecological segregation and Australian phytogeographic elements. *Proceedings of the Royal Society of Queensland* **78**, 101–11.

Hickey, J. E. and Felton, K. C. (1987). Management of Tasmanian cool temperate rainforest. In *Forest management in Australia*, Conference Proceedings of the Institute of Foresters, Australia, Perth. pp. 327–41.

Hickey, J. E., Blakesley, A. J., and Turner, B. (1983). Seedfall and germination of *Nothofagus cunninghamii* (Hook.) Oerst., *Eucryphia lucida* (Labill.) Baill. and *Atherosperma moschatum* Labill. Implications for regeneration practice. *Australian Forest Research* **13**, 21–8.

Hill, R. S. (1982). Rainforest fire in Western Tasmania. *Australian Journal of Botany* **30**, 583–89.

Hill, R. S. (1983). Evolution of *Nothofagus cunninghamii* and its relationship to *N. moorei* as inferred from Tasmanian macrofossils. *Australian Journal of Botany* **31**, 453–65.

Hill, R. S. (1987). Discovery of *Nothofagus* fruits corresponding to an important Tertiary pollen type. *Nature* **327**, 56–8.

Hill, R. S. (1990*a*). The fossil history of Tasmania's rainforest tree species. *Tasforests* **2**, 5–12.

Hill, R. S. (1990*b*). Evolution of the modern high latitude southern hemisphere flora: evidence from the Australian macrofossil record. *Proceedings 3rd International Organisation of Palaeobotany conference, Melbourne*, pp. 31–42.

Hill, R. S. (1991). Tertiary *Nothofagus* (Fagaceae) macrofossils from Tasmania and Antarctica and their bearing on the evolution of the genus. *Botanical Journal of the Linnean Society* **105**, 73–112.

Hill, R. S. and Bigwood, A. J. (1987). Tertiary gymnosperms from Tasmania: Araucariaceae. *Alcheringa* **11**, 225–335.

Hill, R. S. and Gibson, N. (1986). Macrofossil evidence for the evolution of the alpine and subalpine vegetation of Tasmania. In *Flora and fauna of alpine Australasia. Ages and origins* (Ed. B. A. Barlow), pp. 205–17.

Hill, R. S. and Macphail, M. K. (1983). Reconstruction of the Oligocene vegetation at Pioneer, northeast Tasmania. *Alcheringa* **7**, 281–99.

Hill, R. S. and Macphail, M. K. (1985). A fossil flora from rafted Plio-Pleistocene mudstones at Regatta Point, Tasmania. *Australian Journal of Botany* **33**, 497–517.

Hill, R. S. and Read, J. (1987). Endemism in Tasmanian cool temperate rainforest: alternative hypotheses. *Botanical Journal of the Linnean Society* **95**, 113–24.

Hill, R. S., Read, J., and Busby, J. R. (1988). The temperature dependence of photosynthesis of some temperate rainforest trees and its biogeographical significance. *Journal of Biogeography* **15**, 431–49.

Hitchcock, P. (1986*a*). The rainforests of New South Wales. Part 1. The background to their World Heritage Nomination. *Wildlife Australia* **23**, (2), 12–17.

Hitchcock, P. (1986*b*). The rainforests of New South Wales. Part 2. World Heritage Nomination. *Wildlife Australia* **23**, (3), 12–14.

Hnatiuk, R. J. (1990). *Census of Australian vascular plants*. Australian Government Publishing Service, Canberra.

Hodgkinson, C. (1845). *Australia from Port Macquarie to Moreton Bay; with descriptions of the natives, their manners and customs; the geology, natural productions, fertility and resources of that region; first explored and surveyed by order of the Colonial Government by Clement Hodgkinson*. T. and W. Boone, London.

Holland, G. J., McBridge, J. L., and Nicholls, N. (1988). Australian region tropical cyclones and the greenhouse effect. In *Greenhouse—planning for climate change* (ed. G. I. Pearman), pp. 438–55. CSIRO, Melbourne.

Holmes, W. B. K., Holmes, F. M., and Martin, H. A. (1982). Fossil *Eucalyptus* remains from the middle Miocene Chalk Mountain Formation, Warrumbungle Mountains, New South Wales. *Proceedings of the Linnean Society of New South Wales* **106**, 299–310.

Hooker, J. D. (1860). Introductory essay. *Botany of the Antarctic voyage of H.M. Discovery Ships 'Erebus' and 'Terror' in the years 1830–1843. III. Flora Tasmaniae*. Reeve, London.

Hopkins, M. S. (1981). Disturbance and change in rainforests and the resulting problems of functional classification. In *Vegetation classification in Australia* (eds. A. N. Gillison and D. J. Anderson), pp. 42–52. CSIRO/ANU Press, Canberra.

Hopkins, M. S. (1990). Disturbances—the forest transformer. In *Australian tropical rainforests. Science—values—meaning* (eds. L. J. Webb and J. Kikkawa), pp. 40–52. CSIRO, Melbourne.

Hopkins, M. S. and Graham, A. W. (1983). The species composition of soil seed banks beneath lowland biotropical rainforests in north Queensland, Australia. *Biotropica* **15**, 90–9.

Hopkins, M. S. and Graham, A. W. (1984*a*). The role of soil seed banks in regeneration in canopy gaps in Australian tropical lowland rainforest—preliminary field experiments. *Malaysian Forester* **47**, 146–58.

Hopkins, M. S. and Graham, A. W. (1984*b*). Viable soil seed banks in disturbed lowland tropical rainforest sites in North Queensland. *Australian Journal of Ecology* **9**, 71–79.

Hopkins, M. S. and Graham, A. W. (1987*a*). Gregarious flowering in a lowland tropical rainforest: a possible response to disturbance by Cyclone Winifred. *Australian Journal of Ecology* **12**, 25–9.

Hopkins, M. S. and Graham, A. W. (1987*b*). The viability of seeds of rainforest species after experimental soil burials under tropical wet lowland forest

in north-eastern Australia. *Australian Journal of Ecology* **12**, 97–108.
Hopkins, M. S. and Graham, A. W. (1989). Community phenological patterns of a lowland tropical rainforest in north-eastern Australia. *Australian Journal of Ecology* **14**, 399–413.
Hopkins, M. S., Kikkawa, J., Graham, A. W., Tracey, J. G., and Webb, L. J. (1977). An ecological basis for the management of rain forest. In *The Border Ranges. A land use conflict in regional perspective* (eds. R. Monroe and N. C. Stevens), pp. 57–66. Royal Society of Queensland, Brisbane.
Hopkins, M. S., Tracey, J. G., and Graham, A. W. (1990*a*). The size and composition of soil seed banks in remnant patches of three structural rainforest types in north Queensland. *Australian Journal of Ecology* **15**, 43–50.
Hopkins, M. S., Graham, A. W., Hewett, R., Ash, J., and Head, J. (1990*b*). Short note. Evidence of late Pleistocene fires and eucalypt forest from a north Queensland humid tropical rainforest site. *Australian Journal of Ecology* **15**, 345–7.
Hopper, S. D. (1980). Pollination of the rain-forest tree *Syzygium tierneyanum* (Myrtaceae) at Kuranda, northern Queensland. *Australian Journal of Botany* **28**, 223–37.
Horak, E. (1983). Mycogeography in the South Pacific region: Agaricales, Boletales. In *Pacific mycogeography: a preliminary approach*, Australian Journal of Botany Supplementary Series 10 (eds. K. A. Pirozynski and J. Walker), pp. 1–41.
Horne, R. (1981). Recovery of rainforest following logging. In *Proceedings of the rainforest conference* (ed. J. H. Shaw and S. J. Riley), pp. 56–64. Geographical Society of NSW, Sydney.
Horne, R. and Gwalter, J. (1982). The recovery of rainforest overstorey following logging. I. Subtropical rainforest. *Australian Forest Research* **13**, 29–44.
Horne, R. and Gwalter, J. (1987). The recovery of rainforest overstorey following logging. 2. Warm temperate rainforest. *Forest Ecology and Management* **22**, 267–81.
Horne, R. and Hickey, J. (1991). Review. Ecological sensitivity of Australian rainforests to selective logging. *Australian Journal of Ecology* **16**, 119–29.
Horne, R. and Mackowski, C. (1987). Crown dieback, over-storey regrowth and regeneration response in warm temperate rainforest following logging. *Forest Ecology and Management* **22**, 283–9.
Horne, R., King, G., Mackowski, C., Gwalter, J., Mullette, K., Lind, P., and Chapman, W. (1981). Recovery of rainforest following logging. *Research Report 1979 and 1980. Forestry Commission of New South Wales* 69–73.
Horsfall, N. and Hall, J. (1990). People and the rainforest: an archaeological perspective. In *Australian tropical rainforests. Science—values—meaning* (eds. L. J. Webb and J. Kikkawa), pp. 33–9. CSIRO, Melbourne.
Houde, P. (1986). Ostrich ancestors found in the northern hemisphere suggest new hypothesis of ratite origins. *Nature* **324**, 563–5.
House of Representatives Standing Committee on Environment and Conservation (1984). *Protection of the Greater Daintree*. Australian Government Publishing Service, Canberra.
House, S. M. (1989). Pollen movement to flowering canopies of pistillate individuals of three rain forest tree species in tropical Australia. *Australian Journal of Ecology* **14**, 77–93.
Howard, T. M. (1973*a*). Studies on the ecology of *Nothofagus cunninghamii* Oerst. II. Phenology. *Australian Journal of Botany* **21**, 79–92.
Howard, T. M. (1973*b*). Accelerated tree death in mature *Nothofagus cunninghamii* Oerst. forests in Tasmania. *Victorian Naturalist* **90**, 343–5.
Howard, T. M. (1974). *Nothofagus cunninghamii* ecotonal stages. Buried viable seed in north-west Tasmania. *Proceedings of the Royal Society of Victoria* **86**, 137–42.
Howard, T. M. (1981). Southern closed-forests. In *Australian vegetation*. (ed. R. H. Groves), pp. 102–20. Cambridge University Press, Cambridge, UK.
Howe, R. W., Howe, T. D., and Ford, H. A. (1981). Bird distributions on small rainforest remnants in New South Wales. *Australian Wildlife Research* **8**, 637–51.
Hulsman, K., Dale, P., and Jahnke, B. R. (1984). Vegetation and nesting preferences of black noddies at Masthead Island, Great Barrier Reef. II. Patterns at the micro-scale. *Australian Journal of Ecology* **9**, 343–52.
Humphreys, F. R. (1983). *The Division of Wood Technology 1936 to 1975. A personal view of its history and achievements*. Forestry Commission of New South Wales, Sydney.
Hurditch, W. J. (1985). Rainforests in New South Wales: the transition from exploitation to protection. In *Managing the tropical forest*. (ed. K. R. Shepherd

and H. V. Richter), pp. 265–79. Development Studies Centre, Australian National University, Canberra.

Hutchings, P. and Recher, H. F. (1982). The fauna of Australian mangroves. *Proceedings of the Linnean Society of New South Wales* **106**, 83–121.

Hutchings, P. and Saenger, P. (1987). *Ecology of mangroves*. University of Queensland Press, St Lucia.

Hutchings, P. A., van der Velde, J. T., and Keable, S. J. (1987). Guidelines for the conduct of surveys for detecting introductions of non-indigenous marine species by ballast water and other vectors—and a review of marine introductions to Australia. *Occasional reports of the Australian Museum* **3**, 1–147.

Hutton, I. (1986). *Lord Howe Island*. Conservation Press, Canberra.

Huxley, C. R. (1978). The ant-plants *Myrmecodia* and *Hydnophytum* (Rubiaceae) and the relationships between their morphology, ant occupants, physiology and ecology. *New Phytologist* **80**, 231–68.

Huxley, C. R. (1982). Ant-epiphytes of Australia. In *Ant–plant interactions in Australia* (ed. R. C. Buckley), pp. 63–73. Junk, The Hague.

Idriess, I. L. (1959). *The tin scratchers*. Angus and Robertson, Sydney.

Irvine, A. K. and Armstrong, J. (1988). Beetle pollination in Australian tropical rainforests. *Proceedings of the Ecological Society of Australia* **15**, 107–13.

Isaacs, S. (1982). *Report of the Terania Creek inquiry by the Honourable Simon Isaacs, Q.C., Commissioner*. 3 vols. New South Wales Government Printer, Sydney.

Jackson, W. D. (1968). Fire, air, water and earth—an elemental ecology of Tasmania. *Proceedings of the Ecological Society of Australia* **3**, 9–16.

Jackson, W. D. (1978). 'Ecological drift' an argument against the continued practice of hazard reduction burning. In *The South West book* (ed. H. Gee and J. Fenton), pp. 98–101. Australian Conservation Foundation, Hobart, Tasmania.

Jackson, W. D. (1983). Tasmanian rainforest ecology. In *Tasmania's rainforest: what future?* (eds. R. Blackers and P. Robertson), pp. 9–39. Australian Conservation Foundation, Hobart, Tasmania.

Jackson, W. D. and Bowman, D. M. J. S. (1982). Reply: ecological drift or fire cycles in south-west Tasmania. *Search* **13**, 175–6.

Jarman, S. J. and Brown, M. J. (1982). Classification of Tasmanian rainforests. In *Tasmanian rainforests—recent research results* (ed. K. Felton), pp. 35–45. Forestry Commission, Hobart, Tasmania.

Jarman, S. J. and Brown, M. J. (1983). A definition of cool temperate rainforest in Tasmania. *Search* **14**, 81–7.

Jarman, S. J., Brown, M. J., and Kantvilas, G. (1984). *A classification of cool temperate rainforest in Tasmania*. National Parks and Wildlife Service, Tasmania, Hobart.

Jarman, S. J., Brown, M. J., and Kantvilas, G. (1987). The classification, distribution and conservation status of Tasmanian rainforests. In *The rainforest legacy. Australian national rainforests study*, Vol. 1 (eds. G. Werren and A. P. Kershaw), pp. 9–19. Australian Government Publishing Service, Canberra.

Jeans, D. N. (ed.) (1986). *Australia, a geography*. Vol. 1. *The natural environment*. 2nd edn., Sydney University Press, Sydney.

Jennings, J. N. (1986). Introduction. In *Australia, a geography. Vol. 1. The natural environment*, (2nd edn.), (ed. D. N. Jeans), pp. 1–13. Sydney University Press, Sydney.

Jennings, J. N. and Mabbutt, J. A. (1986). Physiographic outlines and regions. In *Australia, a geography*, Vol. 1. *The natural environment*, (2nd edn.), (ed. D. J. Jeans), pp. 80–96. Sydney University Press, Sydney.

Johns, R. J. (1986). The instability of the tropical ecosystem in New Guinea. *Blumea* **31**, 341–71.

Johnson, K. A. (1983). Marsupial Mole *Notoryctes typhlops*. In *The Australian Museum complete book of Australian mammals* (ed. R. Strahan), pp. 88–9. Angus and Robertson, Sydney.

Johnson, L. A. S. and Briggs, B. G. (1981). Three old southern families—Myrtaceae, Proteaceae and Restionaceae. In *Ecological biogeography of Australia* (ed. A. Keast), pp. 427–69. Junk, The Hague.

Johnson, R. W. (1984). Flora and vegetation of the brigalow belt. In *The brigalow belt of Australia* (ed. A. Bailey), pp. 41–59. Royal Society of Queensland, Brisbane.

Johnston, R. D. and Lacey, C. J. (1983). Multi-stemmed trees in rainforest. *Australian Journal of Botany* **31**, 189–95.

Johnston, R. D. and Lacey, C. J. (1984). A proposal for the classification of tree-dominated vegetation in Australia. *Australian Journal of Botany* **32**, 529–49.

Jones, D. L. (1986). *Ornamental rainforest plants in Australia*. Reed Books, Sydney.

Jones, J. G. and Veevers, J. J. (1984). Morphotectonics of the platform regions, focussed on the highlands. (a) Eastern highlands. In *Phanerozoic earth history of Australia* (ed. J. J. Veevers), pp. 115–42, Oxford University Press, Oxford.

Jones, R. E. and Crome, F. H. J. (1990). The biological web–plant/animal interaction in the rainforest. In *Australian tropical rainforests. Science—values—meaning* (eds. L. J. Webb and J. Kikkawa), pp. 74–87, CSIRO, Melbourne.

Jørgensen, P. M. (1983). Distribution patterns of lichens in the Pacific region. In *Pacific mycogeography: a preliminary approach, Australian Journal of Botany Supplementary Series* 10. (eds. K. A. Pirozynski and J. Walker), pp. 43–66.

Kahn, T. P. and Lawrie, B. C. (1987). Vine thickets of the inland Townsville region. In *The rainforest legacy. Australian National rainforests study* Vol. 1 (eds. C. Werren and A. P. Kershaw), pp. 159–99. Australian Government Publishing Service, Canberra.

Kemp, E. M. (1978). Tertiary climatic evolution and vegetation history in the southeast Indian Ocean region. *Palaegeography, Palaeoclimatology, Palaeoecology* **24**, 169–208.

Kemp, E. M. (1981*a*). Tertiary palaeogeography and the evolution of Australian climate. In *Ecological biogeography of Australia* (ed. A. Keast), pp. 31–49. Junk, The Hague.

Kemp, E. M. (1981*b*). Pre-Quaternary fire in Australia. In *Fire and the Australian biota*. (eds. A. M. Gill, R. H. Groves, and I. R. Noble), pp. 3–21. Australian Academy of Science, Canberra.

Kendell, J. and Buivids, E. (1987). *Earth first. The struggle to save Australia's rainforest*. ABC, Sydney.

Kenneally, K. F. and Beard, J. S. (1987). Rainforests of Western Australia. In *The rainforest legacy. Australian national rainforest study*. Vol. 1. pp. 289–304. Australian Government Publishing Service, Canberra.

Kershaw, A. P. (1981). Quaternary vegetation and environments. In *Ecological biogeography of Australia* (ed. A. Keast), pp. 81–101. Junk, The Hague.

Kershaw, A. P. (1985). An extended late Quaternary vegetation record from north-eastern Queensland and its implications for the seasonal tropics of Australia. *Proceedings of the Ecological Society of Australia* **13**, 179–89.

Kershaw, A. P. (1986). Climatic change and Aboriginal burning in north-east Australia during the last two glacial/interglacial cycles. *Nature* **322**, 47–9.

Kershaw, A. P. (1988). Australasia. In *Vegetation History* (eds. B. Huntley and T. Webb), pp. 127–306. Kluwer, Dordrecht, Netherlands.

Kershaw, A. P. and Sluiter, I. R. (1982). Late Cenozoic pollen spectra from the Atherton Tableland, north-eastern Australia. *Australian Journal of Botany* **30**, 279–95.

Kershaw, A. P. and Whiffen, T. (1989). Australia. In *Floristic inventory of tropical countries: The status of plant systematics, collections and vegetation, plus recommendations for the Future*. (ed. D. G. Campbell and H. D. Hammond), pp. 149–65. New York Botanical Garden, New York.

Keto, A. (1985). Tropical rainforests. In *Rainforests of Australia* (ed. P. Figgis), pp. 33–79. Weldon, Sydney.

Keto, A. and Scott, K. (1986). *Tropical rainforests of North Queensland. Their conservation significance. A report to the Australian Heritage Commission by the Rainforest Conservation Society of Queensland*. Australian Government Publishing Service, Canberra.

Keto, A., Scott, K., and Fox, A. (1985). An overview. In *Rainforests of Australia* (ed. P. Figgis), pp. 17–32. Weldon, Sydney.

Kikkawa, J. (1968). Ecological association of bird species and habitats in eastern Australia; similarity analysis. *Journal of Animal Ecology* **37**, 143–65.

Kikkawa, J. (1990). Specialization in the tropical rainforest. In *Australian tropical rainforests. Science—values—meaning* (eds. L. J. Webb and J. Kikkawa), pp. 67–73, CSIRO, Melbourne.

Kikkawa, J., Monteith, G. B., and Ingram, G. (1981*a*). Cape York Peninsula: major region of faunal interchange. In *Ecological biogeography of Australia* (ed. A. Keast), pp. 1695–742. Junk, The Hague.

Kikkawa, J., Webb, L. J., Dale, M. B., Monteith, G. B., Tracey, J. G., and Williams, W. T. (1981*b*). Gradients and boundaries of monsoon forests in Australia. *Proceedings of the Ecological Society of Australia* **11**, 39–52.

Kile, G. A. and Walker, J. (1987). *Chalora australis* n.sp., a vascular pathogen of *Nothofagus cunninghamii* (Fagaceae) in Australia and its relationship to other *Chalora* spp. *Australian Journal of Botany* **35**, 1–32.

King, G. C. and Chapman, W. S. (1983). Floristic

composition and structure of a rainforest area 25 years after logging. *Australian Journal of Ecology* **8**, 415–23.

King, N. J. (1965). Land preparation. In *Manual of cane growing*, (2nd edn.), (eds. N. J. King, R. W. Mungomery, and C. G. Hughes), pp. 34–43. Angus and Robertson, Sydney.

King, R. J., Adam, P., and Kuo, J. (1990). Seagrasses, mangroves and saltmarsh plants. In *Biology of marine plants* (ed. M. N. Clayton and R. J. King), pp. 213–39. Longman Cheshire, Sydney.

Kirkpatrick, J. B. (1986). Tasmanian alpine biogeography and ecology and interpretations of the past. In *Flora and Fauna of Alpine Australasia. Ages and origins* (ed. B. A. Barlow), pp. 231–42. CSIRO, Melbourne.

Kirkpatrick, J. B. and Brown, M. J. (1984). A numerical analysis of Tasmanian higher plant endemism. *Botanical Journal of the Linnean Society* **88**, 165–83.

Kitching, R. L. (1981). The geography of the Australian Papilionoidea. In *Ecological biogeography of Australia* (ed. A. Keast), pp. 977–1005. Junk, The Hague.

Korf, R. P. (1983). *Cyttaria* (Cyttariales): coevolution with *Nothofagus*, and evolutionary relationship to the Boedijnpezizeae (Pezizales, Sarcoscyphaceae). In *Pacific mycogeography: a preliminary approach, Australian Journal of Botany Supplementary Series* 10. (eds. K. A. Pirozynski and J. Walker), pp. 77–87.

Ladd, P. G. (1979). A short pollen diagram from rainforest in highland eastern Victoria. *Australian Journal of Ecology* **4**, 229–37.

Ladd, P. G. (1989). The status of Casuarinaceae in Australian forests. In *Australia's ever changing forests* (eds. K. J. Frawley and N. Semple), pp. 63–86. Australian Defence Force Academy, Canberra.

Lambert, M. J. and Turner, J. (1986). Nutrient concentrations in foliage of species within a New South Wales sub-tropical rainforest. *Annals of Botany* **58**, 465–78.

Lambert, M. J. and Turner, J. (1987). Suburban development and change in vegetation nutritional status. *Australian Journal of Ecology* **12**, 193–6.

Lambert, M. J. and Turner, J. (1989). Redistribution of nutrients in subtropical rainforest trees. *Proceedings of the Linnean Society of New South Wales* **111**, 1–10.

Lambert, M. J., Turner, J., and Kelly, J. (1983). Nutrient relationships of tree species in a New South Wales subtropical rainforest. *Australian Forest Research* **13**, 91–102.

Lange, R. T. (1982). Australian Tertiary vegetation. Evidence and interpretation. In *A history of Australasian vegetation* (ed. J. M. B. Smith), pp. 44–89. McGraw-Hill, Sydney.

Lavarack, P. S. and Godwin, M. (1987). Rainforests of northern Cape York Peninsula. In *The rainforest legacy. Australian national rainforests study*, Vol. 1 (eds. G. Werren, and A. P. Kershaw), pp. 201–26. Australian Government Publishing Service, Canberra.

Lavarack, P. S. and Puniard, D. (1988). The McIlwraith Range. *Wildlife Australia* **25**, (1), 12–17.

Lavarack, P. S. and Stanton, J. P. (1977). Vegetation of the Jardine River catchment and adjacent coastal areas. *Proceedings of the Royal Society of Queensland* **88**, 39–48.

Lear, R. and Turner, T. (1977). *Mangroves of Australia*. University of Queensland Press, St Lucia.

Lee, A. K., Baverstock, P. R., and Watts, C. H. S. (1981). Rodents—the late invaders. In *Ecological biogeography of Australia* (ed. A. Keast), pp. 1521–53. Junk, The Hague.

Le Gay Brereton, J. (1957). Defoliation in rain-forest. *Australian Journal of Science* **19**, 204–5.

Leigh, J., Boden, R., and Briggs, J. (1984). *Extinct and endangered plants of Australia*. Macmillan Australia, Melbourne.

Leighton, M. and Wirawan, N. (1986). Catastrophic drought and fire in Borneo tropical rain forest associated with the 1982–1983 El Niño Southern Oscillation Event. In *Tropical rain forests and the world atmosphere* (ed. G. T. Prance), pp. 75–102. Westview Press, Boulder, Colorado.

Leslie, A. J. (1987). International logic of a high-value policy. In *Prospects for Australian hardwood forests* (eds. J. Dargavel and G. Sheldon), pp. 13–26. Centre for Resource and Environmental Studies, Canberra.

Liew, T. C. (1973). Occurrence of seeds in virgin top soil with particular reference to secondary species in Sabah. *Malaysian Forester* **36**, 185–93.

Lourensz, R. S. (1977). *Tropical cyclones in the Australian region July 1909 to June 1975*. Australian Government Publishing Service, Canberra.

Lovejoy, T. E. (1985). *Rehabilitation of degraded tropical lands*. IUCN Commission on Ecology Occasional Paper, 5.

Loveless, A. R. (1961). A nutritional interpretation of sclerophylly based on differences in the chemical composition of sclerophyllous and mesophytic leaves. *Annals of Botany*, n.s. **25**, 168–84.

Lowe, I. (1990). Scientific objectivity and values. In *Australian tropical rainforests. Science—values—meaning* (eds. L. J. Webb and J. Kikkawa), pp. 133–41. CSIRO, Melbourne.

Lowman, M. D. (1982a). Effects of different rates and methods of leaf area removal on rain forest seedlings of coachwood (*Ceratopetalum apetalum*). *Australian Journal of Botany* **30**, 447–83.

Lowman, M. D. (1982b). Seasonal variation in insect abundance among three Australian rain forests, with particular reference to phytophagous types. *Australian Journal of Ecology* **7**, 353–61.

Lowman, M. D. (1984). An assessment of techniques for measuring herbivory: is rainforest defoliation more intense than we thought? *Biotropica* **16**, 264–68.

Lowman, M. D. (1985a). Temporal and spatial variability in insect grazing of the canopies of five Australian rainforest tree species. *Australian Journal of Ecology* **10**, 7–24.

Lowman, M. D. (1985b). Insect herbivory in Australian rain forests—is it higher than in the Neotropics? *Proceedings of the Ecological Society of Australia* **14**, 109–19.

Lowman, M. D. (1986). Light interception and its relation to structural differences in three Australian rainforest canopies. *Australian Journal of Ecology* **11**, 163–70.

Lowman, M. D. (1987). Relationships between leaf growth and holes caused by herbivores. *Australian Journal of Ecology* **12**, 189–91.

Lowman, M. D. and Box, J. R. (1983). Variation in leaf toughness and phenolic content among five species of Australian rain forest trees. *Australian Journal of Ecology* **8**, 17–25.

Lumholtz, C. (1889). *Among cannibals: account of four years travels in Australia, and of camp life with the aborigines of Queensland*. John Murray, London.

MacArthur, R. O. and Wilson, E. O. (1967). *The theory of island biogeography*. Princeton University Press, Princeton, New Jersey.

Macnae, W. (1966). Mangroves in eastern and southern Australia. *Australian Journal of Botany* **13**, 67–104.

Macnae, W. (1968). A general account of the fauna and flora of mangrove swamps and forests in the Indo-West Pacific region. *Advances in Marine Biology* **6**, 73–270.

Macphail, M. K. (1986). 'Over the top': pollen-based reconstructions of past alpine floras and vegetation in Tasmania. In *Flora and fauna of Alpine Australasia. Ages and origins* (ed. B. A. Barlow), pp. 173–204. CSIRO, Melbourne.

Main, B. Y. (1981). Eco-evolutionary radiation of mygalomorph spiders in Australia. In *Ecological biogeography of Australia* (ed. A. Keast), pp. 853–72. Junk, The Hague.

Malingreau, J. P., Stevens, G. and Fellows, L. (1985). Remote sensing of forest fires: Kalimantan and north Borneo 1982–3. *Ambio* **14**, 314–21.

Martin, H. A. (1977). A history of *Ilex* (Aquifoliaceae) with special reference to Australia. Evidence from pollen. *Australian Journal of Botany* **25**, 655–73.

Martin, H. A. (1978). Evolution of the Australian flora and vegetation through the Tertiary: evidence from pollen. *Alcheringa* **2**, 181–202.

Martin, H. A. (1981). The Tertiary flora. In *Ecological biogeography of Australia* (ed. A. Keast), pp. 391–406. Junk, The Hague.

Martin, H. A. (1984). Australian phytogeography. In *Vertebrate zoogeography and evolution in Australasia* (eds. M. Archer and G. Clayton), pp. 17–30, Hesperian Press, Carlisle, Western Australia.

Martin, H. A. (1987). Cainozoic history of the vegetation and climate of the Lachlan River region, New South Wales. *Proceedings of Linnean Society of New South Wales* **109**, 213–57.

Martin, H. A. (1990). Tertiary climate and phytogeography in southeastern Australia. *Review of Palaeobotany and Palynology* **65**, 47–55.

Matthews, E. G. and Kitching, R. L. (1984). *Insect ecology* 2nd edn. University of Queensland Press, Brisbane.

Maynes, G. (1983). Parma wallaby. In *The Australian Museum complete book of Australian mammals* (ed. R. Strahan), pp. 230–1. Angus and Robertson, Sydney.

McBarron, E. J. (1976). *Medical and veterinary aspects of plant poisons in New South Wales*. Department of Agriculture NSW, Sydney.

McDonald, K. and Winter, J. (1986). Eungella. The land of cloud. *Australian Natural History* **22**, (1), 39–43.

McDonald, W. J. F. and Elsol, J. A. (1984). *Moreton Region vegetation map series. Summary report and species checklist for Coloundra, Brisbane, Beenleigh and Murwillumbah sheets*. Queensland Department of Primary Industries, Brisbane.

McElroy, M. B. and Wofsy, S. C. (1986). Tropical

forests: interactions with the atmosphere. In *Tropical rainforests and the world atmosphere* (ed. G. T. Prance), pp. 33–60. Westview Press, Boulder, Colorado.

McGee, P. A. (1990). Survival and growth of seedlings of coachwood (*Ceratopetalum apetalum*): effects of shade, mycorrhizas and a companion plant. *Australian Journal of Botany* **38**, 583–92.

McIntyre, S., Jenkins, B., and Lott, R. (1989). The Daintree dilemma. *Australian Natural History* **23**, 199–208.

McKenzie, N., Kenneally, K., and Winfield, C. (1987). Western Australia's rainforests. *Landscope* **3**, (2), 16–22.

McMinn, W. G. (1970). *Allan Cunningham. Botanist and explorer*. Melbourne University Press, Melbourne.

Melick, D. R. (1990*a*). Regenerative succession of *Tristaniopsis laurina* and *Acmena smithii* in riparian warm temperate rainforest in Victoria in relation to light and nutrient regimes. *Australian Journal of Botany* **38**, 111–20.

Melick, D. R. (1990*b*). Relative drought resistance of *Tristaniopsis laurina* and *Acmena smithii* from warm temperate rainforest in Victoria. *Australian Journal of Botany* **38**, 361–70.

Melick, D. R. (1990*c*). Flood resistance of *Tristaniopsis laurina* and *Acmena smithii* from riparian warm temperate rainforest in Victoria. *Australian Journal of Botany* **38**, 371–80.

Mellanby, K. (1987). Politics and wildlife in Australia. *Nature* **235**, 112.

Mepham, R. H. (1983). Mangrove floras of the southern continents. Part I. The geographical origin of Indo-Pacific mangrove genera and the development and present status of the Australian mangroves. *South African Journal of Botany* **2**, 1–8.

Mepham, R. H. and Mepham, J. S. (1985). The flora of tidal forests—a rationalization of the use of the term 'mangrove'. *South African Journal of Botany* **51**, 71–99.

Mildenhall, D. M. (1980). New Zealand late Cretaceous and Cenozoic plant biogeography: a contribution. *Palaeogeography, Palaeoclimatology, Palaeoecology* **31**, 197–233.

Mills, K. (1987). The distribution, character and conservation status of the rainforests of the Illawarra district, New South Wales. In *The rainforest legacy. Australian national rainforests study*. Vol. 1. (ed. G. Werren and A. P. Kershaw), pp. 71–94. Australian Government Publishing Service, Canberra.

Mitchell, M. L. and Adam, P. (1989*a*). The relationship between mangrove and saltmarsh communities in the Sydney region. *Wetlands (Australia)* **8**, 37–46.

Mitchell, M. L. and Adam, P. (1989*b*). The decline of saltmarsh in Botany Bay. *Wetlands (Australia)* **8**, 55–60.

Mosley, G. (1988). Lord Howe Island. In *Australia's Wilderness Heritage*. (eds. P. Figgis and G. Mosley), pp. 108–19. Weldon, Sydney.

Mount, A. B. (1979). Natural regeneration processes in Tasmanian forests. *Search* **10**, 180–6.

Mount, A. B. (1982). Fire-cycles or succession in southwest Tasmania. *Search* **13**, 174–5.

Mueller-Dombois, D. and Ellenberg, H. (1974). *Aims and methods of vegetation ecology*. Wiley, New York.

Mulvaney, D. J. and Golson, J. (eds.) (1971). *Aboriginal man and the environment in Australia*. ANU Press, Canberra.

Myers, N. (1980). *Conversion of tropical moist forests*. National Research Council, Washington, DC.

Myers, N. (1986). Tropical forests: patterns of depletion. In *Tropical rain forests and the world atmosphere* (ed. G. T. Prance), pp. 9–22. Westview Press, Boulder, Colorado.

Nadolny, C. (1984). *Nature and conservation of the invertebrate fauna in New South Wales rainforests—a preliminary report*. Report to National Parks and Wildlife Service, Sydney.

Neyland, M. and Hickey, J. (1990). Leatherwood silviculture—implications for apiculture. *Tasforests* **2**, 63–72.

Nicholson, D. I. (1985). The development of silviculture in north Queensland rainforests. In *Managing the tropical forest* (ed. K. R. Shepherd and H. V. Richter), pp. 203–23. Development Studies Centre, Australian National University, Canberra.

Nicholson, D. I., Henry, N. B., and Rudder, J. (1988). Stand changes in north Queensland rainforests. *Proceedings of the Ecological Society of Australia* **15**, 61–80.

Nicholson, D. I., Henry, N. B., and Rudder, J. (1990). Reply. Disturbance regimes in North Queensland rainforest: a re-evalution of their relationship to species richness and diversity. *Australian Journal of Ecology* **15**, 245–6.

Nicholson, N. and Nicholson, H. (1985). *Australian rainforest plants*. Terania Rainforest Nursery, The Channon via Lismore, NSW.

Nix, H. A. (1981). The environment of *Terra Australis*. In *Ecological biogeography of Australia* (ed. A. Keast), pp. 105–33. Junk, The Hague.

Nix, H. A. (1982). Environmental determinants of biogeography and evolution in *Terra Australis*. In *Evolution of the flora and fauna of arid Australia* (eds. W. R. Barker and P. J. M. Greenslade), pp. 47–66. Peacock Publications, Frewville, South Australia.

Nix, H. A. and Kalma, J. D. (1972). Climate as a dominant control in the biogeography of north Australia and New Guinea. In *Bridge and barrier: the natural and cultural history of Torres Strait* (ed. D. A. Walker), pp. 61–91, Research School of Pacific Studies, ANU, Canberra.

Nixon, S. W. (1980). Between coastal marshes and coastal waters—a review of twenty years of speculation and research on the role of salt marshes in estuarine productivity and water chemistry. In *Estuarine and wetland processes with emphasis on modeling* (eds. P. Hamilton and K. B. Macdonald), pp. 437–525, Plenum, New York.

Noble, I. R. and Slatyer, R. O. (1979). The effect of disturbance on plant succession. *Proceedings of the Ecological Society of Australia* **10**, 135–45.

Noble, I. R. and Slatyer, R. O. (1980). The use of vital attributes to predict successional changes in plant communities subject to recurrent disturbances. *Vegetatio* **43**, 5–21.

Noble, I. R. and Slatyer, R. O. (1981). Concepts and models of succession in vascular plant communities subject to recurrent fire. In *Fire and the Australian biota* (ed. A. M. Gill, R. H. Groves, and I. R. Noble), pp. 311–35. Australian Academy of Science, Canberra.

Noble, W. S. (1977). *Ordeal by fire. The week a State burned up*. Hawthorn Press, Melbourne.

Ogden, J. (1978). On the dendrochronological potential of Australian trees. *Journal of Ecology* **3**, 339–56.

Ogden, J. (1981). Dendrochronological studies and the determination of tree ages in the Australian tropics. *Journal of Biogeography* **8**, 405–20.

Ollier, C. D. (1982). The Great Escarpment of eastern Australia: tectonic and geomorphic significance. *Journal of the Geological Society of Australia* **29**, 431–5.

Ollier, C. D. (1986). Early landform evolution. In *Australia. A geography. Vol. 1. The natural environment*, (2nd edn.), (ed. D. N. Jeans), pp. 97–116. Sydney University Press, Sydney.

Olsen, M. and Lamb, D. (1988). Recovery of subtropical rainforest following storm damage. *Proceedings of the Ecological Society of Australia* **15**, 297–301.

Osborne, K. and Smith, T. J. (1990). Differential predation on mangrove propagules in open and closed canopy forest habitats. *Vegetatio* **89**, 1–6.

Pahl, L. T. (1979). *A study of the distributions of selected marsupial species (Families: Phalangeridae, Petauridae, Macropodidae) in a tropical rainforest reduced and fragmented by man*. B.Sc. Hons. Thesis, James Cook University.

Patton, R. T. (1930). The factors controlling the distribution of trees in Victoria. *Proceedings of the Royal Society of Victoria* **42**, 120–54.

Pavlov, P. M. (1983). Pigs. In *The Australian Museum complete book of Australian mammals* (ed. R. Strahan), pp. 494–5. Angus and Robertson, Sydney.

Pearman, G. I. (ed.) (1988*a*). *Greenhouse—planning for climate change*. CSIRO, Melbourne.

Pearman, G. I. (1988*b*). Greenhouse gases: evidence for atmospheric changes and anthropogenic causes. In *Greenhouse—planning for climate change* (ed. G. I. Pearman), pp. 3–21. CSIRO, Melbourne.

Pegg, K. G. and Foresberg, L. I. (1981). *Phytophthora* in Queensland mangroves. *Wetlands (Australia)* **1**, 2–3.

Perrens, S. (1982). Australia's water resources. In *Man and the Australian environment* (ed. W. Hanley and M. Cooper), pp. 24–36. McGraw Hill, Sydney.

Pianka, E. R. and Schall, J. J. (1981). Species densities of Australian vertebrates. In *Ecological biogeography of Australia* (ed. A. Keast), pp. 1675–94. Junk, The Hague.

Pickard, J. (1982). Catastrophic disturbance and vegetation on Little Slope, Lord Howe Island. *Australian Journal of Ecology* **7**, 161–70.

Pickard, J. (1983). Vegetation of Lord Howe Island. *Cunninghamia* **1**, 133–265.

Pidgeon, I. M. (1940). The ecology of the central coast of New South Wales. III. Types of primary succession. *Proceedings of the Linnean Society of NSW* **65**, 221–49.

Plowman, K. P. (1990). Forest-floor decomposers. In *Australian tropical rainforests. Science—values—meaning* (ed. L. J. Webb and J. Kikkawa), pp. 98–104. CSIRO, Melbourne.

Pócs, T. (1982). Tropical forest bryophytes. In *Bryo-

phyte ecology (ed. A. J. E. Smith), pp. 59–104, Chapman and Hall, London.

Podger, F. D. and Brown, M. J. (1989). Vegetation damage caused by *Phytophthora cinnamomi* on disturbed sites in temperate rainforest in western Tasmania. *Australian Journal of Botany* **37**, 443–80.

Poole, A. L. (1987). *Southern beeches*, DSIR, Wellington.

Preston, R. A. and Vanclay, J. K. (1988). *Calculation of timber yields from north Queensland rainforests.* Department of Forestry Queensland Technical Paper 47, Brisbane.

Pulsford, I. F. (1984). Conservation status of brigalow *Acacia harpophylla* (Mimosaceae) in New South Wales. In *The brigalow belt of Australia.* (ed. A. Bailey), pp. 161–75. Royal Society of Queensland, Brisbane.

Pyke, G. (1990). Apiarists versus scientists: a bittersweet case. *Australian Natural History* **23**, 386–92.

Rae, I. D. (1988). Chemists at ANZAAS: Cabbages or Kings? In *The Commonwealth of Science. ANZAAS and the scientific enterprise in Australasia 1888–1988* (ed. R. Macleod), pp. 166–95. Oxford University Press, Melbourne.

Ramsay, H. P., Streimann, H., and Harden, G. (1987). Observations on the bryoflora of Australian rainforests. *Symposia Biologica Hungarica* **35**, 605–20.

Ratcliffe, F. (1938). *Flying fox and drifting sand. The adventures of a biologist in Australia.* Chatto and Windus, London.

Raunkiaer, C. (1934). *The life forms of plants and statistical plant geography.* Clarendon Press, Oxford.

Raven, P. H. and Axelrod, D. I. (1974). Angiosperm biogeography and past continental movements. *Annals of the Missouri Botanical Gardens* **61**, 539–673.

Read, J. (1985). Photosynthetic and growth responses to different light regimes of the major canopy species of Tasmanian cool temperate rainforest. *Australian Journal of Ecology* **10**, 327–34.

Read, J. (1988). Tasmanian rainforest ecology—patterns and processes. In *Tasmania's rainforests: what future*, pp. 7–24. Wilderness Society, Hobart.

Read, J. and Hill, R. S. (1983). Rainforest invasion onto Tasmanian old-fields. *Australian Journal of Ecology* **8**, 149–61.

Read, J. and Hill, R. S. (1985*a*). Dynamics of *Nothofagus*-dominated rainforest on mainland Australian and lowland Tasmania. *Vegetatio* **63**, 67–78.

Read, J. and Hill, R. S. (1985*b*). Photosynthetic responses to light of Australian and Chilean species of *Nothofagus* and their relevance to the rainforest dynamics. *New Phytologist* **101**, 731–42.

Read, J. and Hill, R. S. (1989). The response of some Australian temperate rain forest tree species to freezing temperatures and its biogeographical significance. *Journal of Biogeography* **16**, 21–7.

Read, J., Hope, G., and Hill, R. (1990*a*). The dynamics of some *Nothofagus* dominated rain forests in Papua New Guinea. *Journal of Biogeography* **17**, 185–204.

Read, J., Hope, G., and Hill, R. (1990*b*). Integrating historical and ecophysiological studies in *Nothofagus* to examine the factors shaping development of cool rainforest in southeastern Australia. *Proceedings 3rd International Organisation of Palaeobotany Conference*, Melbourne, 97–106.

Reader's Digest (1977). *Reader's Digest Atlas of Australia.* Reader's Digest Services Pty Ltd. Sydney.

Rich, P. V. (1985). *Genyornis newtoni* Stirling and Zietz 1896—A Mihirung. In *Kadimakara. Extinct vertebrates of Australia* (eds. P. V. Rich and G. F. van Tets), pp. 188–94. Pioneer Design Studio, Lilydale, Victoria.

Richards, B. N. (1967). Introduction of the rain-forest species *Araucaria cunninghamii* Ait. to a dry sclerophyll forest environment. *Plant and Soil* **27**, 201–16.

Richards, B. N. (1968). Effect of soil fertility on the distribution of plant communities as shown by pot cultures and field trials. *Commonwealth Forestry Review* **47**, 200–10.

Richards, P. W. (1943). The biogeographic division of the Indo-Australian Archipelago. 6. The ecological segregation of the Indo-Malayan and Australian elements in the vegetation of Borneo. *Proceedings of the Linnean Society of London, Session* **154**, 154–6.

Richards, P. W. (1952). *The tropical rain forest.* Cambridge University Press, Cambridge.

Richards, P. W., Tansley, A. G., and Watt, A. S. (1940). The recording of structure, life form and flora of tropical communities as a basis for their classification. *Journal of Ecology* **28**, 224–39.

Ridley, W. F. and Gardner, A. (1961). Fires in rain forest. *Australian Journal of Science* **23**, 227–8.

Ritchie, R. (1989). *Seeing the rainforests in 19th century Australia.* Rainforest Publishing, Sydney.

Roberts, R. G., Jones, R., and Smith, M. A. (1990). Thermoluminescence dating of a 50,000-year old human occupation site in northern Australia. *Nature* **345**, 153–6.

Robertson, A. I. (1987). The determination of trophic

relationships in mangrove-dominated systems: areas of darkness. In *Mangrove ecosystems of Asia and the Pacific. Status, exploitation and management* (eds. C. D. Field and A. J. Dartnall), pp. 292–304. AIMS, Townsville, Queensland.

Robertson, A. I. and Duke, N. C. (1987). Insect herbivory on mangrove leaves in North Queensland. *Australian Journal of Ecology* **12**, 1–7.

Rodd, N. W. (1987). Weather. In *The Mount Tomah book*, pp. 111–13. The Mount Tomah Society and the Royal Botanic Gardens, Sydney.

Rogers, R. W. and Barnes, A. (1986). Leaf demography of the rainforest shrub *Wilkiea macrophylla* and its implications for the ecology of foliicolous lichens. *Australian Journal of Ecology* **11**, 341–5.

Routley, R. and Routley, V. (1975). *The fight for the forests*. 3rd edn. Research School of Social Sciences, Australian National University, Canberra.

Russell, R. (1985). *Daintree. Where the rainforest meets the reef*. Weldon/ACF, Sydney.

Russell-Smith, J. (1985). A record of change: studies of Holocene vegetation history in the south Alligator River region, Northern Territory. *Proceedings of the Ecological Society of Australia* **13**, 191–202.

Russell-Smith, J. J. (1988). Monsoon forest. Contributed chapter. In *Top End Native Plants* (ed. J. Brock), pp. 23–35. J. Brock, Winnellie, Northern Territory.

Russell-Smith, J. and Dunlop, C. (1987). The status of monsoon vine forests in the Northern Territory: a perspective. In *The Rainforest Legacy. Australian national rainforests study*, Vol. 1 (ed. G. Werren and A. P. Kershaw), pp. 227–88. Australian Government Publishing Service, Canberra.

Saenger, P. and Moverley, J. (1985). Vegetative phenology of mangroves along the Queensland coast. *Proceedings of the Ecological Society of Australia* **13**, 257–65.

Saenger, P., Specht, M. M., Specht, R. L., and Chapman, V. J. (1977). Mangal and coastal salt-marsh communities in Australasia. In *Wet coastal ecosystems* (ed. V. J. Chapman), pp. 293–345. Elsevier, Amsterdam.

Saenger, P., Hegerl, E. J., and Davie, J. D. S. (1983). Global status of mangrove ecosystems. *The Environmentalist* **3**, Supplement 3, 1–88.

Sands, D. and House, S. (1990). Plant/insect interactions—food webs and breeding systems. In *Australian Tropical Rainforests. Science—values—meaning*. (eds. L. J. Webb and J. Kikkawa), pp. 88–97. CSIRO, Melbourne.

Sattler, J. S. and Webster, R. J. (1984). The conservation status of brigalow (*Acacia harpophylla*) communities in Queensland. In *The brigalow belt of Australia* (ed. A. Bailey), pp. 149–60. Royal Society of Queensland, Brisbane.

Sauer, J. D. (1988). *Plant migration. The dynamics of geographic patterning in seed plant species*. University of California Press, Berkeley.

Saxon, E. C. (1990). Comment. Disturbance regimes in North Queensland rainforests: a re-evaluation of their relationship to species richness and diversity. *Australian Journal of Ecology* **15**, 241–4.

Schimper, A. F. W. (1903). *Plant-geography upon a physiological basis*. (Transl. W. R. Fisher, eds. P. Groom, and I. B. Balfour). Clarendon Press, Oxford.

Schodde, R. (1982). Origin, adaptation and evolution of birds in arid Australia. In *Evolution of the Flora and Fauna of Arid Australia*. (eds. W. R. Barker and P. J. M. Greenslade), pp. 191–224. Peacock Publications, Frewville, South Australia.

Schodde, R. (1986). The origins of Australian birds. In *Reader's Digest complete book of Australian birds* (2nd edn.), (eds. R. Schodde and S. C. Tidemann), pp. 36–9. Reader's Digest, Sydney.

Schodde, R. and Calaby, J. H. (1972). The biogeography of the Australo-Papuan bird and mammal faunas in relation to Torres Strait. In *Bridge and barrier: the natural and cultural history of Torres Strait* (ed. D. Walker), pp. 257–300. ANU Press, Canberra.

Schodde, R. and Tidemann, S. C. (eds.) (1986). *Reader's Digest complete book of Australian birds* (2nd edn.), Reader's Digest, Sydney.

Schodde, R., Mason, I. J., and Gill, H. B. (1982). The avifauna of the Australian mangroves. A brief review of composition, structure and origin. In *Mangrove ecosystems in Australia. Structure, function and management* (ed. B. F. Clough), pp. 141–50. AIMS/ANU Press, Canberra.

Schodde, R., Fullagar, P., and Hermes, N. (1983). *A review of Norfolk Island birds: past and present*. Australian National Parks and Wildlife Service, Canberra.

Scott, G. A. M. (1988). Australian bryogeography: fact, fallacy and fantasy. *Botanical Journal of the Linnean Society* **98**, 203–10.

Seddon, G. (1974). Xerophytes, xeromorphs and sclerophylls: the history of some concepts in ecology. *Biological Journal of the Linnean Society* **6**, 65–87.

Seddon, G. (1984). Characteristics and classification of rainforest. *Landscape Australia* **4/84**, 276–85.

Seddon, G. and Cameron, D. (1985). Temperate rainforest. *Landscape Australia* **2/85**, 141–51.

Selman, B. and Lowman, M. D. (1983). The biology and herbivory rates of *Novacastria nothofagi* Selman (Coleoptera: Chrysomelidae), a new genus and species from *Nothofagus moorei* in Australian cool temperate rain forests. *Australian Journal of Zoology* **31**, 179–91.

Semeniuk, V. (1983). Mangrove distribution in north-western Australia in relationship to regional and local freshwater seepage. *Vegetatio* **53**, 11–31.

Semeniuk, V. (1985). Development of mangrove habitats along ria shorelines in north and north-western tropical Australia. *Vegetatio* **60**, 3–23.

Semeniuk, V. (1987). Threats to, and exploitation and destruction of, mangrove systems in Western Australia. In *Mangrove ecosystems of Asia and the Pacific. Status, exploitation and management* (eds. C. D. Field and A. J. Dartnall), pp. 228–40. AIMS, Townsville, Australia.

Semeniuk, V., Kenneally, K. F., and Wilson, P. G. (1978). *Mangroves of Western Australia*. Western Australian Naturalists Club, Perth.

Shugart, H. H. (1984). *A theory of forest dynamics. The ecological implications of forest succession models.* Springer-Verlag, New York.

Shugart, H. H., Hopkins, M. S., Burgess, I. P., and Mortlock, A. T. (1980). The development of a succession model for subtropical rain forest and its application to assess the effects of timber harvest at Wiangaree State Forest, New South Wales. *Journal of Environmental Management* **11**, 243–65.

Sibley, C. G. and Ahlquist, J. E. (1985). The phylogeny and classification of the Australo-Papuan passerines. *Emu*, **85**, 1–14.

Singh, G. (1982). Environmental upheaval. The vegetation of Australasia during the Quaternary. In *The history of Australasian vegetation* (ed. J. M. B. Smith), pp. 90–108. McGraw-Hill, Sydney.

Singh, G. (1986). Reply. *Forest and Timber* **22**, p. 14.

Singh, G. and Geissler, E. A. (1985). Late Cainozoic history of vegetation, fire, lake levels and climate, at Lake George, New South Wales, Australia. *Philosophical Transactions of the Royal Society of London* **B 311**, 379–447.

Singh, G., Kershaw, A. P., and Clarke, R. (1981). Quaternary vegetation and fire history in Australia. In *Fire and the Australian biota* (eds. A. M. Gill, R. H. Groves, and I. R. Noble), pp. 23–54. Australian Academy of Science, Canberra.

Smith, J. M. B. and Guyer, I. J. (1983). Rainforest–eucalypt forest interactions and the relevance of the biological nomad concept. *Australian Journal of Ecology* **8**, 55–61.

Smith, T. J. (1987). Seed predation in relation to tree dominance and distribution in mangrove forests. *Ecology* **68**, 266–73.

Smith, T. J. and Duke, N. C. (1987). Physical determinants of inter-estuary variation in mangrove species richness around the tropical coastline of Australia. *Journal of Biogeography* **14**, 9–19.

Specht, R. L. (1970). Vegetation. In *The Australian environment*, (4th edn.). (ed. G. W. Leeper), pp. 44–67. CSIRO/Melbourne University Press, Melbourne.

Specht, R. L. (1981*a*). Foliage projective cover and standing biomass. In *Vegetation classification in Australia* (ed. A. N. Gillison and D. J. Anderson), pp. 10–21. CSIRO/ANU Press, Canberra.

Specht, R. L. (1981*b*). Major vegetation formations in Australia. In *Ecological biogeography of Australia* (ed. A. Keast), pp. 165–297. Junk, The Hague.

Specht, R. L. (1988). Origin and evolution of terrestrial plant communities in the wet-dry tropics of Australia. *Proceedings of the Ecological Society of Australia* **15**, 19–30.

Specht, R. L. and Womersley, J. S. (1979). Heathlands and related shrublands of Malesia (with particular reference to Borneo and New Guinea). In *Ecosystems of the World*, Vol. 9a, *Heathlands and related shrublands. Descriptive studies* (ed. R. L. Specht), pp. 321–38. Elsevier, Amsterdam.

Speck, N. H. (1960). Vegetation of the North Kimberley area. W.A. In *Lands and pastoral resources of the North Kimberley, W.A. Land Research Series No. 4*, pp. 41–63. CSIRO, Melbourne.

Stanton, J. P. (1982). The wilderness of Cape York Peninsula. In *Wilderness* (ed. V. Martin), pp. 76–80. Findhorn Press, Findhorn, Scotland.

Steenis, C. G. G. J. van (1950). The delimitation of Malesia and its main biogeographical divisions. *Flora Malesiana* **1**, ixx–ixxv.

Steenis, C. G. G. J. van (1958*a*). Basic principles of rain forest sociology. In *Proceedings of the Kandy Symposium on the Study of Tropical Vegetation*, pp. 159–62. UNESCO, Paris.

Steenis, C. G. G. J. van (1958*b*). Rejuvenation as a factor for judging the status of vegetation types: the biological nomad theory. In *Proceedings of the Kandy Symposium on the Study of Tropical Vegetation 1956*, pp. 212–18, UNESCO, Paris.

Stevens, G. N. (1979). Distribution and related ecology of macrolichens on mangroves on the east Australian coast. *Lichenologist* **11**, 293–305.

Stevens, G., McGlone, M., and McCulloch, B. (1988). *Prehistoric New Zealand*. Heinemann Reed, Auckland.

Stockard, J. and Hoye, G. (1990). Wingham Brush—resuscitation of a rainforest. *Australian Natural History* **23**, 402–9.

Stocker, G. C. (1971). The age of charcoal from old jungle fowl nests and vegetation change on Melville Island. *Search* **2**, 28–30.

Stocker, G. C. (1976). Report on cyclone damage to natural vegetation in the Darwin area after Cyclone Tracey, 25 December 1974. *Forestry and Timber Bureau Leaflet* **127**.

Stocker, G. C. (1981). Regeneration of a North Queensland rain forest following felling and burning. *Biotropica* **13**, 86–92.

Stocker, G. C. and Irvine, A. K. (1983). Seed dispersal by cassowaries (*Casuarius casuarius*) in north Queensland's rainforest. *Biotropica* **15**, 170–6.

Stocker, G. C. and Unwin, G. L. (1989). The rain forests of northeastern Australia—their environment, evolutionary history and dynamics. In *Tropical rain forest ecosystems* (ed. H. Leith and M. J. A. Werger), pp. 241–59, Elsevier, Amsterdam.

Strahan, R. (ed.) (1983). *The Australian Museum complete book of Australian mammals*. Angus and Robertson, Sydney.

Sullivan, S. (1978). Aboriginal diet and food gathering methods in the Richmond and Tweed River Valleys, as seen in early settler records. In *Records of times past. Ethnohistorical essays on the culture and ecology of the New England tribes* (ed. I. Bryde), pp. 101–15. Australian Institute of Aboriginal Studies, Canberra.

Swaine, M. D. and Whitmore, T. C. (1988). On the definition of ecological species groups in tropical rain forests. *Vegetatio* **75**, 81–6.

Tait, N. N., Stutchbury, R. J., and Briscoe, D. A. (1990). Review of the discovery and identification of Onychophora in Australia. *Proceedings of the Linnean Society of New South Wales* **112**, 153–71.

Takhtajan, A. (1969). *Flowering plants, origin and dispersal* (trans. C. Jeffrey). Oliver and Boyd, Edinburgh.

Takhtajan, A. (1986). *Floristic regions of the world*. University of California Press, Berkeley.

Takhtajan, A. (1987). Flowering plant origin and dispersal: the cradle of the angiosperms revisited. In *Biogeographical evolution of the Malay archipelago*. (ed. T. C. Whitmore), pp. 26–31. Oxford University Press, Oxford.

Taylor, J. A. and Dunlop, C. R. (1985). Plant communities of the wet-dry tropics of Australia: the Alligator Rivers region, Northern Territory. *Proceedings of the Ecological Society of Australia* **13**, 83–127.

Tets, G. F. van (1986). Birds of prehistoric Australia. In *Reader's Digest complete book of Australian birds*. 2nd edn. (ed. R. Schodde and S. C. Tidemann), pp. 34–5. Reader's Digest, Sydney.

Thom, B. G. (1982). Mangrove ecology—a geomorphological perspective. In *Mangrove ecosystems in Australia. Structure, function and management.* (ed. B. F. Clough), pp. 3–17. AIMS/ANU Press, Canberra.

Thom, B. G., Wright, L. D., and Coleman, J. M. (1975). Mangrove ecology and deltaic estuarine geomorphology: Cambridge Gulf–Ord River, Western Australia. *Journal of Ecology* **63**, 203–32.

Thomas, D. (1990). *Rainforest conservation status in the Metropolitan and Woronora catchment areas*. Sydney Water Board, Sydney.

Thomas, D. G. (1980). The bird community of Tasmanian temperate rainforests. *Ibis* **122**, 298–306.

Thompson, J. (1989). A revision of the genus *Leptospermum* (Myrtaceae). *Telopea* **3**, 301–448.

Thompson, W. A., Stocker, G. C., and Kriedmann, P. E. (1988). Growth and photosynthetic response to light and nutrients of *Flindersia brayleyana* F. Muell., a rainforest tree with broad tolerance to sun and shade. *Australian Journal of Plant Physiology* **15**, 299–315.

Thorne, R. F. (1986*a*). Antarctic elements in Australasian rainforests. *Telopea* **2**, 611–17.

Thorne, R. F. (1986*b*). Summary statement. *Telopea* **2**, 697–704.

Tomlinson, P. B. (1986). *The botany of mangroves*. Cambridge University Press, Cambridge.

Tracey, J. G. (1969). Edaphic differentiation in some forest types in eastern Australia. I. Soil physical factors. *Journal of Ecology* **57**, 805–16.

Tracey, J. G. (1982). *The vegetation of the humid tropical region of north Queensland*. CSIRO, Melbourne.

Tracey, J. G. and Webb, L. J. (1975). *The vegetation of*

the humid tropical region of North Queensland (Maps and Key). CSIRO, Indooroopilly, Queensland.

Truswell, E. M. (1982). Antarctica: the vegetation of the past and its climatic implications. *Australian Meteorological Magazine* **30**, 169-73.

Truswell, E. M. (1990). Australian rainforests: the 100 million year record. In *Australian tropical rainforests. Science—values—meaning* (eds. L. J. Webb and J. Kikkawa), pp. 7-22. CSIRO, Melbourne.

Truswell, E. M. and Wilford, G. E. (1985). The setting. Changing Australian environments throughout geological time. In *Kadimakara. Extinct vertebrates of Australia* (eds. P. V. Rich, G. F. van Tets, and F. Knight), pp. 59-92. Pioneer Design Studio, Lilyvale, Victoria.

Truswell, E. M., Kershaw, A. P., and Sluiter, I. R. (1987). The Australian-South-east Asian connection: evidence from the palaeobotanical record. In *Biogeographical evolution of the Malay archipelago* (ed. T. C. Whitmore), pp. 32-49. Oxford University Press, Oxford.

Turner, J. (1984). Radiocarbon dating of wood and charcoal in an Australian forest ecosystem. *Australian Forestry* **47**, 79-83.

Turner, J. and Kelly, J. (1981). Relationships between soil nutrients and vegetation in a North Coast forest, New South Wales. *Australian Forest Research* **11**, 201-8.

Turner, J. and Lambert, M. J. (1983). Nutrient cycling within a 27 year old *Eucalyptus grandis* plantation in New South Wales. *Forest Ecology and Management* **6**, 156-68.

Turner, J., Kelly, J., Lambert, M. J., Truman, R., and Booth, T. (1981). Soil and site factors affecting the distribution of native tree species. *Research Report 1979 and 1980, Forest Commission of New South Wales* 57-9.

Turner, J. C. (1976). An altitudinal transect in rain forest in the Barrington Tops area, New South Wales. *Australian Journal of Ecology* **1**, 155-74.

Twilley, R. R. (1988). Coupling of mangroves to the productivity of estuarine and coastal waters. In *Coastal-Offshore Ecosystem Interactions* (ed. B. O. Jansson). pp. 155-80, Springer Verlag, Berlin.

Tyler, M. J. (1979). Herpetofaunal relationships of South America with Australia. In *The South American herpetofauna* (ed. W. F. Duellman). Kansas Museum of Natural History. Monograph 7, reproduced in *Vertebrate Zoogeography and evolution in Australasia (Animals in space and time)*, (eds. M. Archer and G. Clayton), pp. 263-82. Hesperian Press, Carlisle, Western Australia.

Tyler, M. J. (1985). Gastric brooding: a phenomenon unique to Australian frogs. *Search* **16**, 157-9.

Tyler, M. J., Hand, S. J., and Ward, V. J. (1990). Analysis of the frequency of *Lechriodus intergerivus* Tyler (Anura: Leptodactylidae) in Oligo-Miocene local faunas of Riversleigh Station, Queensland. *Proceedings of the Linnean Society of New South Wales* **112**, 105-9.

Unwin, G. L. (1989). Structure and composition of the abrupt rainforest boundary in the Herberton Highland, north Queensland. *Australian Journal of Botany* **37**, 413-28.

Unwin, G. L., Stocker, G. C., and Sanderson, K. D. (1985). Fire and the forest ecotone in the Herberton Highland, north Queensland. *Proceedings of the Ecological Society of Australia* **13**, 215-24.

Unwin, G. L., Applegate, G. B., Stocker, G. C., and Nicholson, D. I. (1988*a*). Initial effects of Tropical Cyclone 'Winifred' on forests in north Queensland. *Proceedings of the Ecological Society of Australia* **15**, 283-96.

Unwin, G. L., Stocker, G. C., and Sanderson, K. D. (1988*b*). Forest succession following European settlement and mining in the Rossville area, north Queensland. *Proceedings of the Ecological Society of Australia* **15**, 303-5.

Vader, J. (1987). *Red Cedar. The tree of Australia's history*. Reed Books, Sydney.

Vanclay, J. K. (1989). A growth model for North Queensland rainforests. *Forest Ecology and Management* **27**, 245-71.

Veblen, T. T. and Ashton, D. H. (1978). Catastrophic influences on the vegetation of the Valdivian Andes, Chile. *Vegetatio* **36**, 149-67.

Veblen, T. T., Schlegel, F. M., and Escobar, B. R. (1980). Structure and dynamics of old-growth *Nothofagus* forests in the Valdivian Andes, Chile. *Journal of Ecology* **68**, 1-31.

Veblen, T. T., Donoso, C. Z., Schlegel, F. M., and Escobar, B. R. (1981). Forest dynamics in south-central Chile. *Journal of Biogeography* **8**, 211-47.

Veevers, J. J. (ed.) (1984). *Phanerozoic earth history of Australia*. Clarendon Press, Oxford.

Walker, J. and Hopkins, M. S. (1984). Vegetation. In *Australian soil and land survey field handbook*, pp. 44-67. Inkata Press, Melbourne.

Ward, J. P. and Kanowski, P. J. (1985). Implementing

control of harvesting operations in north Queensland rainforests. In *Managing the tropical forest* (eds. K. R. Shepherd and H. V. Richter), pp. 156–86. Development Studies Centre, Australian National University, Canberra.

Wardle, P. (1983). Temperate broad-leaved evergreen forests of New Zealand. In *Temperate broad-leaved evergreen forests* (ed. J. D. Ovington), pp. 33–72, Elsevier, Amsterdam.

Warming, E. (1909). *Oecology of plants. An introduction to the study of plant-communities* (Transl. P. Groom and I. B. Balfour), Clarendon Press, Oxford.

Watson, D. (1989). Clearing the scrubs of south-east Queensland. In *Australia's ever changing forests* (ed. K. J. Frawley and N. Semple), pp. 365–92. Australian Defence Force Academy, Canberra.

Watson, I. (1990). *Fighting over the Forests*. Allen and Unwin, Sydney.

Webb, L. J. (1954). Aluminium accumulation in the Australian–New Guinea flora. *Australian Journal of Botany* **2**, 176–96.

Webb, L. J. (1958). Cyclones as an ecological factor in tropical lowland rainforest, north Queensland. *Australian Journal of Botany* **6**, 220–8.

Webb, L. J. (1959). A physiognomic classification of Australian rain forests. *Journal of Ecology* **47**, 551–70.

Webb, L. J. (1966). An ecological comparison of forest-fringe grassland habitats in eastern Australia and eastern Brazil. *Proceedings of the IX International Grassland Congress* 321–30.

Webb, L. J. (1968). Environmental relationships of the structural types of Australian rain forest vegetation. *Ecology* **49**, 296–311.

Webb, L. J. (1969). Australian plants and chemical research. In *The last of lands. Conservation in Australia* (eds. L. J. Webb, D. Whitelock, and J. LeGay Brereton), pp. 82–90. Jacaranda, Brisbane.

Webb, L. J. (1978*a*). A general classification of Australian rainforests. *Australian Plants* **9**, 349–63.

Webb, L. J. (1978*b*). A structural comparison of New Zealand and south-east Australian rain forests and their tropical affinities. *Australian Journal of Ecology* **3**, 7–21.

Webb, L. J. (1984). Nature conservation. In *The brigalow belt of Australia* (ed. A. Bailey), pp. 125–9. Royal Society of Queensland, Brisbane.

Webb, L. J. (1987). Conservation status of the rainforest of north Queensland. In *The rainforest legacy. Australian National rainforests study* Vol. 1 (eds. G. L. Werren and A. P. Kershaw), pp. 153–8. Australian Government Publishing Service, Canberra.

Webb, L. J. and Tracey, J. G. (1967). An ecological guide to new planting areas and site potential for hoop pine. *Australian Forestry* **31**, 224–40.

Webb, L. J. and Tracey, J. G. (1972). An ecological comparison of vegetation communities on each side of Torres Strait. In *Bridge and Barrier: the natural and cultural history of Torres Strait* (ed. D. Walker), pp. 109–29, ANU, Canberra.

Webb, L. J. and Tracey, J. G. (1975). The Cooloola rain forests. *Proceedings of the Ecological Society of Australia* **9**, 317–21.

Webb, L. J. and Tracey, J. G. (1981*a*). Australian rainforests: patterns and change. In *Ecological biogeography of Australia* (ed. A. Keast), pp. 605–94, Junk, The Hague.

Webb, L. J. and Tracey, J. G. (1981*b*). The rainforests of northern Australia. In *Australian vegetation* (ed. R. H. Groves), pp. 67–101. Cambridge University Press, Cambridge.

Webb, L. J., Tracey, J. G., and Haydock, K. P. (1967*c*). A factor toxic to seedlings of the same species associated with living roots of the non-gregarious subtropical rain forest tree *Grevillea robusta*. *Journal of Applied Ecology* **4**, 13–25.

Webb, L. J., Tracey, J. G., and Jessup, L. W. (1986). Recent evidence for autochthony of Australian tropical and subtropical rainforest floristic elements. *Telopea* **2**, 575–89.

Webb, L. J., Tracey, J. G., and Williams, W. T. (1972). Regeneration and pattern in the subtropical rain-forest. *Journal of Ecology* **60**, 675–95.

Webb, L. J., Tracey, J. G., and Williams, W. T. (1976). The value of structural features in tropical forest typology. *Australian Journal of Ecology* **1**, 3–28.

Webb, L. J., Tracey, J. G., and Williams, W. T. (1984). A floristic framework of Australian rainforests. *Australian Journal of Ecology* **9**, 169–98.

Webb, L. J., Tracey, J. G., and Williams, W. T. (1985). Australian tropical forests in a southeast Asian context: a numerical method for site comparison. *Proceedings of the Ecological Society of Australia* **13**, 269–76.

Webb, L. J., Tracey, J. G., Williams, W. T., and Lance, G. N. (1967*a*). Studies in the numerical analysis of complex rain-forest communities. I. A comparison of methods applicable to site/species data. *Journal of Ecology* **55**, 171–91.

Webb, L. J., Tracey, J. G., Williams, W. T., and Lance, G. N. (1967b). Studies in the numerical analysis of complex rain-forest communities. II. The problem of species sampling. *Journal of Ecology* **55**, 525–38.

Webb, L. J., Tracey, J. G., Williams, W. T., and Lance, G. N. (1969). The pattern of mineral return in leaf litter of three subtropical Australian forests. *Australian Forestry* **33**, 99–110.

Webb, L. J., Tracey, J. G., Williams, W. T., and Lance, G. N. (1970). Studies in the numerical analysis of complex rainforest communities. V. A comparison of the properties of floristic and physiognomic-structural data. *Journal of Ecology* **58**, 203–32.

Wellman, P. and McDougall, J. (1974). Cainozoic igneous activity in eastern Australia. *Tectonophysics* **23**, 49–65.

Wells, A. G. (1982). Mangrove vegetation in northern Australia. In *Mangrove ecosystems in Australia* (ed. B. F. Clough), pp. 57–78. AIMS/ANU Press, Canberra.

Wells, A. G. (1983). Distribution of mangrove species in Australia. In *Biology and ecology of mangroves* (ed. H. J. Teas), pp. 57–76. Junk, The Hague.

Wells, A. G. (1985). Grouping of tidal systems in the Northern Territory and Kimberley region of Western Australia on presence/absence of mangrove species. In *Coasts and tidal wetlands of the Australian monsoon region* (eds. K. N. Bardsley, J. D. S. Davie, and C. D. Woodroffe), pp. 167–86. ANU North Australia Research Unit, Darwin, Australia.

Werren, G. L. (1985). A catalogue of Australian rainforests. *Habitat* **13**, (2), 15–25.

Werren, G. L. and Allworth, D. (1982). *Australian rainforests: a review*. Department of Geography, Monash University, Melbourne.

Werren, G. and Kershaw, A. P. (eds.). (1987). *The rainforest legacy. Australian national rainforests study*, Vol. 1. Australian Government Publishing Service, Canberra.

West, R. J. and Thorogood, C. A. (1985). Mangrove dieback in Hunter River caused by caterpillars. *Australian Fisheries* **44**, (9), 27–8.

West, R. J. Thorogood, C. A., and Williams, R. J. (1983). Environmental stress causing mangrove 'dieback' in NSW. *Australian Fisheries* **42**, (8), 16–20.

Westoby, J. (1978). Forestry, foresters and society. *New Zealand Journal of Forestry* **23**, 64–84.

Westoby, M. (1986). Commentary: how to build and repair ecosystems. *Bulletin of the Ecological Society of Australia* **16**, (1), 2–8.

White, A. (1984). Zoogeography of Australian amphibians. In *Vertebrate zoogeography and evolution in Australasia (Animals in space in time)* (eds. M. Archer and G. Clayton), pp. 283–9. Hesperian Press, Carlisle.

White, M. E. (1986). *The greening of Gondwana*. Reed, Sydney.

Whitmore, T. C. (1973). Plate tectonics and some aspects of Pacific plant geography. *New Phytologist* **72**, 1185–90.

Whitmore, T. C. (1978). Gaps in the forest canopy. In *Tropical trees and living systems* (eds. P. B. Tomlinson and M. H. Zimmerman), pp. 639–55. Cambridge University Press, Cambridge.

Whitmore, T. C. (1982). On pattern and process in forests. In *The plant community as a working mechanism* (ed. E. I. Newman), pp. 45–59. Blackwell Scientific Publications, Oxford.

Whitmore, T. C. (1984). *Tropical rain forests of the Far East* (2nd edn.), Oxford University Press, Oxford.

Whitmore, T. C. (ed.) (1987). *Biogeographical evolution of the Malay archipelago*. Oxford University Press, Oxford.

Whitmore, T. C. (1988). The influence of tree population dynamics on forest species composition. In *Plant population ecology* (ed. A. J. Davy, M. J. Hutchings, and A. R. Watkinson), pp. 271–91. Blackwell Scientific Publications, Oxford.

Whitmore, T. C. (1989a). Southeast Asian tropical forests. In *Tropical rain forest ecosystems* (ed. H. Lieth and M. J. A. Werger), pp. 195–218. Elsevier, Amsterdam.

Whitmore, T. C. (1989b). Canopy gaps and the two major groups of forest trees. *Ecology* **70**, 536–8.

Whitmore, T. C. (1990). *An introduction to tropical rainforests*. Oxford University Press, Oxford.

Whitmore, T. C. and Prance, G. T. (eds.) (1987). *Biogeography and Quaternary history in tropical America*. Oxford University Press, Oxford.

Whitten, A. J., Damanik, S. J., Anwar, J., and Hisyam, N. (1984). *The ecology of Sumatra*. Gadjah Mada University Press, Yogyakarta, Indonesia.

Williams, J. G., Harden, G. J., and McDonald, W. J. F. (1984). *Trees and shrubs in rainforests of New South Wales and southern Queensland*. Botany Department, University of New England, Armidale.

Williams, M. A. J. (1984). Quaternary environments. In *Phanerozoic earth history of Australia* (ed. J. A.

Veevers), pp. 42–7. Oxford University Press, Oxford.

Williams, W. T. (1986). Marine Science. In *Tropical plant communities* (ed. H. T. Clifford and R. L. Specht), pp. 199–204. Dept. of Botany, University of Queensland, St Lucia.

Williams, W. T., Lance, G. N., Webb, L. J., Tracey, J. G., and Dale, M. B. (1969a). Studies in the numerical analysis of complex rain-forest communities. III. The analysis of successional data. *Journal of Ecology* **57**, 515–35.

Williams, W. T., Lance, G. N., Webb, L. J., Tracey, J. G., and Connell, J. H. (1969b). Studies in the numerical analysis of complex rain-forest communities. IV. A method for the elucidation of small-scale forest pattern. *Journal of Ecology* **57**, 635–55.

Williams, W. T., Lance, G. N., Webb. L. J., and Tracey, J. G. (1973). Studies in the numerical analysis of complex rain-forest communities. VI. Models for the classification of quantitative data. *Journal of Ecology* **61**, 47–70.

Williams, W. T. and Tracey, J. G. (1984). Network analysis of northern Queensland tropical rainforests. *Australian Journal of Botany* **32**, 109–16.

Willson, M. F. (1988). Spatial heterogeneity of post-dispersal survivorship of Queensland rainforest seeds. *Australian Journal of Ecology*, **13**, 137–45.

Winter, J. W. (1978). The rainforest: where strange things live. In *Exploration North* (ed. H. J. Lavery), pp. 113–46. Richmond Hill Press, Richmond, Victoria.

Winter, J. W. (1984). Conservation studies of tropical rainforest possums. In *Possums and gliders* (eds. A. P. Smith and I. D. Hume), pp. 469–81. Surrey Beatty, Sydney.

Winter, J. W. (1988). Ecological specialization of mammals in Australian tropical and sub-tropical rainforest: refugial or ecological determinism? *Proceedings of the Ecological Society of Australia* **15**, 127–38.

Winter, J. W., Bell, F. C., Pahl, L. I., and Atherton, R. G. (1984). *The specific habitats of selected northeastern Australian rainforest mammals*. Report to the World Wildlife Fund, Australia.

Winter, J. W., Atherton, R. G., Bell, F. C., and Pahl, L. I. (1987a). The distribution of rainforest in northeastern Queensland. In *The rainforest legacy Australian national rainforests study. Vol. 1* (eds. G. Werren and A. P. Kershaw), pp. 223–6. Australian Government Publishing Service, Canberra.

Winter, J. W., Bell, F. C., Atherton, R. G., and Pahl, L. I. (1987b). An introduction to Australian rainforests. In *The rainforest legacy. Australian national rainforests study. Vol. 1* (eds. G. Werren and A. P. Kershaw), pp. 1–7. Australian Government Publishing Service, Canberra.

Winter, J. W., Bell, F. C., Pahl, L. I., and Atherton, R. G. (1987c). Rainforest clearfelling in northeastern Australia. *Proceedings of Royal Society of Queensland* **98**, 41–57.

Wolfe, J. A. (1979). *Temperature parameters of humid to mesic forests of eastern Asia and relation to forests of other regions of the northern hemisphere and Australasia*. Geological Survey Professional Paper **1106**, Washington, DC.

Woodroffe, C. D., Thom, B. G., and Chappel, J. (1985). Development of widespread mangrove swamps in mid-Holocene times in northern Australia. *Nature* **317**, 711–13.

Wright, R. (1986). How old is Zone F at Lake George? *Archaeology in Oceania* **21**, 138–9.

Young, P. A. R. and McDonald, W. J. F. (1987). The distribution, composition and status of the rainforests of southern Queensland. In *The rainforest legacy. Australian national rainforests study*, Vol. 1 (eds. G. Werren and A. P. Kershaw), pp. 119–41. Australian Government Publishing Service, Canberra.

INDEX OF PLACE NAMES

As far as possible names and co-ordinates are those given in Division of National Mapping, Department of Minerals and Energy (1975). *Australia 1:250 000 Map Series Gazeteer.* Australian Government Publishing Service, Canberra, and Reader's Digest (1977). *Reader's Digest Atlas of Australia.* Reader's Digest Services Pty Ltd, Sydney.

Abercrombie 33°58′ 149°19′ 5 (*fig.*)
Abrohlos Islands 28°35′ 113°40′ 202
Adelaide 34°56′ 138°36′ 202
Alice Springs 23°42′ 133°52′ 156
Alligator Rivers Region 13°30′ 133°00′ 38 (*fig.*), 41, 212–13
Anglesea 38°25′ 144°11′ 145
Apsley Gorge 31°06′ 151°15′ 69 (*fig.*)
Archer Bend 13°28′ 142°12′ 46
Archer River 13°35′ 142°09′ 46
Arnhemland 14°50′ 134°40′ 37–8, 214, 218
Atherton 17°16′ 145°29′ 47 (*fig.*)
Atherton Tableland 17°16′ 145°29′ 48, 50, 86, 122, 129, 143, 184, 185 (*fig.*), 186 (*fig.*), 216, 224, 233

Bamaga 10°53′ 142°24′ 45, 46
Banjo Creek 31°23′ 152°07′ 108
Barrington Tops 32°10′ 152°26′ 58, 61, 63, 101, 113, 131, 153, 198
Barron Falls 16°50′ 145°38′ 221
Barron River 16°52′ 145°42′ 170 (*fig.*), 221
Big Scrub 28°49′ 153°15′ 51, 183, 193, 222–3, 224, 236, 262
Billilimbra State Forest 29°13′ 145°32′ 59 (*fig.*), 135 (*fig.*), 255 (*fig.*)
Black Jungle 12°30′ 130°15′ 40
Black Mountain Corridor 145°29′ 16°37′ 47 (*fig.*), 189
Blackall Ranges 26°34′ 152°52′ 234
Bloomfield 15°57′ 145°20′ 47 (*fig.*), 239
Bloomfield River 15°58′ 145°19′ 170 (*fig.*)
Blue Mountains 33°36′ 150°15′ 2 (*fig.*), 78, 93 (*fig.*), 193
Brodribb River Flora & Fauna Reserve, Vic. 37°29′ 148°35′ 65
Bromfield Swamp 17°23′ 145°32′ 170 (*fig.*)
Border Ranges National Park 153°10′ 28°17′ 51, 55 (*fig.*), 58, 61, 222, 224, 245 (*fig.*)

Botany Bay 43°00′ 151°11′ 209
Bougainville Peninsula 13°54′ 126°06′ 35–6
Brisbane 27°28′ 153°01′ 32, 68
Bulga Plateau 31°39′ 151°01′ 61, 223
Bunbury 33°20′ 11°28′ 202
Bundaberg 25°52′ 152°21′ 70
Bunya Mountains 26°53′ 151°37′ 216–17, 217 (*fig.*)
Butchers Creek 17°18′ 145°45′ 151 (*fig.*), 178

Cabbage Tree Creek Flora Reserve 37°42′ 148°43′ 65
Cairns 16°55′ 145°46′ 47 (*fig.*)
Cambridge Plateau 28°41′ 152°16′ 87 (*fig.*)
Canberra 35°17′ 149°13′ 214
Cape Hillsborough 20°53′ 149°03′ 5 (*fig.*)
Cape Tribulation National Park 16°05′ 45°28′ 239
Cape Tribulation 16°04′ 145°27′ 239, 265 (*fig.*)
Cape York 10°41′ 142°32′ 45
Cardwell 18°16′ 146°01′ 47 (*fig.*)
Careys Peak 32°03′ 151°28′ 62 (*fig.*)
Carnarvon 24°53′ 113°40′ 202
Cedar Brush Nature Reserve 31°45′ 150°07′ 88 (*fig.*)
Ceduna 32°08′ 133°41′ 202
Cethana 41°29′ 146°10′ 153 (*fig.*)
Clark Range 18°36′ 143°16′ 234
Clouds Creek State Forest 32°06′ 152°38′ 109 (*fig.*)
Cobourg Peninsula 11°22′ 132°18′ 40
Coffs Harbour 30°18′ 153°08′ 94, 239
Comboyne Plateau 31°36′ 152°28′ 61, 223
Conondale Range 26°45′ 152°37′ 54, 234
Cooktown 15°28′ 145°15′ 1 (*fig.*), 32, 46, 47 (*fig.*), 233
Cooloola 26°05′ 153°07′ 56

Cooma 36°14′ 149°08′ 2
Cowra 33°50′ 148°41′ 147 (*fig.*)

Daintree River 16°08′ 145°17′ 170 (*fig.*)
Daly River 13°50′ 130°20′ 40
Dampier Peninsula 20°34′ 116°47′ 35
Darwin 12°27′ 130°50′ 39, 207 (*fig.*)
Dorrigo 30°20′ 152°43′ 61
Dorrigo National Park 30°21′ 152°50′ 236

East Gippsland 37°12′ 148°10′ 63, 64, 65, 66
Errinundra Plateau 37°17′ 148°55′ 63, 65 (*figs*)
Eungella National Park 20°55′ 148°30′ 234
Eungella Range 21°05′ 148°25′ 32, 46, 51

Florentine Valley 42°35′ 146°25′ 111
Forbes 33°23′ 148°01′ 147 (*fig.*), 186
Franklin River 42°33′ 145°46′ 243
Fraser Island 25°22′ 153°07′ 56, 57, 241

Gibraltar Range 29°35′ 152°15′ 61, 234
Gibraltar Range National Park 29°34′ 152°21′ 240 (*fig.*)
Gilbert River 16°50′ 141°20′ 233
Gladstone 23°51′ 151°16′ 51, 54, 58
Golden Grove 138°44′ 34°47′ 145
Goonmirk Rocks 37°17′ 148°53′ 64
Gradys Creek 28°21′ 153°03′ 55 (*fig.*), 131

Hamilton 28°28′ 148°11′ 184, 185 (*fig.*), 186
Hastings River 31°20′ 152°13′ 198
Hawkesbury River 33′30′ 151′10′ 220, 222, 225
Hayters Hill Nature Reserve 28°48′ 153°15′ 197, 264

INDEX OF PLACE NAMES

Herberton Highland 17°22′ 145°27′ 89 (fig.), 90 (fig.), 170 (fig.)
Herbert River 18°29′ 145°51′ 170 (fig.)
Hillston 33°29′ 145°32′ 147 (fig.)
Hinchinbrook Island 18°22′ 146°15′ 202 (fig.)
Holmes Jungle 12°24′ 130°56′ 40
Homebush Bay 33°51′ 151°04′ 206 (fig.)
Hunter Estuary 32°50′ 151°47′ 208

Iluka 29°24′ 153°21′ 57, 57 (fig.), 118, 236
Ingham 18°39′ 146°10′ 32, 47 (fig.)
Innisfail 17°32′ 146°01′ 8 (fig.)
Inverell 29°46′ 151°07′ 5 (fig.)
Iron Range 12°46′ 143°16′ 43

Jardine River 11°08′ 142°19′ 43, 45 (fig.)
Johnstone River 17°31′ 145°44′ 170 (fig.)
Jones Creek 37°25′ 149°24′ 126–7

Kakadu National Park 12°40′ 132°53′ 39 (fig.), 239, 242, 243
Katherine 14°28′ 132°16′ 40
King River Valley 42°10′ 145°32′ 225
Kinrara Basalt Flow 18°29′ 145°02′ 86
Kroombit Tops 24°22′ 151°01′ 58

Lake George 35°05′ 149°28′ 160, 164, 214–15, 218
Lake Euramoo 17°10′ 145°38′ 151 (fig.), 170 (fig.)
Lamb Range 17°10′ 145°40′ 170 (fig.)
Lamington National Park 28°12′ 153°05′ 121
Lamington Plateau 28°16′ 153°05′ 129, 130 (fig.)
Lightning Ridge 29°26′ 147°59′ 184
Limpinwood Nature Reserve 28°15′ 153°13′ 244, 130 (fig.)
Liverpool Creek 17°44′ 145°50′ 12 (fig.)
Liverpool Range 31°45′ 150°07′ 5 (fig.) 55, 79
Lockerbie Rainforests 10°48′ 142°28′ 45
Lord Howe Island 31°33′ 159°05′ 72–3, 73 (fig.), 74 (fig.), 119, 234, 242
Lower Gordon River Valley 42°26′ 145°35′ 225, 243
Loy-Yang 38°11′ 146°37′ 151 (fig.)

Lynchs Crater 17°24′ 145°40′ 151 (fig.), 164–6, 165 (fig.), 166 (table), 170 (fig.), 218

Macalister Foothills 16°42′ 145°50′ 47 (fig.)
Mackay 21°09′ 141°11′ 32, 46, 51, 54, 68
Macquarie Harbour 42°17′ 145°23′ 225
Maguires Creek 28°50′ 153°26′ 263 (fig.)
Malbon–Thompson Range 17°07′ 145°54′ 47 (fig.), 170 (fig.)
Manning River 31°53′ 152°34′ 262
McBride Region 18°23′ 144°51′ 86
McHenry Uplands 11°38′ 142°42′ 43
McIlwraith Range 13°50° 143°17′ 32, 43, 46
McPherson Range 28°18′ 152°40′ 131
Melbourne 37°49′ 144°58′ 258
Melville Island 11°35′ 131°10′ 38, 41
Mitchell 26°29′ 147°58′ 5 (fig.)
Mitchell Plateau 14°50′ 125°50′ 34–6, 156, 158 (fig.)
Monaro Range 36°18′ 148°55′ 5 (fig.)
Mossmann 16°28′ 145°23′ 47 (fig.)
Mossman River 16°28′ 145°21′ 170 (fig.)
Mount Banda Banda 31°10′ 152°26′ 61, 112, 112 (fig.)
Mount Bartle Frere 17°24′ 145°52′ 48, 152, 170 (fig.)
Mount Bellenden Ker 17°20′ 145°52′ 12, 47 (fig.), 48, 51 (fig.), 170 (fig.)
Mount Boss State Forest 31°09′ 152°23′ 100 (fig.), 102 (table), 126, 228 (fig.), 255 (fig.)
Mount Carbine Tableland 16°27′ 145°15′ 47 (fig.), 170 (fig.)
Mount Donna Buang 37°42′ 145°41′ 152
Mount Dromedary 24°19′ 151°42′ 32, 55, 62, 233
Mount Elliot 19°30′ 146°58′ 47 (fig.)
Mount Finnigan 15°50′ 145°16′ 47 (fig.), 170 (fig.)
Mount Glorious 27°20′ 152°46′ 54
Mount Gower 31°35′ 159°05′ 72, 73 (fig.)
Mount Halifax 146°22′ 19°07′ 47 (fig.), 91
Mount Hyland 30°10′ 152°27′ 61
Mount Kosciusko 36°27′ 148°16′ 2
Mount Lee 18°43′ 145°52′ 47 (fig.)

Mount Lidgbird 31°34′ 159°05′ 72, 73 (fig.)
Mount Moornapa 37°45′ 147°09′ 65
Mount Nothofagus Flora Reserve 28°20′ 152°39′ 244, 245 (fig.)
Mount Nothofagus 28°17′ 152°37′ 112
Mount Spec 18°57′ 146°11′ 47 (fig.)
Mount Spurgeon 20°08′ 144°16′ 189
Mount Surprise 18°09′ 144°19′ 5 (fig.)
Mount Tomah 33°33′ 150°25′ 8, 9 (fig.)
Mount Warning 28°23′ 153°17′ 6 (fig.), 223 (fig.), 236
Mount Windsor Tableland 23°38′ 141°40′ 47 (fig.), 91, 109, 169, 170 (fig.), 253–4
Mulgrave River 17°07′ 145°47′ 170 (fig.)
Murramarrang 35°32′ 150°25′ 56 (fig.)

Nadgee Nature Reserve 37°26′ 149°55′ 62, 63 (fig.)
Nandewar Range 30°18′ 150°32′ 5 (fig.)
Nepean Island 29°04′ 167°58′ 71
New England National Park 30°30′ 152°30′ 91 (fig.), 236
Nightcap Range 28°31′ 153°16′ 23, 236
Noah Creek 16°08′ 145°26′ 145, 170 (fig.)
Norfolk Island 29°05′ 168°00′ 71–2, 172
Normanby River 15°19′ 144°51′ 170 (fig.)
Nulla 28°25′ 149°55′ 5 (fig.)
Numbinbah Nature Reserve 28°15′ 153°21′ 244, 245 (fig.)

Obiri Rock 12°25′ 133°58′ 41 (fig.)
Oenpelli 12°20′ 133°04′ 38 (fig.)
Orange 33°17′ 149°06′ 5 (fig.)
Ord Estuary 15°10′ 128°11′ 205 (fig.)
Osborn Islands 14°21′ 125°59′ 36
Otway Ranges 38°27′ 143°58′ 64
Ourimbah 33°22′ 151°22′ 137

Pacific Palms 32°20′ 152°32′ 265 (fig.), 266 (fig.)
Palmer River 16°01′ 143°13′ 233
Paradise Gully 39°05′ 146°25′ 66
Perth 31°57′ 115°51′ 202
Phillip Island 38°28′ 145°14′ 71
Pitcairn Island 25°04′ 130°06′ 71

INDEX OF PLACE NAMES

Point Lookout 30°29′ 152°25′ 3 (fig.), 198
Port Essington 11°22′ 132°08′ 40

Queenstown 42°05′ 145°33′ 233
Quincan Crater 17°18′ 145°35′ 170 (fig.)

Richmond 33°36′ 150°45′ 222
Riversleigh 19°02′ 138°45′ 184–7, 186 (fig.), 185 (fig.), 195–6
Rockhampton 23°22′ 150°32′ 5 (fig.)
Rossville 15°54′ 145°17′ 233
Royal National Park 34°07′ 151°02′ 236
Russell River 17°22′ 145°58′ 170 (fig.)

Sealers Creek 39°01′ 146°23′ 66
Shoalhaven Gorge 35°02′ 150°02′ 70
Snowy Mountains 36°40′ 148°17′ 5 (fig.), 141
Springsure 24°07′ 148°05′ 5 (fig.)
Strzelecki Ranges 38°20′ 148°55′ 64

Sydney 33°53′ 151°13′ 204, 208, 219, 224, 225, 236

Tennant Creek 19°39′ 134°11′ 38
Terania Creek 28°37′ 153°18′ 96–7, 97 (fig.), 100–1, 237–8
Thornton Peak 16°10′ 145°23′ 47 (fig.), 170 (fig.), 183, 189
Toowoomba 27°34′ 151°57′ 5 (fig.)
Townsville 19°16′ 146°49′ 46
Towra Point 34°01′ 151°09′ 205 (fig.)
Tully River 17°50′ 145°42′ 170 (fig.)
Tweed River 28°18′ 153°27′ 224

Uluru 25°21′ 131°02′ 239, 242

Victoria 11°22′ 132°08′ 40
Victoria Park Nature Reserve 28°48′ 153°15′ 262–3
Victoria River 17°14′ 131°03′ 158 (fig.)

Warrumbungles 31°26′ 149°36′ 5 (fig.)

Washpool 29°20′ 152°22′ 98 (fig.), 238, 252
Washpool National Park 98 (fig.) 240 (fig.)
Watagan Mountains 32°57′ 151°14′ 234
Weipa 12°37′ 141°52′ 45
Wentworth 34°06′ 141°55′ 147 (fig.)
Wiangaree State Forest 28°38′ 152°58′ 108, 114
Willandra Lakes Region 33°12′ 145°07′ 242
Wilsons Promontory 39°05′ 146°25′ 63, 64, 65, 66, 202
Windsor 33°37′ 150°49′ 222
Wingham 31°52′ 152°22′ 221
Wingham Brush 31°52′ 152°22′ 262, 264 (fig.)
Wrights Lookout 30°30′ 152°25′ 61 (fig.)
Wyndham 15°28′ 128°06′ 36 (fig.), 205 (fig.)

Yarrawa Brush 34°35′ 150°34′ 62

VASCULAR PLANTS, SYSTEMATIC INDEX

Nomenclature follows Hnatiuk (1990)

Abrus precatorius L. FABACEAE 40, 171
Acacia MIMOSACEAE 27, 49 (*table*), 51, 72, 73, 84, 94 (*table*), 114, 126, 147, 172, 178
Acacia aulacocarpa Benth. MIMOSACEAE 129
Acacia dealbata Link. MIMOSACEAE 65 (*fig.*), 231 (*table*)
Acacia elata Benth. MIMOSACEAE 178
Acacia harpophylla Benth. MIMOSACEAE 70
Acacia melanoxylon R. Br. MIMOSACEAE 64, 66, 127, 178, 231, 231 (*table*)
Acacia polystachya Benth. MIMOSACEAE 46
Acacia shirleyi Maiden MIMOSACEAE 40
Acalypha eremorum Muell. Arg. EUPHORBIACEAE 70
Acmena MYRTACEAE 147, 166 (*table*), 232
Acmena hemilampra (F. Muell.) Merr. & Perry MYRTACEAE 43, 57
Acmena smithii (Poiret) Merr. & Perry MYRTACEAE 56, 60, 62, 65, 66, 79, 126–7, 127 (*fig.*), 130–1
Acradenia euodiiformis (F. Muell.) T. Hartley RUTACEAE 154 (*table*), 155
Acradenia frankliniae Kippist RUTACEAE 154 (*table*)
Acronychia acidula F. Muell. RUTACEAE 90 (*fig.*)
Acronychia crassipetala T. Hartley RUTACEAE 90 (*fig.*)
Acrostichum speciosum Willd. PTERIDACEAE 201
Adansonia digitata L. BOMBACEAE 36
Adansonia gregorii F. Muell. BOMBACEAE 36–7, 70
Aegiceras corniculatum (L.) Blanco MYRSINACEAE 203
Agastachys PROTEACEAE 67 (*table*)
Agathis ARAUCARIACEAE 19, 27, 143, 153, 225
Agathis atropurpurea B. Hyland ARAUCARIACEAE 49 (*table*), 50
Agathis microstachya J. Bailey & C. White ARAUCARIACEAE 49 (*table*)
Agathis robusta (F. Muell.) Bailey ARAUCARIACEAE 49 (*table*), 50, 56, 124
Ageratum ASTERACEAE 104
Alchornia ilicifolia (J. Smith) Muell. Arg. EUPHORBIACEAE 115–16
Alectryon SAPINDACEAE 172
Alectryon diversifolius (F. Muell.) S. Reyn. SAPINDACEAE 70
Alectryon forsythii Radlk. SAPINDACEAE 69
Alectryon subcinereus (A. Gray) Radlk. SAPINDACEAE 56
Aleurites moluccana (L.) Willd. EUPHORBIACEAE 90 (*fig.*)
Allocasuarina torulosa (Aiton) L. Johnson CASUARINACEAE 90 (*fig.*)
Allophyllus cobbe (L.) Blume SAPINDACEAE 39
Allosyncarpia ternata S. T. Blake MYRTACEAE 38–9, 39 (*fig.*)
Alphitonia excelsa (Fenzl) Benth. RHAMNACEAE 40, 157 (*table*), 230 (*table*)
Alphitonia petriei Braid RHAMNACEAE 114
Alphitonia whitei Braid RHAMNACEAE 90 (*fig.*)
Alstonia actinophylla (Cunn.) Schuman APOCYNACEAE 46
Alstonia scholaris (L.) R. Br. APOCYNACEAE 45, 171
Alstonia spectabilis R. Br. APOCYNACEAE 45
Anacolosa OLACACEAE 145
Angophora MYRTACEAE 225
Anodopetalum biglandulosum J. D. Hook. CUNONIACEAE 24, 64, 67 (*table*), 154 (*table*)
Anopterus glandulosus Labill. GROSSULARIACEAE 24, 67 (*table*), 154 (*table*)
Anopterus macleayanus F. Muell. GROSSULARIACEAE 154 (*table*), 155
Anredera cordifolia (Tenore) Steenis BASELLACEAE 236, 262
Apodytes brachystylis F. Muell. ICACINACEAE 90 (*fig.*)
Araucaria ARAUCARIACEAE 19, 27, 50, 153, 165, 172, 225
Araucaria bidwillii Hook. ARAUCARIACEAE 68, 216, 217 (*fig.*), 226, 261
Araucaria cunninghamii Cunn. ARAUCARIACEAE 43, 54, 55 (*fig.*), 56, 58, 68, 69, 72, 120, 123, 124, 177, 227, 230 (*table*), 261
Araucaria heterophylla (Salisb.) Franco ARAUCARIACEAE 72, 73 (*fig.*)
Archeria EPACRIDACEAE 67 (*table*)
Archidendropsis basaltica (F. Muell.) Nielsen MIMOSACEA 157 (*table*)
Archontophoenix ARECACEAE 49 (*table*)
Archontophoenix alexandrae (F. Muell.) H. A. Wendl. & Drude ARECACEAE 50, 129
Archontophoenix cunninghamiana (H. Wendl.) H. A. Wendl. & Drude ARECACEAE 59, 94 (*table*), 102 (*table*)
Argyrodendron STERCULIACEAE 56, 96 (*fig.*)
Argyrodendron actinophyllum Bailey STERCULIACEAE 54, 102 (*table*)
Argyrodendron trifoliolatum F. Muell. STERCULIACEAE 54, 115, 125, 134 (*fig.*), 230 (*table*)
Aristotelia ELAEOCARPACEAE 67 (*table*)
Asplenium australasicum (J. Smith) Hook. ASPLENIACEAE 55, 134 (*fig.*)
Astelia alpina R. Br. LILIACEAE 24
Atalaya hemiglauca (F. Muell) Benth SAPINDACEAE 157 (*table*)
Atherosperma MONIMIACEAE 18
Atherosperma moschatum Labill. MONIMIACEAE 24, 62, 64, 65 (*fig.*), 66, 67 (*table*), 111, 112, 113, 124, 153, 154, 154 (*table*), 231, 231 (*table*)
Athrotaxis CUPRESSACEAE 18, 66
Athrotaxis cupressoides D. Don CUPRESSACEAE 67 (*table*)
Athrotaxis selaginoides D. Don CUPRESSACEAE 67 (*table*), 114, 154 (*table*), 225
AUSTROBAILEYACEAE 159
Austromyrtus MYRTACEAE 147
Avicennia integra Duke VERBENACEAE 202
Avicennia marina (Forsk.) Vierh. VERBENACEAE 133, 200–1, 203, 205 (*fig.*), 206, 206 (*fig.*), 208, 209, 212, 213

Backhousia MYRTACEAE 147
Backhousia myrtifolia Hook.
    MYRTACEAE 57, 60, 70
Backhousia sciadophora F. Muell.
    MYRTACEAE 69
Baeckea gunniana Schauer
    MYRTACEAE 67 (*table*)
Baloghia lucida Endl.
    EUPHORBIACEAE 72
Bambusa arnhemica F. Muell.
    POACEAE 40
Banksia integrifolia L.f. PROTEACEAE 176
Barringtonia acutangula (L.) Gaertner
    LECYTHIDACEAE 157 (*table*)
Beauprea PROTEACEAE 172
Beilschmiedia LAURACEAE 172
Beilschmiedia bancroftii (Bailey) C.
    White LAURACEAE 216
Blechnum wattsii Tindale
    BLECHNACEAE 24, 67 (*table*)
Blepharocarya involucrigera F. Muell.
    ANACARDIACEAE 43
Bombax ceiba L. BOMBACEAE 40, 43, 45, 46, 50, 51, 157 (*table*), 171
Brachychiton STERCULIACEAE 37, 43
Brachychiton australis (Schott &
    Endl.) Terracino STERCULIACEAE 70
Brachychiton discolor F. Muell.
    STERCULIACEAE 69
Brachychiton diversifolius R. Br.
    STERCULIACEAE 157 (*table*)
Brachychiton rupestris (Lindley)
    Schumann STERCULIACEAE 70
Bruguiera RHIZOPHORACEAE 201, 206
Buchanania arborescens (Blume)
    Blume ANACARDIACEAE 40, 43, 46
Buchanania obovata Engl.
    ANACARDIACEAE 157 (*table*)

Cadellia pentastylis F. Muell.
    SIMAROUBACEAE 70
Calamus ARECACEAE 107 (*fig.*), 232
Calamus australis C. Martius
    ARECACEAE 129
Calamus muelleri H. A. Wendl. &
    Drude ARECACEAE 134 (*fig.*)
Caldcluvia paniculosa (F. Muell.)
    Hoogl. CUNONIACEAE 54, 55, 56, 102 (*table*), 124, 126, 230 (*table*)
Callitris collumellaris F. Muell.
    CUPRESSACEAE 45
Callitris macleayana (F. Muell.) F.
    Muell. CUPRESSACEAE 124
Calophyllum soulattri Burman f.
    CLUSIACEAE 40
Calophyllum australianum Vesque
    CLUSIACEAE 45
Calophyllum sil Lauterb. CLUSIACEAE 38

Calystegia marginata R. Br.
    CONVOLVULACEAE 127
Canarium australianum F. Muell.
    BURSERACEAE 39, 45, 46, 157 (*table*)
Canarium australasicum (Bailey)
    Leenh. BURSERACEAE 90 (*fig.*)
Canthium RUBIACEAE 70
Capparis CAPPARACEAE 43
Capparis lasiantha DC. CAPPARACEAE 157 (*table*)
Cardiospermum halicacabum L.
    SAPINDACEAE 236, 262
Cardwellia sublimis F. Muell.
    PROTEACEAE 90 (*fig.*)
Carissa ovata R. Br. APOCYNACEAE 70
Carpentaria acuminata (H. L. Wendl.
    & Drude) Becc. ARECACEAE 38, 40
Cassine australis (Vent.) Kuntze
    CELASTRACEAE 70
Castanospermum australe Hook.
    FABACEAE 60, 60 (*fig.*) 118, 122, 216, 232
Casuarina CASUARINACEAE 14, 73, 94 (*table*), 160, 164
CELASTRACEAE 68
Celtis paniculata (Endl.) Planchon
    ULMACEAE 72
Celtis philippinensis Blanco
    ULMACEAE 39
Cenarrhenes nitida Labill.
    PROTEACEAE 67 (*table*), 111
Ceriops RHIZOPHORACEAE 206
Ceratopetalum CUNONIACEAE 27
Ceratopetalum apetalum D. Don.
    CUNONIACEAE 22, 56, 57, 58, 59 (*fig.*), 61, 78, 82, 94 (*table*), 96 (*fig.*), 102 (*table*), 107, 108, 110, 112, 124, 126, 134, 134 (*fig.*), 154 (*table*), 155, 177, 222, 227–9, 228 (*fig.*), 230 (*table*), 238, 251, 255 (*fig.*)
Ceratopetalum succirubrum C. White
    CUNONIACEAE 78
Chrysanthemoides monilifera (L.) T.
    Norlindh ssp. *rotundata* (DC.) T.
    Norlindh ASTERACEAE 235
Chusquea POACEAE 113
Cinnamomum camphora T. Nees &
    C. Eberm LAURACEAE 107 (*fig.*), 136, 223, 236
Cissus hypoglauca A. Gray VITACEAE 65, 127
Cissus opaca F. Muell VITACEAE 70
Citriobatus spinescens (F. Muell.)
    Druce PITTOSPORACEAE 70
Citronella moorei (Benth.) R.
    Howard ICACINACEAE 55
Citronella smythii (F. Muell.) R.
    Howard ICACINACEAE 90 (*fig.*)
Clerodendron cunninghamii Benth.
    VERBENACEAE 157 (*table*)

Cochlospermum gillvraei Benth.
    BIXACEAE 43, 45, 51
Codonocarpus attenuatus (Hook.) H.
    Walter GYROSTEMONACEAE 119
Commersonia bartramia (L.) Merr.
    STERCULIACEAE 263
Coprosma quadrifida (Labill.)
    Robinson RUBIACEAE 67 (*table*), 111
Cordyline AGAVACEAE 166 (*table*)
Corynocarpus CORYNOCARPACEAE 172
Croton EUPHORBIACEAE 39
Croton insularis Baillon
    EUPHORBIACEAE 70
Cryptocarya corrugata C. White &
    Hyland LAURACEAE 90 (*fig.*)
Cryptocarya erythroxylon Maiden &
    Betche LAURACEAE 54, 230 (*table*)
Cryptocarya foveolata C. White &
    Francis LAURACEAE 61
Cryptocarya glaucescens R. Br.
    LAURACEAE 55, 102 (*table*), 230 (*table*)
Cryptocarya nova-anglica Hyland &
    Floyd LAURACEAE 61
Cryptocarya obovata R. Br.
    LAURACEAE 94 (*table*)
Cryptocarya rigida Meissner
    LAURACEAE 90 (*fig.*)
CUNONIACEAE 78, 166 (*table*)
Cyathea CYATHEACEAE 72, 164
Cyathodes EPACRIDACEAE 111

Dacrycarpus PODOCARPACEAE 145
Dacrydium PODOCARPACEAE 166
Daphnandra micrantha (Tul.) Benth.
    MONIMIACEAE 56, 94 (*table*), 102 (*table*)
Decaspermum MYRTACEAE 147
Dendrobium biggibum Lindley
    ORCHIDACEAE 43
Dendrocnide cordata (Winkl.) Chew
    URTICACEAE 227
Dendrocnide excelsa (Wedd.) Chew
    URTICACEAE 55, 56, 134, 134 (*fig.*), 135 (*fig.*)
Dendrocnide moroides (Wedd.) Chew
    URTICACEAE 107, 227
Dicksonia DICKSONIACEAE 164
Dicksonia antarctica Labill.
    DICKSONIACEAE 24, 64
Dietes robinsoniana (C. Moore & F.
    Muell.) Klatt IRIDACEAE 73
Diospyros ferrea (Willd.) Bakh.
    EBENACEAE 71
Diospyros humilis (R. Br.) F. Muell.
    EBENACEAE 39
Diploglottis australianus Leenh.
    SAPINDACEAE 94 (*table*)
Diploglottis cunninghamii (Hook.)
    J. D. Hook. SAPINDACEAE 56

DIPTEROCARPACEAE 168
Diselma CUPRESSACEAE 18, 67 (*table*)
Doryphora aromatica (Bailey) L. S. Smith MONIMIACEAE 78, 90 (*fig.*)
Doryphora sassafras Endl. MONIMIACEAE 22, 55, 56, 58, 61, 62, 78, 102 (*table*), 112, 115, 124, 126, 134 (*fig.*), 153, 154 (*table*), 155, 228, 230 (*table*)
Dracophyllum sayeri F. Muell. EPACRIDACEAE 50
Drypetes lasiogyna (F. Muell.) Pax & K. Hoffm. EUPHORBIACEAE 39
Duboisia myoporoides R. Br. SOLANACEAE 232
Duboisia hopwoodii (F. Muell.) F. Muell. SOLANACEAE 157 (*table*)
Dysoxylum MELIACEAE 172
Dysoxylum fraseranum (Adr. Juss.) Benth. MELIACEAE 54, 115, 225, 230 (*table*)
Dysoxylum oppositiflorum F. Muell. MELIACEAE 45, 46
Dysoxylum patersonianum F. Muell. MELIACEAE 72

EBENACEAE 69
Ehretia acuminata R. Br. BORAGINACEAE 56
Ehretia membranifolia R. Br. BORAGINACEAE 70
Ehretia saligna R. Br. BORAGINACEAE 157 (*table*)
Elaeocarpus ELAEOCARPACEAE 166 (*table*)
Elaeocarpus angustifolius Blume ELAEOCARPACEAE 38, 90 (*fig.*), 118
Elaeocarpus holopetalus F. Muell. ELAEOCARPACEAE 61, 62, 63, 64
Elaeocarpus obovatus G. Don ELAEOCARPACEAE 94 (*table*)
Elaeocarpus reticulatus Smith ELAEOCARPACEAE 65, 115
Elaeodendron curtipendulum Endl. CELASTRACEAE 72
Elattostachys nervosa (F. Muell.) Radlk. SAPINDACEAE 134 (*fig.*)
Endiandra palmerstonii (Bailey) C. White & Francis LAURACEAE 216, 227
Endiandra pubens Meissner LAURACEAE 90 (*fig.*)
Endiandra sieberi Nees LAURACEAE 126
Erythrina vespertilio Benth. FABACEAE 157 (*table*)
ESCALLONIACEAE 69
Eucalyptus MYRTACEAE 14, 51, 70, 72, 73, 84, 92, 100, 147, 164, 165, 172, 176, 177, 215, 225

Eucalyptus acmenoides Schauer MYRTACEAE 95, 96 (*fig.*)
Eucalyptus andrewsii Maiden MYRTACEAE 98 (*fig.*)
Eucalyptus botryoides Smith MYRTACEAE 59
Eucalyptus cypellocarpa L. Johnson MYRTACEAE 127
Eucalyptus deanei Maiden MYRTACEAE 93 (*fig.*)
Eucalyptus fastigata Deane & Maiden MYRTACEAE 127
Eucalyptus grandis Maiden MYRTACEAE 89, 90 (*fig.*), 91, 94, 95, 96 (*fig.*), 239
Eucalyptus gummifera (Gaertner) Hochr. MYRTACEAE 94 (*table*)
Eucalyptus intermedia R. Baker MYRTACEAE 89, 90 (*fig.*)
Eucalyptus marginata Smith MYRTACEAE 131
Eucalyptus microcorys F. Muell. MYRTACEAE 94 (*table*), 96 (*fig.*), 98 (*fig.*), 102 (*table*)
Eucalyptus pilularis Smith MYRTACEAE 91 (*fig.*), 94 (*table*), 95, 96, 96 (*fig.*), 97 (*fig.*), 237
Eucalyptus propinqua Deane & Maiden MYRTACEAE 94 (*table*), 98 (*fig.*)
Eucalyptus resinifera Smith MYRTACEAE 91
Eucalyptus robusta Smith MYRTACEAE 59
Eucalyptus saligna Smith MYRTACEAE 96 (*fig.*), 98 (*fig.*), 101, 102 (*table*), 109 (*fig.*)
Eucalyptus siderophloia Benth. MYRTACEAE 98 (*fig.*)
Eucalyptus umbra R. Baker MYRTACEAE 96 (*fig.*), 98 (*fig.*)
Eucryphia EUCRYPHIACEAE 28, 67 (*table*)
Eucryphia lucida (Labill.) Baillon EUCRYPHIACEAE 111, 124, 152, 231 (*table*), 232
Eucryphia milliganii J. D. Hook. EUCRYPHIACEAE 152, 154 (*table*)
Eucryphia moorei F. Muell. EUCRYPHIACEAE 62–3, 63 (*fig.*), 124, 152, 154, 154 (*table*), 155, 233
Euodia elleryana F. Muell. RUTACEAE 90 (*fig.*)
Euodia micrococca F. Muell. RUTACEAE 56
Eupatorium ASTERACEAE 104
EUPHORBIACEAE 68
EUPOMATIACEAE 159
Eupomatia laurina R. Br. EUPOMATIACEAE 90 (*fig.*), 116, 127

Euroschinus falcata J. D. Hook. ANACARDIACEAE 102 (*table*)
Excoecaria agallocha L. EUPHORBIACEAE 206, 208
Excoecaria parvifolia Muell. Arg. EUPHORBIACEAE 157 (*table*)

Falcatifolium PODOCARPACEAE 145
Ficus MORACEAE 39, 55, 100, 232
Ficus albipila (Miq.) King MORACEAE 43, 45, 50, 51
Ficus columnaris C. Moore & F. Muell. MORACEAE 74 (*fig.*)
Ficus leptoclada Benth. MORACEAE 90 (*fig.*)
Ficus nodosa Teijsm. & Binnend. MORACEAE 43
Ficus opposita Miq. MORACEAE 157 (*table*)
F. platypoda (Miq.) Miq. MORACEAE 45
Ficus racemosa L. MORACEAE 40, 171
Ficus stenocarpa F. Muell. MORACEAE 94 (*table*)
Ficus virens Aiton MORACEAE 40, 45, 50
Flindersia RUTACEAE 69, 70
Flindersia australis R. Br. RUTACEAE 123
Flindersia bourjotiana F. Muell. RUTACEAE 90 (*fig.*)
Flindersia brayleyana F. Muell. RUTACEAE 90 (*fig.*), 105, 226, 261
Flindersia collina Bailey RUTACEAE 70
Flindersia ifflaiana F. Muell. RUTACEAE 43
Flindersia maculosa (Lindley) Benth. RUTACEAE 157 (*table*)
Flindersia pimenteliana F. Muell. RUTACEAE 90 (*fig.*)
Francisodendron laurifolia (F Muell.) Steenis & Hyland STERCULIACEAE 90 (*fig.*)
Freycinetia PANDANACEAE 166 (*table*), 172

Ganophyllum falcatum Blume SAPINDACEAE 45, 46
Gardenia vilhelmii Domin RUBIACEAE 157 (*table*)
Geijera parviflora Lindl. RUTACEAE 70
Geissois benthamii F. Muell. CUNONIACEAE 124, 230 (*table*)
Gleichenia GLEICHENIACEAE 164
Glochidion ferdinandi (Muell. Arg.) Bailey EUPHORBIACEAE 115
Gmelina leichardtii (F. Muell.) Benth. VERBENACEAE 56, 225
Grevillea parallela J. Knight PROTEACEAE 157 (*table*)

VASCULAR PLANT, SYSTEMATIC INDEX

Grevillea robusta R. Br. PROTEACEAE 69, 123, 232
Grewia breviflora Benth. TILIACEAE 40
Gymnoschoenus sphaerocephalus (R. Br.) J. D. Hook. CYPERACEAE 98
Gyrocarpus americanus Jacq. HERNANDIACEAE 40, 45, 50, 51, 157 (*table*), 171

Halfordia scleroxyla F. Muell. RUTACEAE 90 (*fig.*)
Hedycarya MONIMIACEAE 172
Hedyscepe canterburyana (C. Moore & F. Muell.) H. A. Wendl. & Drude ARECACEAE 73
Helicia australasica F. Muell. PROTEACEAE 90 (*fig.*)
Histiopteris DENNSTAEDTIACEAE 164
Horsfieldia australiana S. T. Blake MYRISTICACEAE 38, 40
Howea belmoreana (C. Moore & F. Muell.) Becc. ARECACEAE 73
Howea forsteriana (C. Moore & F. Muell.) Becc. ARECACEAE 73, 74 (*fig.*), 119
Hydnophytum formicarium Jack RUBIACEAE 202
Hymenosporum flavum (Hook.) F. Muell. PITTOSPORACEAE 45

IDIOSPERMACEAE 159
Idiospermum australiense (Diels) S. T. Blake IDIOSPERMACEAE 119
Ilex AQUIFOLIACEAE 144
Ilex arnhemensis (F. Muell.) Loes. AQUIFOLIACEAE 38, 149
Imperata cylindrica L. (P. Beauv.) POACEAE 91, 106, 107 (*fig.*)

Jasminum didymum G. Forster subsp. *racemosum* (F. Muell.) P. Green 70

Lagarostrobos PODOCARPACEAE 150
Lagarostrobos franklinii (J. D. Hook.) Quinn PODOCARPACEAE 18, 67 (*table*), 125, 154 (*table*), 225, 231, 231 (*table*), 251
Lagerostroemia archeriana Bailey LYTHRACEAE 43, 45
Lagunaria patersonia (Andr.) G. Don MALVACEAE 72
LAURACEAE 153
Lantana camara L. VERBENACEAE 72, 89, 106, 107 (*fig.*), 136, 235
Lepidozamia hopei Regel ZAMIACEAE 143
Lepoidorrhacis mooreana (F. Muell.) Cook ARECACEAE 73
Leptospermum MYRTACEAE 63, 67 (*table*)

Leptospermum flavescens auct. MYRTACEAE 51
Leptospermum javanicum Blume MYRTACEAE 51
Leptospermum riparium D. Morris MYRTACEAE 66
Leptospermum rupestre J. D. Hook MYRTACEAE 67 (*table*)
Leptospermum wooroonooran Bailey MYRTACEAE 50, 51
Leucopogon maccraei F. Muell. EPACRIDACEAE 64
Licuala ARECACEAE 49 (*table*)
Licuala ramsayi (F. Muell.) Domin. ARECACEAE 50
Ligustrum lucidum Aiton OLEACEAE 136, 236
Ligustrum sinense Lour. OLEACEAE 136, 236
Litsea leefeana (F. Muell.) Merr. LAURACEAE 90 (*fig.*)
Litsea reticulata (Meissner) F. Muell. LAURACEAE 230 (*table*)
Livistona australis (R. Br.) C. Martius ARECACEAE 56 (*fig.*), 59, 66, 265 (*fig.*), 266 (*fig.*)
Livistona benthamii Bailey ARECACEAE 40
Livistona eastonii C. Gardner ARECACEAE 156, 158 (*fig.*)
Livistona humilis R. Br. ARECACEAE 156
Livistona mariae F. Muell. ARECACEAE 156
Livistona muelleri Bailey ARECACEAE 157 (*table*)
Lomatia fraseri R. Br. PROTEACEAE 64
Lophostemon MYRTACEAE 147
Lophostemon confertus (R. Br.) Peter G. Wilson & Waterhouse MYRTACEAE 56, 94 (*table*), 96, 97 (*fig.*), 98 (*fig.*), 100, 100 (*fig.*), 102 (*table*), 125, 237
Lophostemon suaveolens (Gaertner) Peter G. Wilson & Waterhouse 90 (*fig.*)
Lycopodium LYCOPODIACEAE 164
Lysiphyllum cunninghamii (Benth.) de Wit CAESALPINIACEAE 40

Macadamia ternifolia F. Muell. PROTEACEAE 231
Macadamia tetraphylla L. Johnson PROTEACEAE 231
Macaranga tanarius (L. Muell. Arg. EUPHORBIACEAE 171
Macfadyena unguis-cati (L.) A. Gentry BIGNONIACEAE 236, 261, 262
Macropteranthes keckwickii F. Muell. COMBRETACEAE 40

Macropteranthes leichardtii F. Muell. COMBRETACEAE 70
Mallotus phillippensis (Lam.) Muell. Arg. EUPHORBIACEAE 90 (*fig.*), 171
Marsdenia rostrata R. Br. ASCLEPIADACEAE 127
Melaleuca MYRTACEAE 45
Melaleuca leucadendra (L.) L. MYRTACEAE 34
MELIACEAE 69
Melia azeredach L. MELIACEAE 70, 171, 219
Microstrobos PODOCARPACEAE 67 (*table*)
Mimusops elengi L. SAPOTACEAE 45, 46
MONIMIACEAE 69, 78
Myristica muelleri Warb. MYRISTICACEAE 45
MYRTACEAE 147, 149, 159, 168, 176
Myrmecodia RUBIACEAE 202

Nauclea orientalis (L.) L. RUBIACEAE 43
Neofabricea myrtifolia (Gaertner) J. Thompson, MYRTACEAE 45
Neolitsea dealbata (R. Br.) Merr. LAURACEAE 90 (*fig.*), 115
Nestegis apetala (Vahl) L. Johnson OLEACEAE 72
Notelaea microcarpa R. Br. OLEACEAE 69, 157 (*table*)
Notelaea venosa F. Muell. OLEACEAE 63
Nothofagus FAGACEAE 18, 28, 113, 145, 147, 149–50, 152–3, 164, 171, 178
Nothofagus cunninghamii (Hook.) Oersted FAGACEAE 24, 63, 64, 66, 67 (*table*), 83, 111–14, 124, 132–3, 144, 152–3, 153 (*fig.*), 154, 154 (*table*), 231, 231 (*table*)
Nothofagus gunnii (J. D. Hook.) Oersted FAGACEAE 66, 114, 144, 155
Nothofagus moorei (F. Muell.) Krasser FAGACEAE 61, 63, 83, 112, 112 (*fig.*), 113, 124, 126, 134, 144, 152–3, 153 (*fig.*), 154, 154 (*table*), 155, 223, 224, 227, 228, 255 (*fig.*)
Nothofagus pumilio (Poepp & Endl) Krasser 155
Nypa ARECACEAE 145

Omalanthus populifolius Graham EUPHORBIACEAE 115, 119, 136
Opisthiolepis heterophylla L. S. Smith PROTEACEAE 90 (*fig.*)
Orites excelsa R. Br. PROTEACEAE 56, 61, 78, 102 (*table*), 112, 177, 230 (*table*)

Owenia acidula F. Muell. MELIACEAE 157(*table*)
Owenia cepiodora F. Muell. MELIACEAE 222

Pandanus aquaticus F. Muell. PANDANACEAE 34
Pandorea pandorana (Andrews) Steenis BIGNONIACEAE 65
Paraserianthes toona (Bailey) I. Nielsen MIMOSACEAE 50
Parsonsia APOCYNACEAE 70
Parsonsia straminea (R. Br.) F. Muell. APOCYNACEAE 154(*table*), 155
Pennantia ICACINACEAE 172
Pennisetum clandestinum Chiov. POACEAE 263
Peperomia PIPERACEAE 172
Persoonia silvatica L. Johnson PROTEACEAE 64.
Phyllocladus PHYLLOCLADACEAE 18, 67(*table*), 166
Phyllocladus aspleniifolius (Labill.) J. D. Hook. PHYLLOCLADACEAE 83, 154(*table*), 225, 231(*table*)
Pinus PINACEAE 109(*fig.*)
Pinus radiata D. Don PINACEAE 229
Pisonia grandis R. Br. NYCTAGINACEAE 71, 71(*fig.*)
Pittosporum bicolor Hook. PITTOSPORACEAE 64
Pittosporum phillyraeoides DC. PITTOSPORACEAE 157(*table*)
Pittosporum undulatum Vent. PITTOSPORACEAE 56, 65, 136
Planchonella SAPOTACEAE 172
Planchonella cotinifolia (A. DC.) Dubard SAPOTACEAE 70
Planchonella obovata (R. Br.) Pierre SAPOTACEAE 71
Planchonella pohlmaniana (F. Muell.) Dubard SAPOTACEAE 157(*table*)
Planchonella xerocarpa (Benth) H. J. Lam. SAPOTACEAE 38
Pleiogynium timorense (DC.) Leenh. ANACARDIACEAE 50
Podocarpus PODOCARPACEAE 45, 67(*table*), 191
Podocarpus lawrencei J. D. Hook. PODOCARPACEAE 64
Podocarpus neriifolius Lamb. PODOCARPACEAE 45
Polyscias ARALIACEAE 115
Polyscias elegans (C. Moore & F. Muell.) Harms ARALIACEAE 102(*table*), 114, 134(*fig.*)
Polystichum proliferum (R. Br.) Presl. ASPIDIACEAE 24, 65
Pongamia pinnata (L.) Pierre FABACEAE 45, 171

Prostanthera lasianthos Labill. LAMIACEAE 64
Prostanthera ovalifolia R. Br. LAMIACEAE 63
PROTEACEAE 69, 73, 144, 149, 168, 176
Prumnopitys PIODOCARPACEAE 145
Psidium guajava L. MYRTACEAE 72
Pteris PTERIDACEAE 164
Pullea stutzeri (F. Muell.) Gibbs CUNONIACEAE 90(*fig.*)

Quintinia GROSSULARIACEAE 154, 171, 172
Quintinia quatrefagesii F. Muell. GROSSULARIACEAE 155
Quintinia sieberi A. DC. GROSSULARIACEAE 124, 154, 154(*table*)

Rapanea MYRSINACEAE 166(*table*)
Rapanea howittiana Mez. MYRSINACEAE 65
Rhopalostylis baueri Wendl. & Drude ARECACEAE 72
Richea EPACRIDACEAE 67(*table*)
Richea pandanifolia J. D. Hook EPACRIDACEAE 67(*table*)
RHIZOPHORACEAE 206, 207
Rhizophora RHIZOPHORACEAE 201, 206, 207, 213
Rhizophora stylosa Griffith RHIZOPHORACEAE 202(*fig.*)
Rhodamnia MYRTACEAE 147
Rhodamnia argentea Benth. MYRTACEAE 94(*table*)
Rhodamnia trinervia (Smith) Bl. MYRTACEAE 94(*table*)
Rhododendron lochae F. Muell. ERICACEAE 50
RUBIACEAE 69
Rubus hillii F. Muell. ROSACEAE 127
RUTACEAE 68, 168

SAPINDACEAE 166(*table*)
Sarcopetalum harveyanum F. Muell. MENISPERMACEAE 127
Schefflera actinophylla (Endl.) Harms ARALIACEAE 232
Schizomeria ovata D. Don CUNONIACEAE 55, 57, 58, 78, 94(*table*), 102(*table*), 110, 124, 230(*table*)
Schizomeria whitei Mattf. CUNONIACEAE 78
Sclerostegia arbuscula (R. Br.) Paul G. Wilson CHENOPODIACEAE 200–1
Scolopia braunii (Klotzsch) Sleumer FLACOURTIACEAE 90(*fig.*)
Senecio ASTERACEAE 104

Siphonodon pendulus Bailey CELASTRACEAE 157(*table*)
Sloanea australis (Benth.) F. Muell. ELAEOCARPACEAE 102(*table*)
Sloanea woollsii F. Muell. ELAEOCARPACEAE 54, 56, 102(*table*), 112, 112(*fig.*), 115, 230(*table*)
Smilax australis R. Br. SMILACACEAE 127
Solanum mauritianum Scop. SOLANACEAE 104, 108, 235, 263
Stenocarpus reticulatus C. White PROTEACEAE 90(*fig.*)
Stenocarpus sinuatus (Loudon) Endl. PROTEACEAE 232
Stenotaphrum secundatum (Walter) Kuntze POACEAE 263
Sterculia quadrifida R. Br. STERCULIACEAE 39, 40
Streblus MORACEAE 172
Streblus pendulinus (Endl.) F. Muell. MORACEAE 72
Strychnos lucida R. Br. LOGANIACEAE 39, 40
Swietenia MELIACEAE 219
Syncarpia glomulifera (Smith) Niedenzu MYRTACEAE 100–1, 102(*table*)
Syncarpia hillii Bailey MYRTACEAE 56
Synoum glandulosum (Smith) Adr. Juss. MELIACEAE 94(*table*), 102(*table*)
Syzygium MYRTACEAE 38, 147, 166(*table*), 232
Syzygium eucalyptoides (F. Muell.) B. Hyland MYRTACEAE 157(*table*)
Syzygium cormiflorum (F. Muell.) B. Hyland MYRTACEAE 116
Syzygium fibrosum (Bailey) T. Hartley & Perry MYRTACEAE 45
Syzygium forte (F. Muell.) B. Hyland MYRTACEAE 45
Syzygium leuhmannii (F. Muell.) L. Johnson MYRTACEAE 57, 90(*fig.*)
Syzygium nervosum DC. MYRTACEAE 40

Tamarindus indica L. CAESALPINACEAE 40
Tasmannia WINTERACEAE 67(*table*), 119
Tasmannia sp. aff. *xerophila* WINTERACEAE 64
Tasmannia lanceolata (Poiret) A. C. Smith WINTERACEAE 24, 64, 111
Tecomanthe BIGNONIACEAE 172
Telopea oreades F. Muell. PROTEACEAE 64
Telopea truncata (Labill.) R. Br. PROTEACEAE 67(*table*)

Terminalia COMBRETACEAE 45
Terminalia canescens (DC.) Radlk COMBRETACEAE 157 (*table*)
Terminalia sericocarpa F. Muell. COMBRETACEAE 40, 45
Tetrameles nudiflora R. Br. DATISCACEAE 43
Toona australis (F. Muell.) Harms MELIACEAE 56, 70, 94 (*table*), 122, 124–5, 133, 134 (*fig.*), 171, 219–22, 220 (*fig.*)
Tradescantia albiflora Kunth COMMELINACEAE 236
Trema tomentosa subsp. *aspera* (Brongn.) Hewson ULMACEAE 115
Tristaniopsis collina Peter G. Wilson & Waterhouse MYRTACEAE 58
Tristaniopsis laurina (Smith) Peter G. Wilson & Waterhouse MYRTACEAE 60, 65, 94 (*table*), 130–1

Trochocarpa sp. nov. aff. *laurina* EPACRIDACEAE 61
Trochocarpa laurina R. Br. EPACRIDACEAE 102 (*table*), 126

Ventilago viminalis Hook. RHAMNACEAE 70
Vesselowskya rubifolia (F. Muell.) Pampan CUNONIACEAE 61, 154, 154 (*table*)
Vitex VERBENACEAE 172

Waterhousea floribunda (F. Muell.) B. Hyland MYRTACEAE 60
Welchiodendron longivalve (F. Muell.) Peter G. Wilson & Waterhouse MYRTACEAE 45, 46
Wilkiea macrophylla (Cunn.) A. DC. MONIMIACEAE 133
WINTERACEAE 144, 159

Wrightia pubescens R. Br. APOCYNACEAE 39
Wrightia saligna (R. Br.) Benth. APOCYNACEAE 157 (*table*)

Xanthophyllum octandrum (F. Muell.) Domin XANTHOPHYLLACEAE 90 (*fig.*)
Xanthostemon MYRTACEAE 39
Xanthostemon eucalyptoides F. Muell. MYRTACEAE 38

Zanthoxylum pinnatum (Forster & G. Forster) Druce RUTACEAE 72
Ziziphus oenoplia (L.) Miller RHAMNACEAE 43
Ziziphus quadrilocularis F. Muell. RHAMNACEAE 35
Zygogynum WINTERACEAE 172

GENERAL INDEX

An 'equals' sign indicates that a cross-reference is to be made to the Vascular Plants Index

Aborigines
  and fire 36, 98, 99, 126, 185, 186, 215, 217, 218
  history of occupation of Australia 214–15
  in rainforest 215–19
  use of mangrove 208
  use of rainforest resources 216–17
Acanthiza ewingii (Tasmanian thornbill) 192
Acanthiza katherina (mountain thornbill) 191
acrobatids 185 (*fig.*), 186
Africa 1 (*table*)138, 156, 159, 258
agriculture, clearing of rainforest for 222–5, 249
Ailuroedus dentirostris (tooth-billed catbird) 253
Albert's lyrebird 191
Albian stage 142, 144, 159
Alectura lathami (brush turkey) 120, 191, 193, 234
algae 201, 211
allelopathy 123
alliances, floristic 15, 22–4
aluminium
  accumulation by plants 78
  in soil 78, 82
Amazonia 258
amenochory 110
America, N 1 (*table*)
America, S. 1 (*table*), 2, 138, 144, 159, 171, 178, 179
amphibia 195–6
Andes 84
Angophora-bloodwood pollen type 147
Anodopetalum–Anopterus alliance 24
Anous minotus (black noddy) 71, 72 (*fig.*)
Antarctica 1 (*table*), 138, 139, 142, 149, 159, 171
  icecap 142
Antarctic biogeographic element 15, 137, 148, 149, 166
Antechinus 189
Antechinus flavipes (yellow-footed antechinus) 182 (*table*)
Antechinus godmani (Atherton antechinus) 181 (*table*)
Antechinus leo (cinnamon antechinus) 181 (*table*)

Antechinus stuartii (brown antechinus) 180, 182 (*table*)
Antechinus swainsonii (dusky antechinus) 182 (*table*)
ant plants 202
ants 119, 202
Apis mellifera (honey bee) 116–17
Aplonis metallica (metallic starling) 191
apomixis 115–16
araucarian notophyll vine forest 27, 29 (*table*), 49 (*table*), 68–9, 76 (*table*), 80 (*fig.*), 81 (*fig.*), 224, 227, 228
area of rainforest in Australia 83 (*fig.*), 84, 84 (*table*)
Argentina 84
aridity, development of 142, 146 (*fig.*)
ash bed effect 93
Asia 1 (*table*), 188; *see also* South-east Asia
Asian water buffalo 41, 43, 235
Ash Wednesday bushfires 95
Assa darlingtonii (pouched frog) 196
association analysis 22
associations 14, 23
Atherton antechinus 181 (*table*)
Atherton scrubwren 191
Atrichornis clamosus (noisy scrub-bird) 192
Atrichornis rufescens (rufous scrub-bird) 61, 192
Australian Constitution 239
Australian Federation 239
Australian Heritage Commission 239, 241, 246
Australian Conservation Foundation (ACF) 237
Australian High Court 243, 250, 258
autochthonous biogeographic element 148, 149, 166
avifauna 190–5
avifauna, evolution of 194–5

ballast water 209
bamboo 113
Banks, Sir Joseph 219
bark, conferring protection from fire 78, 126, 127
basal area, trends of variation in 83
Bass Strait 154, 212

bats 187–9
  insectivorous 109
  fruit 117, 124, 188
bauxite 35, 36
beech = *Nothofagus* spp.
beech, white = *Gmelina leichardtii*
beech forest 29 (*table*); *see also* cool temperate rainforest
Bennett's tree kangaroo 181 (*table*), 190
bitou bush = *Chrysanthemoides monilifera* ssp. *rotundata*
black bean = *Castanospermum australe*
Black Bean forest type 59, 60
blackbutt = *Eucalyptus pilularis*
black noddy 71, 72 (*fig.*)
black striped wallaby 182 (*table*)
blady grass = *Imperata cylindrica*
Bligh, Captain 71
boab = *Adansonia gregorii*
booyong = *Argyrodendron* spp.
Booyong forest type 54, 58
Booyong–Coachwood forest type 56
bottle tree = *Brachychiton* spp.
bottle tree scrub 29 (*table*)
boundaries to rainforest 55, 56, 86–95, 87 (*fig.*), 88 (*figs*), 90 (*fig.*)
  stability of 87–91, 95–9
Bounty, HMS 71
Bower's shrikethrush 191
bracken 94
brackish water forest 200
Brazil 84
breeding systems of plants 115
bridled honeyeater 116, 191
brigalow = *Acacia harpophylla*
brigalow belt 70, 224
brown antechinus 180, 182 (*table*)
brown gerygone 191
brush 14
brushbox = *Lophostemon confertus*
brush turkey 120, 191, 193, 234
bryophytes 60, 73, 108, 109, 178, 179, 202
Bubalis bubalis (Asian water buffalo) 41, 43, 235
budgerigar 180
buffalo, Asian water 41, 43, 235
Bufo marinus (cane toad) 190, 196
Bukit Timah 264
Bunya pine = *Araucaria bidwillii*
Burma 139

bush rat 182 (*table*)
bush tucker 23
butterflies 196–7
button grass = *Gymnoschoenus sphaerocephalus*
buttresses 25, 50, 112, 112 (*fig.*)

Cainozoic 141, 156
California 159, 212
callidendrous rainforest 66, 67 (*table*), 68 (*fig.*), 132
camphor laurel = *Cinnamomum camphora*
cane toad 190, 196
Canis familliaris dingo (dingo) 182 (*table*), 187, 218, 235
canthrophily 116
Cape York melomys 181 (*table*)
Cape York Peninsula 2, 22, 43–6, 44 (*fig.*), 46 (*fig.*), 57, 116, 137, 176, 181 (*table*), 183, 190
Cape York rat 181 (*table*), 190
cats, feral 190, 235
carabeen, yellow = *Sloanea woollsii*
cassowary 118, 118 (*fig.*), 191, 193, 194
Casuarius casuarius (cassowary) 118, 118 (*fig.*), 191, 193, 194
cattle 46, 122
cedar-getters 220 (*fig.*), 220–2, 224
cedar, onion = *Owenia cepiodora*
cedar, red = *Toona australis*
cedar, silviculture of 226
cedar, uses of timber 219–21, 225, 250
cedar, white = *Melia azeredach*
cedar tip moth 133, 226
celery top pine = *Phyllocladus aspleniifolius*
Ceratopetalum apetatalum–Doryphora sassafras suballiance 22
Cercartetus caudatus (long-tailed pygmy possum) 181 (*table*)
Cercartetus nanus (eastern pygmy possum) 182 (*table*)
Chalara australis (cause of myrtle wilt) 132–3
charcoal 97, 147, 148, 164–5, 165 (*fig.*), 169, 214–15
China 85
Chile 62, 110, 113, 155
cinnamon antechinus 181 (*table*)
chowchilla 191
classification, of vegetation 13–14
Clavatipollenites (pollen type) 159
climatic change 142, 143
climax species 104, 105, 106, 110, 113, 114, 115, 120, 127
cloud stripping 8, 55

coachwood = *Ceratopetalum apetalum*
coachwood forest 29 (*table*)
Coachwood–Crabapple forest type 101
Coachwood–Sassafras forest type 58
Collins, Robert 236
Collocalia esculenta (glossy swiftlet) 191
Collurincla boweri (Bower's shrike-thrush) 191
common brushtail possum 182 (*table*)
common ringtail possum 182 (*table*)
Commonwealth Scientific and Industrial Research Organisation (CSIRO) 33
compensatory mechanisms 123
conservation, history in Australia 236–9
constitution of Australia 258
continental drift 137–43, 148
Cook, Captain 222
cool temperate rainforest 18, 60–5, 79, 80 (*fig.*), 81 (*fig.*), 110–14, 124, 126, 233
coppice growth 106, 107, 113, 121, 127
coral cays 71
coriaceous leaves 19
corkwood = *Caldcluvia paniculosa*
Corkwood–Sassafras–Crabapple–Silver Sycamore forest type 55
Costa Rica 115
crabapple = *Schizomeria ovata*
crabs 211
Crateroscelis gutteralis (fernwren) 191
crayfish 198
Cretaceous 139, 141, 142, 143, 144, 149, 150, 158, 159, 184
crocodiles 211
Crocodylus johnstonii (freshwater crocodile) 211
Crocodylus porosus (saltwater crocodile) 211
crown cover 17
Cunningham, Allan 33
cycads 143
cycadophytes 143
cyclones 12, 83, 89, 106, 128–9, 160, 206–8, 260
  cyclone Agnes 128–9
  cyclone Kathy 206, 208
  cyclone Peter 12
  cyclone Tracey 129
  cyclone Winifred 129, 253
Cyttaria 168, 178

Dactylopsila trivirgata (striped possum) 181 (*table*)
dasyurids 186

Dasyurus maculatus (spotted-tailed quoll) 182 (*table*)
day length 144
defoliation by insects 133–5, 134 (*fig.*), 135 (*fig.*)
  measurement of 133–4
dendrochronology 124–5
Dendrolagus 186, 189
Dendrolagus bennettianus (Bennett's tree kangaroo) 181 (*table*), 190
Dendrolagus lumholtzi (Lumholtz's tree kangaroo) 181 (*table*), 190
Dicksonia–Polystichum alliance 24
dieback, crown 228–9
dieback, jarrah 131
dingo 182 (*table*), 187, 218, 235
diprotodontids 187
disturbance 18, 23
  agents of 125–36
  role in temperate rainforest 113–14
diversity, maintenance of 123
DNA hybridization 194
dogs, feral 190
Dromornithidae 194
drought 8, 89
dry rainforest 14, 19, 21 (*fig.*), 29 (*table*); *see also* ecofloristic provinces C_1, C_2 and vine thicket
Ducula spilorrhoa (Torres Strait pigeon) 117
duricrust 35
dusky antechinus 182 (*table*)

Eastern Highlands, *see* Great Dividing Range
eastern pygmy possum 182 (*table*)
eclectus parrot 191
Eclectus rotatus (eclectus parrot) 191
ecofloristic regions 30–2, 31 (*fig.*)
ecofloristic provinces
  A_1 30, 51–63, 75, 78, 84
  A_2 31, 51–63, 75, 78, 84
  A_3 31, 63–6
  B_1 31, 36, 40, 46, 51, 84
  B_2 31, 46, 51, 84
  B_3 31, 36, 40, 46, 51, 84
  C_1 31, 66, 68–70, 75
  C_2 31, 66, 68–70, 75
echidna 182 (*table*)
Echymipera rufescens (rufous spiny bandicoot) 181 (*table*)
ecological drift 98–9
ecological sifting model 149
Elleschodes hamiltoni, pollinating *Eupomatia* 116
emu 117, 180, 194
endemism 156, 168
environmental impact statements 238
Environmental Planning and Assessment Act (1979) 238

Eocene 142, 144, 145, 147 (*fig.*), 149, 184, 212
epiphytes 15, 20–1 (*fig.*), 23, 24, 25, 50, 202
epiphytic establishment 123–4
Euastacus aqullus 198
Euastacus polyetosus 198
Euastacus reductus 198
Euastacus spinosus 198
Eungella honeyeater 191
Europe 1 (*table*), 2, 142, 159
Euschemon rafflesia (regent skipper butterfly) 197

fauna
  aquatic 197–8
  community organization 198–9
  composition 180, 181 (*table*)
  factors affecting 183
  invertebrate 196–7
  of Riversleigh 184, 185 (*fig.*), 186 (*fig.*)
  of Hamilton 184, 185 (*fig.*)
  of New Guinea 181 (*table*), 184, 186 (*fig.*)
fawn-footed melomys 182 (*table*)
Federation, of Australia 239
feral animals 41, 42, 46
fernwren 191
fig = *Ficus* spp.
Fig–Giant Stinger forest type 55
fire 18, 36, 41, 78, 86, 87, 92–9, 113, 120, 125–8, 164, 169, 214–15, 260, 261–2, 254
  and Aborigines 36, 98, 99, 126, 185, 186, 217, 218
  and rainforest boundaries 86–9, 95–9
  and regeneration of tall open forest 92–9
  effects on soil temperature 120
  in Kalimantan 125
fire climax model 95
First Fleet 218
fish 197, 211
flowering 115
flooded gum = *Eucalyptus grandis*
flooding 18, 130–1
floristics 14–15, 19, 25
  classification of rainforest vegetation 22–4
forestry 225–33, 248–57
Forestry Commission of NSW 100, 227, 229, 237, 238, 239, 249
  forest classification 19, 22
  indigenous forest policy 229
Forestry Department of Queensland 131, 226, 227, 249, 256
Forestry Revocation and National Parks Reservation Acts (1983, 1984) 238

formations, vegetation 15
fossil record, interpretation of 143
fox 235
Fraser, Malcolm 243
Friends of the Earth 237
frogs 195–6, 195 (*fig.*)
  gastric brooding 234–5
frost 11 (*fig.*), 12, 48, 109–10, 109 (*fig.*), 129–30
fruit bats 117, 124, 188
fungi 143, 178

gallery forest 20 (*fig.*), 40, 45, 59, 60, 118, 146, 216
gap-filling model 102–14, 105 (*fig.*), 107 (*fig.*)
gastric brooding frogs 234–5
geocols 157
Geoffroyus geoffroyi (red cheeked parrot) 191
germination 111
  trigger for 120–2
Gerygone mouki (brown gerygone) 191
glacial processes 4
glaciation, Quaternary 4, 7, 153
glaciation, Permian 4
glossy swiftlet 191
Glycichaera fallax (green-backed honeyeater) 191
golden bowerbird 253
goldmining 233
Gondwana 36, 138, 139, 142, 149, 159, 167, 188, 171, 178, 179, 194, 195
graceful honeyeater 116, 191
granite 39, 50, 86
Great Barrier Reef 70, 71
Great Dividing Range 1–2, 7, 8, 43, 61, 139–41, 140 (*fig.*), 148, 156, 159 (*fig.*), 169
Great Escarpment 1, 3 (*fig.*), 48, 139–41, 140 (*fig.*)
green-backed honeyeater 191
greenhouse effect 259–60
Greenpeace 237
green ringtail possum 181 (*table*)
grey cuscus 181 (*table*), 190
grey headed fruit bat 117
guano 71
Gulf of Carpentaria 2
Gulf Stream 142
Gunei 218

hail 8, 129, 130 (*fig.*)
Hastings region 238
Hawke, R. J. L. 243
Hawaii 172, 212, 231
hazard reduction burning 128
Headland Brushbox forest type 57
heath forest 45, 57, 85

Heidewald 45, 45 (*fig.*), 57
helmeted friarbird 191
Hemibelideus 189
Hemibelideus lemuroides (lemuroid ringtail possum) 181 (*table*), 189
Herbert River ringtail possum 181 (*table*), 189
Herbert, Xavier 257
herbicides
  to control stinging trees 107–8
  in regeneration projects 282–3
herbivory by insects 133–5, 206–8
Heritage Act (1977) 211
herpetofauna 195–6
High Court of Australia 243, 250, 258
honey 116–17, 232
honey-bees 116–17, 232
honeyeaters 116–17, 191, 210
hoop pine = *Araucaria cunninghamii*
Hoop Pine forest type 69
horticultural use of rainforest species 232–3
Hulitherium 187
Hunter Valley 78
Huon pine = *Lagarostrobos franklinii*
hydroelectricity power generation schemes 231, 234, 243, 248
Hydromys chrysogaster (water rat) 183 (*table*)
Hylidae 195
hypersalinity 204
Hypsiprymnodon 186, 189, 190
Hysiprymnodon moschatus (musky rat kangaroo) 181 (*table*), 186, 189–90
Hypsipyla robusta (cedar tip moth) 133, 226

Illawarra 61, 62, 137, 221
implicate rainforest 66, 67 (*table*), 68 (*fig.*), 132
Indo-Australian plate 139
India 138, 156, 168
Indo-Malayan biogeographic element 15, 137, 148, 149, 166
Indonesia 187, 208
International Union for the Conservation of Nature 242
introduced species 235–6
invasion model for rainforest 148–9
invertebrates 196–7
Irian avifaunal element 191
Iridomyrmex 202
Isaacs, Justice S. 100–1, 238
Israel 159

Japan 84
jarrah = *Eucalyptus marginata*
jungle fowl 40–1, 193
Jurassic 138, 139, 142–4, 149

Kalimantan 125
kangaroos 180
kangaroos, tree 181 (*table*), 186
Kawau Island 234
KIAMBRAM model of forest regeneration 108
Kimberley region 30, 33–7, 34 (*fig.*), 35 (*fig.*), 36 (*fig.*), 38, 46, 158 (*fig.*)
King Billy Pine = *Athrotaxis selaginoides*
King, Governor 225
knee roots 201
koala 180, 187

lace bark tree = *Brachychiton discolor*
Lachlan Valley 147–8, 147 (*fig.*)
Lahey, R. 236
landbridges 149
LANDSAT 33
landslips 99, 130, 133
large billed scrubwren 191
laterite 46
Laos 85
Laura Basin 43
Laurasia 149, 164
leaf longevity 133, 134
leaf margins 83
leaf size 20–1 (*fig.*), 25, 27 (*table*)
 trends of variation in 82 (*fig.*), 83, 171
 use in physiognomic classification 26–9, 27 (*table*)
leagues in NSW forest classification 19
Lechriodus intergerius 196
legislation
 Environmental Planning and Assessment Act (1979) 238
 Forestry Revocation and National Parks Reservation Acts (1983, 1984) 49
 Heritage Act (1977) 221
 National Parks and Wildlife Conservation Act (1975) 243
 World Heritage properties Conservation Act (1983) 243, 246, 248, 258
lemuroid ringtail possum 181 (*table*), 189
lesser sooty owl 191
Lewin's honeyeater 116
Lichenostomus frenatus (bridled honeyeater) 116, 191
Lichenostomus hindwindi (Eungella honeyeater) 191
lichens 59 (*fig.*), 178, 202
light-demanding species 103
lightning 18, 98, 103, 128, 164, 208
lignotubers 92, 124, 126–7, 127 (*fig.*)

lilly pilly = *Acmena*, *Syzygium* spp.
litter fall 8, 212
littoral rainforest 33, 40, 56–8, 57 (*fig.*), 58 (*fig.*), 85, 216, 234, 262
logging 105–6, 108, 250–5
 effects on avifauna 108–9, 253
 effects on microclimate 109–10
 models of effects 108, 253
 recovery from 108–9
 see also forestry
logrunner 191
long-nosed bandicoot 182 (*table*)
long-nosed potoroo 182 (*table*)
long-tailed pygmy possum 181 (*table*)
Lopholaimus antarcticus (topnot pigeon) 191
Lord Howe Rise 139, 173
Lumholtz's tree kangaroo 181 (*table*), 190

macadamia nut 231
Macassans 40, 219
Macleay's honeyeater 116
macrofossils 43
macropodids 186
Macropus dorsalis (black-striped wallaby) 182 (*table*)
Macropus parma (parma wallaby) 182 (*table*), 234
Madagascar 36, 138, 168
Magela floodplain 213
mahogany, white = *Eucalyptus acmenoides*
Malaya 115, 139, 198
Malaysia 168, 208
Malesia 45, 57, 85, 150, 156, 172, 173, 173 (*fig.*)
mangroves 145, 200–13
 development pressure on 208–9, 210 (*fig.*)
 distribution 202–3
 ecophysiology of 204–6
 effects of disturbance 206–8
 fauna of 209–11
 and fisheries 210–12
 insect herbivory on 206–8
 pathogens of 208
 pollution impacts 209
 succession in 204
 zonation 204
Manucodia keraudrenii (trumpet manucode) 191
maple, Queensland = *Flindersia brayleyana*
mariculture 208
marsupial mole 187
megafauna 217
Megapodes freycinetia (jungle fowl) 40

Megapodius reinwardt (jungle fowl) 193
megatherm 30, 167 (*fig.*)
Meliphagidae 116, 210
Meliphaga gracilis (graceful honeyeater) 116, 191
Meliphaga lewinii (Lewin's honeyeater) 116
Meliphaga notata (yellow spotted honeyeater) 116, 191
Melomys 187, 188, 189
Melomys capensis (Cape York melomys) 186 (*table*)
Melomys cervinipes (fawn-footed melomys) 182 (*table*)
Melomys hadrourus (Thornton Peak melomys) 181 (*table*), 183
Menura 193
Menura albertii (Albert's lyrebird) 191
mesophyll vine forest 26, 27, 29 (*table*), 34, 38, 80 (*fig.*), 81 (*fig.*), 123, 160, 164, 193, 259
mesotherm 30
metallic starling 191
Mexico 84
Microhylidae 196
microphyll fern forest 18, 28, 29 (*table*), 80 (*fig.*), 81 (*fig.*), 152, 171, 192
microphyll moss forest 18, 28, 29 (*table*), 81 (*fig.*), 152, 171, 259
microtherm 30, 167 (*fig.*)
mining 46, 233–4
Miocene 142, 146–8, 146 (*fig.*), 147 (*fig.*), 150, 151 (*fig.*), 166, 168, 172, 176–7, 178, 184, 187, 214
mistletoe 201
mixed forest 18, 50
molluscs, terrestrial 196
monsoon forest 15, 29 (*table*), 76–7 (*table*), 79, 155–7
montane forest 50–1, 65, 73
mountain brushtail possum 182 (*table*)
mountain thornbill 191
Mountain Walnut forest type 61
Muridae 187
murids 186
Murinae 187
Murray Basin 2, 141
musky rat kangaroo 181 (*table*), 186, 189–90
Myobatrachidae 195
myrmecochory 116
Myrtaceidites eucalyptoides (pollen type) 147
myrtle = *Nothofagus cunninghamii*
myrtle wilt 132–3

National Conservation Strategy for Australia 250

National Estate 241, 246
National Parks and Wildlife
 Conservation Act (1975) 243
Nectarinia jugularis (yellow bellied
 sunbird) 191
New Caledonia 71, 172
New Guinea 84, 113, 117, 125, 144,
 148-9, 159, 171-3, 174-5 (*fig.*),
 181-3 (*table*), 184, 187, 188-9,
 190-1, 208, 212, 219, 234
New South Wales forest classification
 19-22, 20-1 (*fig.*)
Neozeylandic phytogeographic
 region 71-2
New Zealand 71-3, 84, 113, 139,
 144-5, 149, 171-2, 178-9
noisy scrub bird 192
nomad species 99, 104
Norfolk Island pine = *Araucaria
 heterophylla*
Norfolk Ridge 172
Nothofagus brassii (pollen type) 147
Nothofagus fusca (pollen type) 147
Nothofagus johnstonii (fossil leaf)
 153 (*fig.*)
Nothofagus menziesii (pollen type)
 147
Nothofagus muelleri (fossil leaf)
 153 (*fig.*)
Nothofagus ninnisiana (fossil leaf)
 153 (*fig.*)
Notomys 187
notophyll vine forest 27, 28,
 29 (*table*), 43, 45, 49 (*table*), 78-
 9, 95, 105 (*fig.*), 107-8,
 109 (*fig.*), 110-11, 117, 121,
 124, 130 (*fig.*), 132, 145, 160,
 164, 171, 193, 228, 228 (*fig.*),
 233, 255 (*figs.*)
Notoryctes typhlops (marsupial mole)
 187
notoryctids 187
Nullarbor Plain 141
nutrient budgets 254
nutrient distribution following
 disturbance 114

oil pollution, effects on mangroves
 209
Oligocene 142, 145, 146-7,
 147 (*fig.*), 151 (*fig.*), 189
olive whistler 192
onion cedar = *Owenia cepiodora*
Onycophora 196
Ornithoptera richmondia (Richmond
 birdwing butterfly) 196, 197
Ornithorhynchus anatinus (platypus)
 180, 183 (*table*)
Orthonyx 194
Orthonyx spaldingii (chowchilla) 191

Orthonyx temminckii (logrunner)
 191
oxygen isotope dating 142

Palaeocene 144, 145, 155, 172, 184
pale yellow robin 191
palm cockatoo 191
palorchestids 187
Pangaea 168, 196
Parastacidae 198
Paris 242-3
parma wallaby 182 (*table*), 234
pathogens 131-3
Peloridae 197
Pentoxylon australica 143
Perameles nasuta (long-nosed
 bandicoot) 182 (*table*)
Permian 139
peroryctids 7
Petaurus breviceps (sugar glider)
 181 (*table*)
Petroica rodinogaster (pink robin)
 192
Phanerozoic 142
pharmacological products of
 rainforest 232
phascaloarctids 187
Phellinus 133
Philemon buceroides (helmeted
 friarbird) 191
Philippines 208
phosphorus 7, 71, 114
photosynthesis, rates of in cool
 temperate species 111-13, 152
physiognomic classification 29-30
 key to 26-8
physiognomic features, trends of
 variation in 25, 79-83, 79 (*fig.*),
 80 (*fig.*), 81 (*fig.*), 82 (*fig.*)
phytogeographic analysis of rainforest
 floras 30-2, 31 (*fig.*)
Phytophthora 131-3, 208
Phytophthora cinnamomi 131-3
phytosociological classification of
 rainforest 24
pigeons 117, 193
 as seed dispersers 114
pigs 41, 46, 119, 122, 131, 190, 219,
 235
pilkipildrids 185 (*fig.*), 186
pine, Bunya = *Araucaria bidwillii*
pine, celery top = *Phyllocladus
 aspleniifolius*
pine, hoop = *Araucaria cunninghamii*
pine, Huon = *Lagarostrobos franklinii*
pine, King Billy = *Athrotaxis
 selaginoides*
pine, Norfolk Island = *Araucaria
 heterophylla*
pine, radiata = *Pinus radiata*
piners 225

pink robin 192
pioneer species 99, 104-5, 110-14,
 120, 134
Pitcairn Island 71
Pitta erythrogaster (red bellied pitta)
 191
plate tectonics 137-9, 138 (*fig.*), 149
Platycephala olivacea (olive whistler)
 192
platypus 180, 183 (*table*)
Platypus subgranosus (Platypus
 beetle) 132-3
Pleistocene 146 (*fig.*), 148, 151 (*fig.*),
 151 (*fig.*), 172, 187, 195, 214
Pliocene 142, 146 (*fig.*), 147,
 147 (*fig.*), 148, 151 (*fig.*), 172,
 184, 187
plywood 227, 229-30, 239
pneumatophores 201, 206, 206 (*fig.*)
Pogomys 189, 190
Pogonomys, cf *molipillosus*
 (prehensile tailed rat)
 181 (*table*), 190
pollen analysis 150, 151 (*fig.*)
pollination 115-17
 by bats 116
 by birds 116
 by beetles 116
polychaetes 209
Potoroidae 190
potoroids 185 (*fig.*)
Potorus tridactylus (long nosed
 potoroo) 182 (*table*)
pouched frog 196
prehensile tailed rat 181 (*table*), 190
primitive angiosperms 78, 116, 157-
 60
Prionodura newtoniana (golden
 bowerbird) 253
Probosciger aterrimus (palm
 cockatoo) 191
projective foliage cover 15
Pseudocheirus 189
Pseudocheirus cinereus
Pseudocheirus herbertensis (Herbert
 River ringtail possum)
 181 (*table*), 189
Pseudocheirus peregrinus (common
 ringtail possum) 182 (*table*)
Pseudochirops 186
Pseudochirops archeri (green ringtail
 possum) 181 (*table*)
Pseudomys 187
Psittacidae 210
Pteropodidae 189
Pteropus poliocephalus (grey headed
 fruit bat) 117
Ptilonorhynchus violaceus (satin
 bowerbird) 253
Puffinus pacificus (wedge-tailed
 shearwater) 71

Quaternary 4, 153, 154, 160–6, 161 (*fig.*), 162 (*fig.*), 163 (*fig.*), 168, 189, 213, 215
Queensland Forestry Department 131, 226–7, 249, 256
Queensland maple = *Flindersia brayleyana*
Queensland walnut = *Endiandra palmerstonii*

radiocarbon dating 41, 97, 125, 215
rainfall 7 (*fig.*), 8 (*fig.*), 9 (*fig.*)
  under greenhouse scenario 259–60
  in the Kimberley 34, 35 (*fig.*), 39 (*fig.*)
  on the Liverpool Range 55–6
  at Lynchs Crater 164–5, 165 (*fig.*)
  in the Miocene 147
  at Mount Tomah 8, 10 (*fig.*)
  in northeast Queensland 48, 48 (*table*)
  in northern NSW 54
  in the Northern Territory 37 (*fig.*), 38 (*fig.*)
  in provinces C_1 and C_2 68–8
  relationship to physiognomy 79, 82 (*fig.*)
  relationship to rainforest boundaries 86–7
  seasonality 7–8, 79
  in tall open forest 92
  in Victoria 64–5
Rainforest Conservation Society of Queensland 241
Rattus 118, 119, 187
Rattus fuscipes (bush rat) 182 (*table*)
Rattus leucopus (Cape York rat) 181 (*table*), 190
Rattus rattus 119, 122
red bellied pitta 191
red cedar = *Toona australis*
red cedar, silviculture of 226
red cheeked parrot 191
red-legged pademelon 182 (*table*), 190, 217
red-necked pademelon 182 (*table*), 217
refugia 40, 148, 154, 156, 159, 160, 169–70, 170 (*fig.*)
regeneration
  characteristics of species 18, 108, 114
  diffuse 103, 106
  gap filling model of 103–14, 105 (*fig.*), 107 (*fig.*)
  phases in 105–6, 105 (*fig.*), 107 (*fig.*)
  spotwise 103
regent skipper butterfly 197
reptiles 195, 195 (*fig.*)

Rheobatrachus silus (gastric brooding frog) 234–5
Rheobatrachus vitellinus (gastric brooding frog) 234–5
Richmond birdwing butterfly 196, 197
road construction 137, 254, 256
rodents 187–8
rodents as seed dispersers 119, 122
roots in mangroves 201
rosewood = *Dysoxylum fraseranum*
rufous scrub bird 61, 192
rufous spiny bandicoot 181 (*table*)

Sahul shelf 148, 160
salt-flats 204, 205 (*fig.*)
salt-marsh 200, 201, 204, 205 (*fig.*)
salt manufacture 208
salt spray 57, 58, 66, 71
sassafras = *Atherosperma moschatum* or *Doryphora sassafras*
satin bowerbird 253
sclerophyll 18–19, 57, 91, 100, 176–8
sclerophyll forest 18, 79, 89
sclerophyll forest, wet *see* tall open forest
sclerophyll woodland 109, 116
Schimper, A. F. W., classification of vegetation 13–14, 18–19, 75, 85, 101
scrub 14
scrubtit 192
sea-level 141, 159, 160, 212, 213, 215
secondary succession 103–4, 110
seed bank, in soil 105, 119–22, 121 (*fig.*)
seed dispersal 110–11, 117–19, 199
seed production 110
seed sources 105–6
seed viability 105, 110, 120
seedling establishment 92, 111
seedling mortality 122–3
seedlings stored in soil 122
selective logging 252–4
Sericornis beccarii (tropical scrubwren) 191
Sericornis citreogularis (yellow throated scrubwren) 191
Sericornis keri (Atherton scrubwren) 191
Sericornis magnirostris (large billed scrubwren) 191
Sericornis magnus (scrubtit) 192
Seychelles 71
shade tolerance 104, 111
silviculture 226–7
Sminthopsis leucopus (white footed dunnart) 182 (*table*)
snails 36

snow 8, 130
soap manufacture 208
softwood scrub 37
soil fertility 4 (*fig.*), 7, 50, 54–8, 78, 79, 83
  boundaries of rainforest stands and 86–9, 88 (*fig.*), 95–7, 96 (*fig.*), 135
  fire and 86, 177
  physiognomy and 79, 82–3, 81 (*fig.*)
  secondary succession and 106
sooty owl 191
South Africa 73
South-east Asia 50, 57, 117, 139, 148, 149, 187, 200
South Tasman Rise 142
southern beech = *Nothofagus*
Specht scheme 15, 16 (*table*), 17, 24, 91, 92, 201
Spilocuscus 189
Spilocuscus maculatus (spotted cuscus) 181 (*table*), 190
spotted cuscus 181 (*table*), 190
spotted-tailed quoll 182 (*table*)
Steropodon galmani 184
stilt roots 201, 202 (*fig.*), 212
stinger = *Dendrocnide* spp.
stinger, giant = *Dendrocnide excelsa*
Strigocuscus 186, 189
Strigocuscus minicus (grey cuscus) 181 (*table*), 190
striped possum 181 (*table*)
structure of vegetation as basis for classification 14
subtropical rainforest subformation 15, 54–5, 61, 78, 121
succession in oldfields 110
succession, secondary 103–4, 110
successional processes 101–2, 204
sugar farming 89, 128
sugar glider 181 (*table*)
Sumatra 139, 208
supernomads 171
sustainable yield forestry 251
Sydney Opera House 250

Tachyglossus aculeatus (echidna) 182 (*table*)
Tahiti 172
tall open forest 88–101, 90 (*fig.*), 91 (*fig.*), 93 (*fig.*), 96 (*fig.*), 97 (*fig.*), 98 (*fig.*), 184, 198, 240 (*fig.*)
  distribution of 91–2, 92 (*fig.*)
  evolution of 99
  in southwest Australia 99, 172
Talbragar Fish Beds 143
tamarind = *Tamarindus indica*
Tasmanian thornbill 192
Tasman Sea 171, 172

Tasmannia–Astelia order 24
Taudactylus 196
tawny breasted honeyeater 116, 191
teak = *Flindersia australis*
temperate rainforest subformation 15
temperature 9, 10 (*fig.*), 11 (*fig.*), 12 (*fig.*), 38, 38 (*fig.*)
Terania Creek Inquiry 100–1, 238, 250
termites 197
Tertiary 5 (*fig.*), 114, 141–2, 148, 150, 152–4, 158, 160, 166, 168, 197
Tethys Ocean 139
Thailand 139, 208
thamnic rainforest 66, 67 (*table*), 68 (*fig.*), 132
thermal reponse groups 30
thermoluminescence dating 214
Thornton Peak melomys 181 (*table*), 183
thylacine 187
thylacinids 187
thylacoleonids 187
Thylogale stigmatica (red-legged pademelon) 182 (*table*), 190, 217
Thylogale thetis (red-necked pademelon) 182 (*table*), 217
tidal regimes 202–3
tooth-billed catbird 253
topnot pigeon 191
Torres Strait pigeon 117
tourism 257
Tregellasia capito (pale yellow robin) 191
trepang 40, 219
Trichosurus caninus (mountain brush-tailed possum) 182 (*table*)
Trichosurus vulpecula (common brushtailed possum) 182 (*table*)
tropical rainforest (tropische Regenwald), definition of 13
tropical rainforest (tropische Regenwald), subformation in Beadle & Costin classification 15
tropical scrubwren 191
trumpet manucode 191
Tuckeroo forest type 57
tulipwood, yellow = *Brachychiton discolor*
Tumbunan avifaunal element 191, 195
turpentine = *Syncarpia glomulifera*
Tyto multipunctata (lesser sooty owl) 191
Tyto tenebricosa (sooty owl) 191

UNESCO 241
UNESCO classification of vegetation 76–7 (*table*)
Uromys 187, 189
Uromys caudimaculatus (white tailed rat) 181 (*table*), 190

vagility of taxa 169
Vietnam 85
vine thickets 35, 40, 43, 51, 57, 69–70, 87, 122, 155
volcanoes 5 (*fig.*), 7, 128, 141–2

wallabies 124
walnut, Queensland = *Endiandra palmerstonii*
warm temperate rainforest subformation 75–8, 79, 126–8
water availability 8
water rat 183 (*table*)
wedge-tailed shearwater 71
white beech = *Gmelina leichardtii*
white cedar = *Melia azeredach*
white footed dunnart 182 (*table*)
White's thrush 191
white tailed rat 181 (*table*), 190
white mahogany = *Eucalyptus acmenoides*
wilderness 256–7

Wilderness Society 237
windthrow 18
World Conservation Strategy 250
World Heritage Committee 241–4, 246
World Heritage Convention 240, 242
World Heritage Fund 243–4
World Heritage list 6 (*fig.*), 57 (*fig.*), 72, 73 (*fig.*), 241–4, 246, 257–8, 264
 Australian properties on 242
 criteria for inclusion 241–2
 NSW rainforest nomination for 240 (*fig.*), 244–6, 245 (*fig.*)
 wet tropics nomination 246–8, 247 (*fig.*)
World Heritage Properties Conservation Act (1983) 243, 246, 248, 258
Wynyardidae 186

Xanthotis flaviventer (tawny breasted honeyeater) 116, 191
Xanthotis macleayana (Macleay's honeyeater) 116

yalkaparidontids 186
yingabalanarids 186
yellow bellied sunbird 191
yellow carabeen = *Sloanea woollsii*
Yellow Carabeen forest type 54
yellow footed antechinus 182 (*table*)
yellow spotted honeyeater 116, 191
yellow throated scrubwren 191
yellow tulipwood = *Brachychiton discolor*
Yellow Tulipwood forest type 89

zonation of mangroves 204
Zone F at Lake George 164, 215
Zoothera dauma (White's thrush) 191
Zyzomys 187